The Evolution of Beauty

The Evolution of Beauty

HOW DARWIN'S FORGOTTEN THEORY
OF MATE CHOICE SHAPES THE ANIMAL WORLD—
AND US

Richard O. Prum

Doubleday

New York   London   Toronto   Sydney   Auckland

All rights reserved. Published in the United States by Doubleday,
a division of Penguin Random House LLC, New York, and distributed
in Canada by Random House of Canada,
a division of Penguin Random House Canada Limited, Toronto.

www.doubleday.com

DOUBLEDAY and the portrayal of an anchor with a dolphin are
registered trademarks of Penguin Random House LLC.

Book design by Maria Carella
Jacket design by John Fontana
Front-of-jacket photographs (clockwise from top left): © Tim Laman/National
Geographic Creative; © Juan Carlos Vinda/Minden Pictures; © Glenn Bartley/
All Canada Photos/Superstock; © NHPA/Photoshot/Superstock; © Tim Laman/
National Geographic Creative
Pen-and-ink drawings by Michael DiGiorgio
Charts by Rebecca Gelernter

Library of Congress Cataloging-in-Publication Data
Names: Prum, Richard O., author.
Title: The evolution of beauty : how Darwin's forgotten theory of mate choice
shapes the animal world and us / Richard O. Prum.
Description: First edition. | New York : Doubleday, 2017. | Includes
bibliographical references and index.
Identifiers: LCCN 2016050808 (print) | LCCN 2016059440 (ebook) |
ISBN 9780385537216 (hardcover) | ISBN 9780385537223 (ebook)
Subjects: LCSH: Sexual selection in animals. | Sexual selection. | Courtship in
animals. | Human evolution.
Classification: LCC QL761 .P744 2017 (print) | LCC QL761 (ebook) |
DDC 591.56/2—dc23
LC record available at https://lccn.loc.gov/2016050808

MANUFACTURED IN THE UNITED STATES OF AMERICA

First Edition

TO ANN,
for inspiring and tolerating my many flights of fancy

MOTHER GOOSE: What is the secret Nature knows?
TOM RAKEWELL: What Beauty is and where it grows.

—*The Rake's Progress,*
an opera in three acts by Igor Stravinsky
Fable libretto by W. H. Auden and
Chester Kallman

Contents

The Evolution of Beauty

Introduction

I started bird-watching and studying birds at the age of ten, and I never really considered doing anything else in my life. Which is fortunate, because I am now unfit for any other sort of employment.

It all started with glasses. I got my first pair of eyeglasses during fourth grade, and within six months I was a bird-watcher. Before glasses, I spent a lot of time memorizing facts out of the *Guinness Book of World Records* and asking my siblings to quiz me on them. I was especially interested in the records of extreme human "achievement," like the tallest and heaviest men, and the now suppressed category of "gastronomical" records, like the greatest number of whelks eaten in five minutes. But after glasses, the outer world came into focus. Soon, my amorphous nerdiness found something to organize around, something to run with—birds.

The next catalyst was a book. My family lived in Manchester Center, Vermont, a small town nestled in a beautiful valley between the Taconics and the Green Mountains. As I was browsing in a small, local bookstore one day, my eyes landed on Roger Tory Peterson's *A Field Guide to the Birds*. I was transfixed by the paintings of the Cardinal, the Evening Grosbeak, and the

Atlantic Puffin on the book's cover. The book was a pleasing and efficient pocket size. Thumbing through its pages, I immediately began imagining all the places I would have to go to see all these birds—with the book, of course, in my back pocket. I showed the book to my mother with a not so subtle pitch that I would love to take it home. "Well," she responded encouragingly, "you do have a birthday coming up!" About a month later, for my tenth birthday, I did indeed receive a bird guide, but it was the *other* one, Chandler Robbins's *Birds of North America,* with the text and range maps opposite the color plates. It was a great book with a really bad binding, and I would trash several more copies before I was out of elementary school.

Starting with a clunky old pair of family binoculars, I began to scour our rural neighborhood looking for birds. Within a year or so I had bought myself a new pair of Bausch & Lomb Custom 7x35s, paying for them with money I'd earned from mowing lawns and working a paper route. On my next birthday, I received a bird song record, and I began to learn them. My initial curiosity grew into obsession and then into a consuming passion. On a good day of birding, my pulse would race with excitement. Sometimes, it still does.

Many people cannot understand *what* there is about birds to become obsessed about. What are bird-watchers actually doing out there in the woods, swamps, and fields? The key to comprehending the passion of birding is to realize that bird-watching is really a hunt. But unlike hunting, the trophies you accumulate are in your mind. Of course, your mind is a great place to populate with trophies because you carry them around with you wherever you go. You don't leave them to gather dust on a wall or up in the attic. Your birding experiences become part of your life, part of who you are. And because birders are human, these birding memories—like most human memories—improve over time. The colors of the plumages become more saturated, the songs sweeter, and those elusive field marks more vivid and distinct in retrospect.

The exciting buzz of birding creates the desire to see more

birds, to see the earliest arrivals and the latest departures, the biggest and the smallest, and to know their habits. Most of all birding creates the desire to see *new* birds—birds you have never seen before—and to keep records of your sightings. Many birders keep a "Life List" of all the bird species they have seen in their lives; each new bird they add is called a lifer.

Most kids are probably not thinking about what they will be doing for the rest of their lives, but I was very sure. By the time I was twelve, I knew I would be birding. Birding was an open invitation to adventures straight out of the gorgeously illustrated pages of *National Geographic* magazine. I soon found myself lusting after ever more remote and exotic habitats and locales. In 1976, I was again browsing in a bookstore, this time with my father, and I came across the gorgeous new *Guide to the Birds of Panama* by Robert Ridgely. It cost $15, which was more than I had. My parents were usually game for going fifty-fifty on such worthy purchases, so I asked Dad if he would be willing to split it with me. He looked at me incredulously and asked, "But, Ricky, when are *you* going to *Panama*?" My adolescent voice probably cracked as I responded, "But don't you see, Dad, you *get* the book, and *then* you go!" I guess I was pretty convincing, because I brought the book home, and it initiated my lifelong fascination with neotropical birds.

Of course, the ultimate goal of birding is to know *all* the birds of the world. All ten thousand plus species. But I don't mean *know* the birds in the same sense that one can know the laws of gravity, the height of Everest, or the fact that Robert Earl Hughes was the heaviest human in the world at 1,070 pounds. Birding is about knowing the birds in a more intimate, profound way.

To understand what I mean, let's imagine what it's like for a bird-watcher to see a bird. Not just any bird, but a particular bird—for example, a male Blackburnian Warbler (*Setophaga fusca*) (color plate 1). I can remember exactly my first sighting of a male Blackburnian Warbler, which was perched in a thinly leaved white birch tree in my front yard in Manchester Center on a bright May morning in about 1973. In the years since,

I have seen Blackburnian Warblers many times, and in many places, from their breeding grounds in the boreal forests along the Allagash River in northern Maine to their wintering distribution in the Andean cloud forests of Ecuador. I *know* Blackburnian Warbler.

Of course, no one who sees a male Blackburnian Warbler can fail to observe its crisp black body plumage, brilliantly orange throat and face patterns, and white wing bars, belly, and tail spots. The sight of a Blackburnian Warbler would create a truly stunning and memorable sensory impression on anyone. But birding is about more than just seeing a bird and taking in the visual experience of it. Birding is about recognizing all the physical characteristics of the bird *and* being able to attach the correct name, or proper noun, to that observation.

When a bird-watcher sees a male Blackburnian Warbler or any other bird she has identified, she has a neurological experience distinct from the mere sensory perception of its bold pattern of black, orange, and white plumage. We know this is true because functional magnetic resonance imaging studies of the brains of bird-watchers have shown that unlike untrained human observers, birders use the face recognition module in the visual cortex of the brain to recognize and identify bird species and plumages. In other words, when a birder identifies a Blackburnian Warbler, she uses the same parts of the brain that people use to recognize familiar faces—like those of Jennifer Aniston, Abraham Lincoln, and your Aunt Lou. Birding trains your brain to transform a stream of natural history perceptions into encounters with identifiable individuals. This is like the difference between walking along a city street amid a sea of strangers and walking the halls of your old high school, where you recognize every individual instantly. The key difference between what a bird-watcher experiences and a simple walk in the woods is what's happening in your brain.

The English language falls short in communicating this distinction, because English provides us with only one verb for "to know." Many other languages, however, have two distinct verbs. One means knowing a fact or understanding a concept, and the other means being familiar with someone or something through personal experience. In Spanish, to know or understand a fact is *saber,* but to be familiar with someone or something through experience is *conocer;* in French, these verbs are *savoir* and *connaître,* and in German *wissen* and *kennen.* The key difference between birding and mere observation is that birding is really about building a bridge between these two kinds of knowing— connecting familiarity and personal experience to facts and understanding. It's about accumulating knowledge about the natural world through your own personal experience. That's why, to a birder, it always matters whether or not you have actually seen the bird in real life and not just on the page! Knowing that the bird exists without seeing it for yourself is merely knowledge without experience—*savoir* without *connaissance*—which is never enough.

When I got to college, I discovered that evolutionary biology was the field of science that was about the aspect of birds that I found most fascinating—their tremendous diversity and endless, exquisite differences. Evolution was the explanation of how all ten thousand species of birds came to be the way they are. I realized that my birding—all that cognitive stamp collecting—had laid the foundation for a much grander intellectual project: a lifelong engagement in scientific research on the evolution of birds.

In more than forty years of birding and thirty years of studying avian evolution, I have had the joy and good fortune to research an enormous range of topics in science. Along the way, I have been given the opportunity to watch birds on all continents and to see more than one-third of the bird species of the world, though I have no doubt that my twelve-year-old self would be sorely disappointed at how slowly I have progressed at the impossible task of seeing them all. I have worked in the rain forests of South America discovering the previously unknown display behaviors of manakins (Pipridae). I have dissected the syringes of birds—the tiny, avian vocal organs—in order to use this anatomical feature to reconstruct the evolutionary relationships of species. I have worked on avian biogeography (the study of the distribution of species around the globe), on the development and evolution of feathers, and on the origin of avian feathers in theropod dinosaurs. I have investigated the physics and chemistry of avian plumage coloration and the four-color vision of birds.

During such forays, my research has taken many surprising turns, directing me to topics I would never have imagined studying—such as the shockingly violent sex lives of ducks. Sometimes, my various investigations turned up connections that were entirely unexpected. For example, separate research initiatives on the coloration of bird feathers and the evolution of dinosaur feathers ultimately led to a collaborative discovery of the dramatic colors in the plumage sported by a 150-million-year-old feathered dinosaur—*Anchiornis huxleyi* (color plate 15).

For a long time I thought that my research was just an eclectic grab bag of "stuff Rick is into." In recent years, however, I have realized that a large portion of my research is really about one big issue—the evolution of beauty. I don't mean beauty as *we* experience it. Rather, I am interested in the beauty of birds to *themselves*. In particular I am fascinated by the challenge of understanding how the social and sexual choices of birds have driven so many aspects of avian evolution.

In various social contexts, birds observe each other, they evaluate what they've observed, and they make social decisions—real choices. They choose which birds to flock with, which baby bird mouths to feed, and whether or not to incubate a given clutch of eggs. And, of course, the most crucial social decision that birds make is whom to mate with.

Birds use their preferences for particular plumages, colors, songs, and displays to choose their mates. The result is the evolution of sexual ornaments. And birds have a lot of them! Scientifically speaking, sexual beauty encompasses all of the observable features that are desirable in a mate. Over millions of years and among thousands of avian species, mate choice has resulted in an explosive diversity of sexual beauty in birds.

Ornaments are distinct in function from other parts of the body. They do not function solely in ecological or physiological interactions with the physical world. Rather, sexual ornaments function in interactions with *observers*—through the way in which sensory perceptions and cognitive evaluations by other individual organisms create a subjective experience in those organisms. And by subjective experience, I mean the unobservable, internal mental qualities produced by a flow of sensory and cognitive events: like the sight of the color red, the smell of a rose, or the feeling of pain, hunger, or desire. Crucially, the function of sexual ornaments is to inspire the qualities of desire and attachment in the observer.

What can we possibly know about the subjective experience of desire in animals? Subjective experience is, almost by definition, unmeasurable and unquantifiable. As Thomas Nagel has written in his classic paper "What Is It Like to Be a Bat?," subjective experience encompasses the "what it is like" for a given organism—be it a bat, a flounder, or a person—to have a perceptual or cognitive event. But if you are not a bat, you will never be able to grasp the experience of perceiving the three-dimensional "acoustic structure" of the world through sonar. Although we can imagine that our individual subjective experiences are similar in quality to those of other individuals, perhaps even to those of other species, we can never confirm this, because we cannot actually share the qualities of our internal mental experiences with one another. Even among humans who can express their thoughts and experiences in words, the actual content and quality of our internal sensory experiences are ultimately unknowable by anyone else and inaccessible to scientific measurement and reduction.

Most scientists have therefore been allergic to the idea of making a scientific study of subjective experiences, or even to admitting that they exist. If we cannot measure them, many biologists think that such phenomena cannot be an appropriate subject of science. For me, however, the concept of subjective experience is absolutely critical to understanding evolution. I will argue that we need an evolutionary theory that encompasses the subjective experiences of animals in order to develop an accurate scientific account of the natural world. We ignore them at our intellectual peril, because the subjective experiences of animals have critical and decisive consequences for their evolution. If subjective experience is not reducible to measurement, then how can we study it scientifically? I think we can take a lesson from physics. In the early twentieth century, Werner Heisenberg proved that we cannot simultaneously know the position and the momentum of an electron. Although Heisenberg's Uncertainty Principle proved the electron could not be reduced to Newtonian mechanics, physicists did not abandon or ignore the problem of the electron. Rather, they devised new methods to approach it. Similarly, biology needs to develop new methods to investigate the subjective experiences of animals. We cannot measure or know what these experiences are like in any detail, but we can sneak up on them, and as with the electron we can learn fundamental things about them indirectly. For example, as we will see, we can investigate how subjective experience evolves by tracing the evolution of ornaments and the sexual preferences for them among closely related organisms.

I call the evolutionary processes that are driven by the sensory judgments and cognitive choices of individual organisms aesthetic evolution. The study of aesthetic evolution requires engaging with both sides of sexual attraction: the object of desire and the form of desire itself, which biologists refer to as display traits and mating preferences. We can observe the consequences of sexual desire by studying which mates are preferred. More powerfully, perhaps, we can also study the evolution of sexual desire by studying the evolution of the objects of that desire—the ornaments that are particular to a given species and how those ornaments have evolved among multiple species.

What emerges from an understanding of the workings of sexual selection is the startling realization that desire and the object of desire coevolve with each other. As I will discuss later, most examples of sexual beauty are the results of coevolution; in other words, the form of the display and the mating preference do not accidentally correspond to each other, but have shaped each other over evolutionary time. It is through this coevolutionary mechanism that the extraordinary aesthetic diversity of the natural world comes into being. This book, then, is ultimately a natural history of beauty and desire.

How does aesthetic evolution differ from other modes of evolution? To explore the difference, let's compare "normal," adaptive evolution by natural selection—the evolutionary mechanism famously discovered by Charles Darwin—with aesthetic evolution by mate choice, another amazing discovery of Darwin's. In the bird world, the beaks of the Galápagos Finches are one of Darwin's best-known examples of adaptive evolution. The approximately fifteen different species of Galápagos Finches evolved from a single common ancestor, and they differ from each other mainly in the size and shape of their beaks. Certain beak shapes and sizes are particularly effective at handling and opening certain kinds of plant seeds; large beaks are better at cracking larger, harder seeds, while smaller beaks are more efficient at handling smaller, finer seeds. Because the environment of the Galápagos varies in the size, hardness, and abundance of the plant seeds available in different areas and times, some finches will survive better in certain environments than do others. Because beak size and shape are highly heritable traits, differential survival of beak shapes *within* one generation of Galápagos Finches will result in evolutionary change in beak shape *among* generations. This evolutionary mechanism—called natural selection—leads to *adaptation* because subsequent generations will have evolved beak shapes that function better in their environment, contributing directly to improvements in individual survival and fecundity (that is, individual capacity for reproduction and energy and resources to lay lots of eggs, to lay bigger eggs, and to raise lots of healthy offspring).

By contrast, let's imagine the evolution of an avian ornament, such as the song of the thrush or the iridescent plumage of the hummingbird. These features evolve in response to criteria very different from those involved in natural selection on beak shape. Sexual ornaments are aesthetic traits that have evolved as a result of mate choices based on subjective evaluations. They function through the perception and evaluations of other individuals through mate choice. The cumulative effect of many individual mating decisions shapes the evolution of ornament. In other words, members of these species act as agents in their own evolution.

As Darwin himself realized, evolution by natural selection and aesthetic evolution by mate choice produce profoundly different patterns of variation in nature. For example, there are a limited number of ways to crack open a seed with a bird beak and therefore a limited number of variations in beak size and shape to do it. Consequently, seed-eating birds from more than a dozen different bird families have independently and convergently evolved very similar, robust, finchy beak shapes in order to perform this particular physical task. But the task of attracting a mate is an infinitely more open-ended, unconstrained, and dynamic challenge than opening a seed. Each species evolves its own solution to the challenge of intersexual communication and attraction— what Darwin called independent "standards of beauty." Thus, it is no surprise that each of the world's ten thousand plus bird species has evolved its own, unique aesthetic repertoire of ornaments and preferences to accomplish this task. The result is the earth's nearly unfathomable variety of biological beauty.

Now, I have a problem—a scientific problem. Although doing research in evolutionary biology has been a real joy for me, the community of science is not without diversity of opinion, disagreement, and intellectual conflict. And as it turns out, my ideas about aesthetic evolution run counter to the main flow of ideas in evolutionary biology—not just for the last few decades, but for nearly a century and a half, indeed, since the time of Darwin himself. Most evolutionary biologists, then and now, think that sexual ornaments and displays—they generally avoid using the word "beauty"—evolve because such ornaments provide specific, honest information about the quality and condition of potential mates. According to this "honest signaling" paradigm, the extraordinary electric-blue smiley face display on the erectable breast feathers of a male Superb Bird of Paradise (*Lophorina superba*) (color plate 2) functions like a birdie Internet dating profile, providing multiple pieces of information that a discerning female bird of paradise needs to know. Who are his "people"? Does he come from a good egg? Was he raised in a good nest? Does he have a good diet? Does he take care of himself? Does he have sexually transmitted diseases? In species of birds that form enduring pairs, such courtship displays may communicate additional information: Will he or she energetically defend our territory from competitors? Will he or she help feed and shelter me, be a good parent to our offspring, and be faithful to me?

According to this BioMatch.com theory of ornament, beauty is all about utility. In this view, the subjective mating preferences of individuals are shaped by the objective quality of their available mates. Beauty is only desirable because it brings other, real-world benefits, like vigor, health, or good genes. Although sexual beauty may indeed be sensually pleasing, according to this view, sexual selection is just another form of natural selection; there is no fundamental difference between the evolutionary forces acting on the beaks of Galápagos Finches and those shaping the courtship displays of the birds of paradise. Beauty is merely the handmaiden of natural selection.

This is very different from my own view of beauty and how it arises. Although I am rather hesitant to admit it, I think that the process of adaptation by natural selection is sort of boring. Of course, as an evolutionary biologist I am well aware that it is a fundamental and ubiquitous force in nature. I do not deny its immense importance. But the process of adaptation by natural selection is *not* synonymous with evolution itself. A lot of evolutionary process and evolutionary history cannot be explained by natural selection alone. Throughout this book, I will argue that evolution is frequently far quirkier, stranger, more historically contingent, individualized, and less predictable and generalizable than adaptation can explain.

Evolution can even be "decadent," in the sense of its resulting in sexual ornaments that not only fail to signal anything about objective mate quality but actually lower the survival and fecundity of the signaler and the chooser. In short, in pursuit of their subjective preferences, individuals can make mating choices that are *maladaptive*—resulting in a worse fit between the organism and its environment. This is something that quite a few evolutionary biologists would argue is impossible, but I beg to differ, and this book is my explanation of why. In the larger sense, I hope to communicate to my readers that natural selection alone cannot possibly explain the diversity, complexity, and extremity of the sexual ornaments we see in nature. Natural selection is not the only source of design in nature.

It seems to me that the kinds of scientific questions one likes to ask, and the kinds of scientific answers one finds satisfying, are deeply personal. For some reason, I have always been more fascinated by those aspects of evolutionary process that defy simplistic adaptive explanations. Somehow, the way my personal, lifelong engagement with birds connected to the science of their evolution led me to a different view. However, as I will document in these pages, this aesthetic theory of evolution was first proposed and championed by Charles Darwin himself and roundly criticized at the time. Indeed, Darwin's aesthetic theory of mate choice has been so marginalized in evolutionary biology that it has been nearly forgotten. Contemporary "neo-Darwinism"—which posits that sexual selection is merely another form of natural selection—is highly popular yet not Darwinian at all. Rather, the adaptationist view comes down to us from Darwin's intellectual acolyte and subsequent antagonist Alfred Russel Wallace. Aesthetic evolution, I will argue, restores the real Darwin to Darwinism, by showing how the subjective mate choice decisions of animals play a critical and often decisive role in evolution. But can we really talk about beauty as a quality that animals respond to? The concept of beauty is so fraught with people's preconceptions, expectations, and misunderstandings that perhaps it would be wiser to continue to shun any scientific use of the term. Why use such a problematic and loaded word? Why not continue with the sanitized and nonaesthetic language that most biologists prefer?

I have thought a lot about this. I have decided to embrace beauty as a scientific concept because, like Darwin, I think it captures in ordinary language exactly what is involved in biological attraction. By recognizing sexual signals as *beautiful* to those organisms that prefer them—whether they are Wood Thrushes, bowerbirds, butterflies, or humans—we are forced to engage with the full implications of what it means to be a sentient animal making social and sexual choices. We are forced to entertain the Darwinian possibility that beauty is not merely utility shaped by adaptive advantage. Beauty and desire in nature can be as irrational, unpredictable, and dynamic as our own personal experiences of them.

This book aspires to bring beauty back into the sciences—to reanimate Darwin's original aesthetic conception of mate choice and elevate beauty to a mainstream subject of scientific concern.

Darwin's concept of mate choice has another controversial element that I will also champion in these pages. In proposing the mechanism of evolution by mate choice, Darwin hypothesized that female preferences can be a powerful and independent force in the evolution of biological diversity. Not surprisingly, Victorian scientists ridiculed Darwin's revolutionary idea that females had either the cognitive ability or the opportunity to make autonomous decisions about their choice of mates. But the concept of freedom of sexual choice—or sexual autonomy—needs to be revived. In this book, we will do some long-overdue work—140 years overdue—on the evolution of sexual autonomy and its implications for both nonhuman and human traits and behaviors.

As my research on the often violent sexual behavior of waterfowl has taught me, the primary challenge to female sexual autonomy is male sexual coercion via sexual violence and social control. Through investigations of ducks and other birds, we will explore the diverse evolutionary responses to male sexual coercion. We will see that mate choice can evolve in ways that specifically *enhance* female freedom to choose. In short, we will discover that reproductive freedom of choice is not merely a political ideology invented by modern suffragettes and feminists. Freedom of choice matters to animals, too.

Leaping from birds to people, I will explore the ways in which sexual autonomy is fundamental to understanding the evolution of many of the unique and distinctive features of human sexuality, including the biological roots of female orgasm, the boneless human penis, and same-sex sexual desire and preference. Aesthetic evolution and sexual conflict are also likely to have played a critical role in the origins of human intelligence, language, social organization, and material culture and the diversity of human beauty.

In short, the evolutionary dynamics of mate choice are essential to understanding ourselves.

I have been interested in the theory of aesthetic evolution for my entire career, and over the years I have become accustomed to its marginal status in the discipline of evolutionary biology. But I remember the exact moment when I realized how strong the resistance to aesthetic evolution really is and how the strength of this resistance is really a measure of the threat this idea poses to mainstream adaptationist evolutionary thought. At that moment, I realized how necessary it was to write this book.

The epiphany came during a visit to an American university a few years ago as I described my views on the evolution of sexual ornaments to a fellow evolutionary biologist over lunch. After every few sentences, my host interrupted me with an objection or two, each of which I answered before I got back to outlining my view. Toward the end of the lunch, when I had finally managed to give a full explanation of my views on evolution by mate choice, he cried out, "But that's *nihilism*!" Somehow, what I thought of as a powerful and awe-inspiring explanation of the diversity of ornament in the natural world, my evolutionary colleague saw as a bleak worldview that, should he adopt it, would deprive him of any sense of purpose or meaning in life. After all, if mate choice results in the evolution of ornaments that are merely beautiful, rather than being indicators of mate quality, doesn't that mean that the universe is not rational? At this moment, I realized why it was necessary to embrace Darwin's aesthetic perspective on evolution and explain it to a wider audience.

My scientific view has grown directly from my experience of the natural world as a bird-watcher and natural historian and from my work as a scientific researcher—*connaissance* and *savoir.* This work has given me enormous intellectual and personal pleasure. Never in my career have I been more excited and inspired to do science. I get goose bumps just thinking about the evolution of avian beauty. But this same worldview would seem to deny some of my professional colleagues any reason to get out of bed in the morning. In this book, I will try to explain why I think this more subtle, less deterministic view of evolution provides a richer, more accurate, and more scientific understanding of nature than the common adaptationist view. When we look at evolution through sexual selection, we see a world of freedom and choice that is deeply thrilling—a world of greater beauty than can possibly be accounted for without it.

CHAPTER 1

Darwin's *Really* Dangerous Idea

Adaptation by natural selection is among the most successful and influential ideas in the history of science, and rightly so. It unifies the entire field of biology and has had a profound influence on many other disciplines, including anthropology, psychology, economics, sociology, and even the humanities. The singular genius behind the theory of natural selection, Charles Darwin, is at least as famous as his most famous idea.

You might think that my contrarian view of the limited power of adaptation by natural selection would mean that I am "over" Darwin, that I am ready to denigrate the cultural/scientific personality cult that surrounds Darwin's legacy. Quite to the contrary. I hope to celebrate that legacy but also to transform the popular understanding of it by shedding new light on Darwinian ideas that have been neglected, distorted, ignored, and almost forgotten for nearly a century and a half. It's not that I'm interested in doing a Talmudic-style investigation of Darwin's every word; rather, my focus is on the science of today, and I believe that Darwin's ideas have a value to contemporary science that has yet to be fully exploited.

Trying to communicate the richness of Darwin's ideas puts me in the unenviable position of having to convince people that we don't actually know the real Darwin and that he was an even greater, more creative, and more insightful thinker than he has been given credit for. I am convinced that most of those who think of themselves as Darwinians today—the neo-Darwinists— have gotten Darwin all wrong. The real Darwin has been excised from modern scientific hagiography.

The philosopher Daniel Dennett referred to evolution by natural selection—the subject of Darwin's first great book, *On the Origin of Species by Means of Natural Selection*—as "Darwin's dangerous idea." Here I propose that Darwin's *really* dangerous idea is the concept of aesthetic evolution by mate choice, which he explored in his second great book, *The Descent of Man, and Selection in Relation to Sex*.

Why is the idea of Darwinian mate choice so dangerous? First and foremost, Darwinian mate choice really *is* dangerous—to the neo-Darwinists—because it acknowledges that there are limits to the power of natural selection as an evolutionary force and as a scientific explanation of the biological world. Natural selection cannot be the only dynamic at work in evolution, Darwin maintained in *Descent,* because it cannot fully account for the extraordinary diversity of ornament we see in the biological world.

It took Darwin a long time to grapple with this dilemma. He famously wrote, "The sight of a feather in a peacock's tail, whenever I gaze at it, makes me sick!" Because the extravagance of its design seemed of no survival value whatsoever, unlike other heritable features that are the result of natural selection, the peacock's tail seemed to challenge everything that he had said in *Origin*. The insight he eventually arrived at, that there was another evolutionary force at work, was considered an unforgivable apostasy by Darwin's orthodox adaptationist followers. As a consequence, the Darwinian theory of mate choice has largely been suppressed, misinterpreted, redefined, and forgotten ever since.

Aesthetic evolution by mate choice is an idea so dangerous that it had to be laundered out of Darwinism itself in order to preserve the omnipotence of the explanatory power of natural selection. Only when Darwin's aesthetic view of evolution is restored to the biological and cultural mainstream will we have a science capable of explaining the diversity of beauty in nature.

Charles Darwin, a member of England's nineteenth-century rural gentry, led a privileged life within the most elite class of an expanding global empire. Yet Darwin was no idle member of the upper class. A man of careful habits and a steady, hardworking disposition, he used his privilege (and his generous independent income) to support the searching of a stubbornly relentless intellect. By following where his interests took him, he ultimately discovered the fundamentals of modern evolutionary biology. He thus delivered a fatal blow to the hierarchical Victorian worldview, which put man on a pedestal above, and totally removed from, the rest of the animal kingdom. Charles Darwin became a radical despite himself. Even today the full creative impact of his intellectual radicalism—its implications for science and for the culture at large—has yet to be appreciated.

The traditional image of Darwin as a young man portrays him as an indifferent and undisciplined student who mostly liked to roam around outside collecting beetles. He dropped out of his original course of medical education and bounced aimlessly among various interests with little outward commitment to any of them until he was offered the opportunity to go on his famous *Beagle* voyage. According to legend, Darwin was transformed by his world travels and became the revolutionary scientist we remember today.

I think it more likely that Darwin had the same voracious, quiet, but stubborn intellect as a young man that he displayed later in life, an intellect that would have given him an instinctive sense of what good science looked like. Just prior to publishing *On the Origin of Species* in 1859, Darwin characterized the giant creationist masterwork of the world-famous Harvard professor Louis Agassiz, the *Essay on Classification,* as "utterly impracticable rubbish!" As a medical student, Darwin, I think, likely came to the same conclusion about most of his biological education.

And he would have been right. Most of what was taught as medicine in the 1820s *was* impracticable rubbish. There was no central mechanistic understanding of the workings of the body and no broader scientific concept of the causes of disease. Medical treatments were a grab bag of irrelevant placebos, powerful poisons, and dangerous quackery. It would be hard to identify more than a handful of professional medical treatments from that time that would be recognized today as being likely to do any patient any good whatsoever. Indeed, in his autobiography Darwin describes his experience of attending lectures at the Royal Medical Society in Edinburgh: "Much rubbish was talked there." I suspect that it was only when Darwin went all the way to the unexplored reaches of the Southern Hemisphere that he found an intellectual space free enough from the hidebound dogmas of his day to allow him the full play of his far-reaching, brilliant, and ever-curious mind.

Once he could make his own unfiltered observations, what he saw led him to the two great biological discoveries he revealed in *Origin:* the mechanism of evolution by natural selection, and the concept that all organisms are historically descended from a single common ancestor and thus related to one another in a "great Tree of Life." The enduring debates in some corners over whether these ideas should be taught in public schools give us some sense of how profoundly they must have challenged Darwin's readers a century and a half ago.

In confronting the fierce attacks that were mounted against *Origin* after its publication, Darwin had three gnawing problems. The first problem was the absence of any working theory of genetics. Not knowing the work of Mendel, Darwin struggled and failed to develop a functioning theory of inheritance, which was fundamental to the mechanism of natural selection. Darwin's second problem was the evolutionary origin of human beings, human nature, and human diversity. When it came to human evolution, Darwin pulled his punches in *Origin* and evasively concluded only that "light will be thrown on the origin of man and his history."

Darwin's third big problem was the origin of impracticable beauty. If natural selection was driven by the differential survival of heritable variations, what could explain the elaborate beauty of that peacock's tail that troubled him so much? The tail obviously did not help the male peacock to survive; if anything, the huge tail would be a hindrance, slowing him down and making him much more vulnerable to predators. Darwin was particularly obsessed with the eyespots on the peacock's tail. He had argued that the perfection of the human eye could be explained by the evolution of many incremental advances over time. Each evolutionary advance would have produced slight improvements in the ability of the eye to detect light, to distinguish shadows from light, to focus, to create images, to differentiate among colors, and so on, all of which would have contributed to the animal's survival. But what purpose could the intermediate stages in the evolution of the peacock's eyespots have served? Indeed, what purpose do the "perfect" eyespots of a peacock serve today? If the problem of explaining the evolution of the human eye was an intellectual challenge, the problem of explaining the peacock's eyespot was an intellectual nightmare. Darwin lived this nightmare. It was in that context that in 1860 he wrote that oft-quoted line to his American friend the Harvard botanist Asa Gray: "The sight of a feather in a peacock's tail, whenever I gaze at it, makes me sick!"

In 1871, with the publication of *The Descent of Man, and Selection in Relation to Sex,* Darwin boldly addressed both the problem of human origins and the evolution of beauty. In this book he proposed a second, independent mechanism of evolution—sexual selection—to account for armaments and ornaments, battle and beauty. If the results of natural selection were determined by the differential survival of heritable variations, then the results of sexual selection were determined by their differential sexual success—that is, by those heritable features that contribute to success at obtaining mates.

Within sexual selection, Darwin envisioned two distinct and potentially opposing evolutionary mechanisms at work. The first mechanism, which he called the law of battle, was the struggle between individuals of one sex—often male—for sexual control over the individuals of the other sex. Darwin hypothesized that the battle for sexual control would result in the evolution of large body size, weapons of aggression like horns, antlers, and spurs, and mechanisms of physical control. The second sexual selection mechanism, which he called the taste for the beautiful, concerned the process by which the members of one sex—often female—choose their mates on the basis of their own innate preferences. Darwin hypothesized that mate choice had resulted in the evolution of many of those traits in nature that are so pleasing and beautiful. These ornamental traits included everything from the songs, colorful plumages, and displays of birds to the brilliant blue face and hindquarters of the Mandrill (*Mandrillus sphinx*). In an exhaustive survey of animal life from spiders and insects to birds and mammals, Darwin reviewed the evidence for sexual selection in many different species. Using the law of battle and the taste for the beautiful, he proposed to explain the evolution of both armament and ornament in nature.

In *The Descent of Man*, Darwin finally presented the explicit theory of the evolutionary origins of humans that he had avoided articulating in *Origin*. The book begins with a long discussion of the continuity between human beings and other animals, slowly and incrementally chipping away at the edifice of human uniqueness and exceptionalism. Because of the obvious cultural sensitivity of the subject, Darwin proceeded at a very deliberate pace to build the argument for this evolutionary continuity. He put off until his final chapter, "General Summary and Conclusion," the incendiary conclusion to which all this was leading: "We thus learn that man is descended from a hairy quadruped."

Then, after discussing how sexual selection worked in the animal world, Darwin analyzed its impact on human evolution. From our furless bodies, to the enormous geographic, ethnic, and tribal diversity in human appearance, to our highly social character, to language and music, Darwin made a powerful case that sexual selection had played a critical role in the shaping of the human species:

> Courage, pugnacity, perseverance, strength and size of body, weapons of all kinds, musical organs, both vocal and instrumental, bright colors, stripes and marks, and ornamen-

tal appendages, have all been indirectly gained . . . through the influence of love and jealousy, through the appreciation of the beautiful . . . and through the exertion of a choice.

Although tackling two subjects as complex and controversial as the evolution of beauty and the origins of humankind in one volume was an intellectually daring feat, *Descent* is generally considered a difficult, or even flawed, work. By building his argument so slowly and incrementally, writing in such dry, discursive prose, and citing so many learned authorities in support of the ideas he was advancing, Darwin might have thought he could draw any reasonable reader to accept the inevitability of his radical conclusions. But his rhetorical tactics failed, and in the end *Descent* was criticized by both creationist opponents of the very concept of evolution and fellow scientists who accepted natural selection but were adamantly opposed to sexual selection. To this day, *Descent* has never had the same intellectual impact as *Origin*.

The most notable and revolutionary feature of Darwin's theory of mate choice is that it was explicitly *aesthetic*. He described the evolutionary origin of beauty in nature as a consequence of the fact that animals had evolved to be beautiful to *themselves*. What was so radical about this idea was that it positioned organisms—especially female organisms—as active agents in the evolution of their own species. Unlike natural selection, which emerges from external forces in nature, such as competition, predation, climate, and geography, acting on the organism, sexual selection is a potentially independent, self-directed process in which the organisms themselves (mostly female) were in charge. Darwin described females as having a "taste for the beautiful" and an "aesthetic faculty." He described males as trying to "charm" their mates:

> With the great majority of animals . . . the taste for the beautiful is confined to the attractions of the opposite sex.*
> The sweet strains poured forth by many male birds during

the season of love are certainly admired by the females, of which fact evidence will hereafter be given. If female birds had been incapable of appreciating the beautiful colours, the ornaments, and voices of their male partners, all the labour and anxiety by the latter in displaying their charms before the females would have been thrown away; and this is impossible to admit . . .

On the whole, birds appear to be the most aesthetic of all animals, excepting of course man, and they have nearly the same taste for the beautiful as we have . . . [Birds] charm the female by vocal and instrumental music of the most varied kinds.

From the scientific and cultural perspectives of today, Darwin's choice of aesthetic language may seem quaint, anthropomorphic, and possibly even embarrassingly silly. And that may help to account for why Darwin's aesthetic view of mate choice is treated today like the crazy aunt in the evolutionary attic; she is not to be spoken of. Clearly, Darwin did not have our contemporary fear of anthropomorphism. Indeed, because he was vitally engaged in breaking down the previously unquestioned barrier between humans and other forms of life, his use of aesthetic language was not just a curious mannerism or a quaint Victorian affectation. It was an integral feature of his scientific argument about the nature of evolutionary process. Darwin was making explicit claims about the sensory and cognitive abilities of animals and the evolutionary consequences of those abilities. Having put humans and all other organisms on different branches of the same great Tree of Life, Darwin used ordinary language to make an extraordinary scientific claim: that the subjective sensory experiences of humans can be compared scientifically to those of the animals.

The first implication of Darwin's language was that animals are choosing among their prospective mates on the basis of judgments about their aesthetic appeal. To many Victorian readers, even those sympathetic to evolution, this was patently absurd. It seemed impossible that animals could make fine aesthetic judgments. Even if they were able to *observe* differences in the color of their suitors' plumage or the musical notes of their songs, the notion that they could cognitively distinguish among them, and then demonstrate a specific preference for one or another variation, was considered ludicrous.

These Victorian-era objections have been definitively rejected. Darwin's hypothesis that animals are able to make sensory evaluations and exercise mate preferences is now supported by volumes of evidence and is universally accepted. There have been numerous experiments across the animal kingdom—from birds to fishes, grasshoppers to moths—showing that animals have the capacity to make sensory evaluations that influence their mate choices.

Although Darwin's proposal of animal cognitive choice is now the accepted wisdom, the second implication of his aesthetic theory of sexual selection remains as revolutionary today, and as controversial, as when he first proposed it. By using the words "beauty," "taste," "charm," "appreciate," "admire," and "love," Darwin was suggesting that mating preferences could evolve for displays that had no utilitarian value at all to the chooser, only aesthetic value. In short, Darwin hypothesized that beauty evolves primarily because it is *pleasurable* to the observer.

Darwin's views on this issue developed over time. In an early discussion of sexual selection in *Origin,* Darwin wrote, "Amongst many animals, sexual selection will give its aid to ordinary [natural] selection by assuring to the most vigorous and best adaptive of all males the greatest number of offspring."

In other words, in *Origin,* Darwin saw sexual selection as simply the handmaiden of natural selection, another means of guaranteeing the perpetuation of the most vigorous and best-adapted mates. This view still prevails today. By the time he wrote *Descent,* however, Darwin had embraced a much broader concept of sexual selection that may have nothing to do with a potential mate's being more vigorous or better adapted per se, but only with being aesthetically appealing, as he stated clearly for the mesmerizing example of the Argus Pheasant: "The case of the male Argus pheasant is eminently interesting, because it affords good evidence that the most refined beauty may serve as a sexual charm, *and for no other purpose* [emphasis added]."

Moreover, in *Descent,* Darwin viewed sexual selection and natural selection as two distinct and frequently independent evolutionary mechanisms. Thus, the concept of two distinct but potentially interacting and even conflicting sources of selection is a fundamental and vital component of an authentically Darwinian vision of evolutionary biology. As we will see, however, this view has been rejected by most modern evolutionary biologists in favor of Darwin's earlier view of sexual selection as just another variant on natural selection.

Another distinctive feature of Darwin's theory of mate choice was that it was *coevolutionary.* Darwin hypothesized that the specific display traits and the "standards of beauty" used to select a mate evolved together, mutually influencing and reinforcing each other—as demonstrated again by the Argus Pheasant:

> The male Argus Pheasant acquired his beauty gradually through the preference of the females during many generations for the more highly ornamented males; the aesthetic capacity of females advanced through exercise or habit just as our own taste is gradually improved.

Here, Darwin envisions an evolutionary process in which each species coevolves its own, unique, cognitive "standards of beauty" in concert with the elaboration of the display traits that meet those standards. According to this hypothesis, behind every biological ornament is an equally elaborate, coevolved cognitive preference that has driven, shaped, and been shaped by that ornament's evolution. By modern scientific criteria, Darwin's description of the coevolutionary process in the Argus Pheasant is rather hazy, but it is no less substantive than his explanations of the mechanism of natural selection, which are viewed today as being brilliantly prescient, despite his ignorance of genetics.

Within Darwin's argument for mate choice in *Descent* was another revolutionary idea: that animals are not merely subject to the extrinsic forces of ecological competition, predation, climate, geography, and so on that create natural selection. Rather, animals can play a distinct and vital role in their *own* evolution through their sexual and social choices. Whenever the opportunity evolves to enact sexual preferences through mate choice, a new and distinctively *aesthetic* evolutionary phenomenon occurs. Whether it occurs within a shrimp or a swan, a moth or a human, individual organisms wield the potential to evolve arbitrary and useless beauty completely independent of (and sometimes in opposition to) the forces of natural selection.

In some species—like penguins and puffins—there is mutual mate choice, and both sexes exhibit the same displays and coevolved mating preferences. In polyandrous species, like the phalaropes (*Phalaropus*) and lily-trodding jacanas (Jacanidae), successful females may take multiple mates. These females are larger and brighter than the males, and they're the ones who perform courtship displays and sing songs to attract mates, while the males are the ones who exhibit mate choice, build the nests, and care for the young. But Darwin observed that in many of the most highly ornamented species the evolutionary force of sexual selection acted predominantly through female mate choice, which is why this book focuses largely on female mate choice. If *female* aesthetic preferences drove the process, then *female* sexual desire was responsible for creating, defining, and shaping the most extreme forms of sexual display that we see in nature. Ultimately, it is female sexual autonomy that is predominantly responsible for the evolution of natural beauty. This was a very unsettling concept in Darwin's time—as it is to many today.

Because the concept sexual autonomy has not been well explored in evolutionary biology, it is worthwhile to define it and understand its far-reaching implications. Whether in ethics, political philosophy, sociology, or biology, autonomy is the capacity of an individual agent to make an informed, independent, and uncoerced decision. So, sexual autonomy is the capacity for an individual organism to exercise an informed, independent, and uncoerced *sexual* choice about whom to mate with. The individual elements of the Darwinian concept of sexual autonomy—that is, sensory perception, cognitive capacities for sensory evaluation and mate choice, the potential for independence from sexual coercion, and so on—are all common concepts in evolutionary biology today. Yet few evolutionary biologists since Darwin have aligned these dots as clearly as he did.

In *Descent,* Darwin presented his hypothesis that female sexual autonomy—the taste for the beautiful—is an independent and transformative evolutionary force in the history of life. He also hypothesized that it can sometimes be matched, counterbalanced, or even overwhelmed by an independent force of male sexual control: the law of battle, the combat among members of one sex for control over mating with the other sex. In some species, one evolutionary mechanism or the other may dominate the outcome of sexual selection, but in other species—ducks, for example, as we shall see—female choice and male competition and coercion will both be operative and can give rise to an escalating process of sexual conflict. Darwin did not have the intellectual framework to fully describe the dynamics of sexual conflict, but he clearly understood that it existed—in humans and in other animals.

In short, *Descent* was as mechanistically innovative and analytically thoughtful as *Origin,* but to most of Darwin's contemporaries it was a bridge too far.

Upon publication in 1871, Darwin's theory of sexual selection was swiftly and brutally attacked. Or more precisely, part of it was. Darwin's concept of male-male competition—the law of battle—was immediately and almost universally accepted. Clearly, the notion of male-male competition for dominance over female sexuality was not a hard sell in the patriarchal Victorian culture of Darwin's time. For example, in an initially anonymous review of *Descent* that appeared soon after the book was published, the biologist St. George Mivart wrote,

Under the head of sexual selection, Darwin put two very distinct processes. One of these consists in the action of superior strength or activity, by which one male succeeds in obtaining possession of mates and in keeping away rivals. This is, undoubtedly, a *vera causa;* but may be more conveniently reckoned as one kind of "natural selection" than as a branch of "sexual selection."

In these few words, Mivart established an intellectual gambit that is still operative today. He took the element of Darwin's sexual selection theory that he agreed with—male-male competition—and declared it to be just another form of natural selection, rather than an independent force, in direct opposition to Darwin's own view. But at least he acknowledged that it existed. Not so the other aspect of Darwin's sexual selection theory.

When he came to a consideration of female mate choice, Mivart launched an all-out attack: "The second process consists in the alleged preference or choice, exercised freely by the female in favour of particular males on account of some attractiveness or beauty of form, colour, odor, or voice which the males may possess."

By referring to "choice, exercised freely," Mivart documents that Darwin's theory implied female sexual autonomy to his Victorian readers. However, the notion of an animal's exercising any kind of choice was a complete impossibility to Mivart:

Even in Mr. Darwin's specially-selected instances, there is not a tittle of evidence tending, however slightly, to show that any brute possesses the representative reflective faculties . . . It cannot be denied that, looking broadly over the whole animal kingdom, there is no evidence of advance in mental power on the part of brutes.

Mivart asserts that animals lack the requisite sensory powers, cognitive capacity, and free will necessary to make sexual choices based on display traits. Therefore, they could not possibly be active players, or selective agents, in their own evolution. Moreover, in discussing the role of the peahen in the evolution of the peacock's tail, Mivart found the idea of choice being exercised by female "brutes" particularly preposterous: "such is the instability of *vicious feminine caprice*, [emphasis added] that no constancy of coloration could be produced by its selective actions."

To Mivart, female sexual whims were so malleable—that is, fickle females preferring one thing one minute, and another the next—that they could never lead to the evolution of something as marvelously complex as the peacock's tail.

We need to take a closer look at Mivart's language, because the meanings of some of his words have changed in common English usage over the past 140 years. Today, the word "vicious" means intentionally violent or ferocious, but its original meaning was immoral, depraved, or wicked—literally, characterized by vice. Likewise, today "caprice" refers to a delightful, lighthearted whim, but in Victorian times it had the less appealing meaning of an arbitrary "turn of mind made without apparent or adequate motive." Thus, to Mivart, the concept of female mate choice and autonomy had overtones not just of fickleness but of unjustifiable immorality and sin.

Mivart did concede that display might play a role in sexual arousal: "The display of the male may be useful in supplying the necessary degree of stimulation to her nervous system, and to that of the male. Pleasurable sensations, perhaps very keen in intensity, may thence result to both."

Mivart's evocation of "stimulation" that creates "pleasurable sensations" reads like advice for a fulfilling sex life from a Victorian marriage manual. In this view, females merely require sufficient stimulation in order to elicit an appropriate sexual response and coordinate their sexual behavior with that of the male.

But if the purpose of sexual display is simply to supply "the necessary degree of stimulation," then females do not have their own individual, autonomous sexual desires. Rather, females should inevitably, and in due time, respond to the workmanlike stimulatory efforts of their suitors. This autonomy-denying conception of female sexual desire would reverberate throughout the next century, reaching its apogee in Freud's theory of human sexual response (see chapter 9). According to this physiological interpretation of female sexual pleasure, men need never entertain the possibility that "maybe she's just not that into you." Absence of female sexual response always means that there is something wrong with *her* physiology—in short, that she's frigid. As we will see, it is probably not an accident that the rediscovery of the biological theory of evolution by mate choice, the broad acknowledgment in Western culture of female autonomy, and the collapse of the Freudian conception of female sexuality all occurred during a short period of time that coincided with the advent of the women's liberation movement in the 1970s.

Mivart's review of *Descent* also established another enduring intellectual trend. He was the very first person to portray Darwin as a traitor to his own great legacy—a traitor to *true* Darwinism: "The assignment of the law of 'natural selection' to a subordinate position is virtually an abandonment of the Darwinian theory; for the one distinguishing feature of that theory was *the all-sufficiency of 'natural selection'* [emphasis added]."

Mere weeks after publication of *Descent,* Mivart mounted an attack against it that is still in use—citing *Origin* to argue against *Descent.* To Mivart, Darwin's signature achievement had been the creation of a single, "all-sufficient" theory of biological evolution. By diluting the theory of natural selection with a mechanism that rested largely on the power of aesthetic subjective experiences—*vicious feminine caprice*—Darwin had gone beyond the pale of what was acceptable. Many evolutionary biologists would still agree.

Mivart's attacks on sexual selection set many others in motion. But the most consistent, relentless, and effective critique of sexual selection came from Alfred Russel Wallace. Wallace was famous as the co-discoverer of the theory of natural selection. In 1859, he sent Darwin a manuscript from the jungles of Indonesia in which he set down a theory quite similar to Darwin's, and he asked for his advice and assistance with the manuscript. Fearful of being preempted by the younger man after decades of private work on his theory of natural selection, Darwin quickly published Wallace's article along with a short article summarizing his own theory. Then he rushed the full manuscript of *On the Origin of Species* into publication. By the time Wallace returned to England, Darwin and his theory were world famous.

There is no evidence that Wallace ever held this against Darwin, nor could he. Darwin had been working away on the idea of natural selection for more than twenty years, while Wallace was just beginning to think it through. But Darwin and Wallace never agreed on the subject of mate choice, and Wallace soon mounted a relentless attack on it. The two men debated their opposing views in a series of publications and in private letters that continued until Darwin's death in 1882, with neither man ever changing his mind. In what turned out to be his last scientific publication, Darwin wrote, "I may perhaps be here permitted to say that, after having carefully weighed to the best of my ability the various arguments which have been advanced against the principle of sexual selection, I remain firmly convinced of its truth."

In contrast to Darwin's always polite and understated expression of his views, Wallace's attack on evolution by mate choice grew ever more strident after Darwin's death and continued until his own death in 1913. Ultimately, Wallace was so successful that the subject of sexual selection was almost completely marginalized and forgotten within evolutionary biology until the 1970s.

Wallace expended an enormous amount of energy arguing that the "ornamental" differences between the sexes that Darwin described were not ornaments at all and that Darwin's theory of mate choice was unnecessary to explain animal diversity. Like Mivart, Wallace was skeptical about the possibility that animals had sensory and cognitive capacities to make mate choices. Wallace believed that humans had been specially created by God and divinely endowed with cognitive capacities that animals lacked. Thus, Darwin's concept of mate choice violated Wallace's spiritual theory of human exceptionalism.

However, faced with overwhelming evidence in the form of

elaborate ornaments and displays, especially among birds, Wallace was never able to reject evolution by mate choice entirely. But when forced to admit the possibility, he insisted that sexual ornaments could only have evolved because they had an adaptive, utilitarian value. Thus, in his 1878 book, *Tropical Nature, and Other Essays,* under the heading "Natural Selection as Neutralizing Sexual Selection," Wallace wrote, "The only way in which we can account for the observed facts is by supposing that colour and ornament are strictly correlated with health, vigor, and general fitness to survive."

Here, Wallace articulates the idea that sexual displays constitute "honest" indicators of quality and condition—an entirely orthodox view in sexual selection today. But how can it be that Wallace, the man justly credited with having destroyed sexual selection theory for over a century, actually wrote a statement that would be entirely at home in any modern biology textbook, or practically any contemporary paper on mate choice? The answer is that today's mainstream views of mate choice are as stridently anti-Darwinian as Wallace's critiques.

Wallace was the first to propose the now exceedingly popular BioMatch.com hypothesis, which holds that all beauty provides a rich profile of practical information about the adaptive qualities of potential mates. This view of evolution has become so pervasive that it even found its way into the 2013 Princeton University graduation speech by the Federal Reserve chairman, Ben Bernanke, who admonished the graduates to "remember that physical beauty is evolution's way of assuring us that the other person doesn't have too many intestinal parasites."

Today, most researchers agree with Wallace that all of sexual selection is simply a form of natural selection. But Wallace went further than they do and rejected the term "sexual selection" entirely. In that same passage, he continued,

> If there is (as I maintain) such a correlation [between ornament and health, vigor, and fitness to survive], then the sexual selection of color or ornament, for which there is little or no evidence, becomes needless, because natural selection,

which is the admitted *vera causa,* will itself produce all the results . . . Sexual selection becomes as unnecessary as it would certainly be ineffective.

Of course, it was the arbitrary and aesthetic components of Darwin's theory of sexual selection that Wallace rejected as "needless," "unnecessary," and "ineffective." Today, most evolutionary biologists would still agree.

Like Mivart, Wallace, who saw Darwin's aesthetic heresy as a threat to their shared intellectual legacy, took steps to fix what he perceived as Darwin's error. In the introduction to his 1889 book *Darwinism,* Wallace wrote,

> In rejecting that phase of sexual selection depending on female choice, I insist on the greater efficacy of natural selection. This is pre-eminently the Darwinian doctrine, and I therefore claim for my book the position of being the advocate of pure Darwinism.

Here, Wallace claims to be more Darwinian than Darwin! After wrangling unsuccessfully over mate choice with the living Darwin, within just a few years of Darwin's death Wallace has begun to reshape Darwinism in his *own* image.

In these passages, we witness the birth of adaptationism—the belief that adaptation by natural selection is a universally strong force that will *always* be predominant in the evolutionary process. Or, as Wallace put it in a strikingly absolutist statement, "Natural selection acts perpetually and on an enormous scale"—so enormous that it would "neutralise" any other evolutionary mechanisms.

Wallace set in motion the transformation of Darwin's fertile, creative, and diverse intellectual legacy into the monolithic and intellectually impoverished theory with which he is almost universally associated today. Notably, Wallace also invented the characteristic style of adaptationist argument—mere stubborn insistence.

This is kind of a big deal. The Darwin we have inherited, through the filter of Wallace's outsized influence on evolutionary biology in the twentieth century, has been laundered, retailored, and cleaned up for ideological purity. The true breadth and creativity of Darwin's ideas, especially his aesthetic view of evolution, have been written out of history. Alfred Russel Wallace might have lost the battle for credit over the discovery of natural selection, but he won the war over what evolutionary biology and Darwinism would become in the twentieth century. More than one hundred years later, I am still pissed about it.

In the century following the publication of Darwin's *Descent of Man*, the theory of sexual selection was almost entirely eclipsed. Despite a few scattered attempts to revive the topic, Wallace's hatchet job on mate choice was so successful that generation after generation would turn exclusively to natural selection to account for sexual ornament and display behavior.

During the century-long dark age of mate choice theory, however, one man did make a fundamental contribution to the field. In a 1915 paper and a 1930 book, Ronald A. Fisher proposed a genetic mechanism for the evolution of mate choice that built on and extended Darwin's aesthetic view. Unfortunately, however, Fisher's ideas on sexual selection would be mostly ignored for the next fifty years.

Fisher was a gifted mathematician who had a huge effect on the sciences through his fundamental work developing both the basic tools and the intellectual structure underlying modern statistics. However, he was first and foremost a biologist, and his statistical research grew directly from his desire for a more rigorous understanding of the workings of genetics and evolution in nature, agriculture, and human populations. His interest in genetics and evolution was motivated in part by his ardent support for eugenics—the now disgraced theory and social movement that advocated the use of social, political, and legal regulation of reproduction in order to genetically improve the human species and maintain "racial purity." Appalling as his beliefs were, Fisher's investigations led him to some brilliant scientific conclusions—conclusions that, in the end, conflicted with his eugenic beliefs.

Fisher permanently reframed the sexual selection debate with a critical observation: Explaining the evolution of sexual ornaments is easy; all other things being equal, display traits should evolve to match the prevailing mating preferences. The more critical scientific question is, why and how do mating preferences evolve? This insight remains fundamental to all contemporary discussions of evolution by sexual selection.

Fisher actually proposed a two-stage evolutionary model: one phase for the initial origin of mating preferences, and a second, subsequent phase for the coevolutionary elaboration of trait and preference. The first phase, which is solidly Wallacean, holds that preferences initially evolve for traits that are honest and accurate indices of health, vigor, and survival ability. Natural selection would ensure that mate choice based on these traits would lead to objectively better mates and to genetically based mating preferences for these better mates. But then, after the origin of mating preference, Fisher hypothesized in his second-phase model, the very existence of mate choice would *unhinge* the display trait from its original honest, quality information by creating a new, unpredictable, aesthetically driven evolutionary force: sexual attraction to the trait itself. When the honest indicator trait becomes disconnected from its correlation with quality, that doesn't make the trait any less attractive to a potential mate; it will continue to evolve and to be elaborated merely because it is preferred.

In the end, according to the Fisher phase-two model, the force that drives the subsequent evolution of mate choice is mate *choice* itself. In an exact reversal of the Wallacean view of natural selection as neutralizing sexual selection, arbitrary aesthetic choices (per Darwin) trump choices made for adaptive advantage (per Wallace), because the trait that was originally preferred for some adaptive reason has become a source of attraction in its own right. Once the trait is attractive, its attractiveness and popularity become ends in themselves. According to Fisher, mating preference is like a Trojan horse. Even if mate choice originally acts to enhance traits that carry adaptive information, desire for the preferred trait will eventually undermine the ability of natural selection to dictate the evolutionary outcome. Desire for beauty will endure and undermine the desire for truth.

How does this happen? Fisher hypothesized that a positive feedback loop between the sexual ornament and the mating preference for that ornament will evolve through genetic covariation (that is, correlated genetic variation) between the two. To understand how this could work, imagine a population of birds with genetic variation for a display trait—say tail length—and for mating preferences for different tail lengths. Females who prefer males with long tails will find mates with those longer tails. Likewise, females who prefer males with shorter tails will find mates with shorter tails. The action of mate choice means that variation in genes for traits and preferences will no longer be found randomly in the population. Rather, most individuals will soon carry genes for correlated traits and preferences—that is, genes for long tails *and* preferences for long tails, or genes for short tails *and* preferences for short tails. Likewise, there will be fewer and fewer individuals who carry genes for short tails and preferences for long tails, or vice versa. The very action of mate choice will distill and concentrate genetic variation for trait and preference into correlated combinations. To Fisher, this observation was merely a mathematical fact. This outcome is what mating preference means.

As a consequence of genetic covariation, genes for a given trait and the preference for that trait will coevolve with each other. When females exercise their mate choices based on particular displays—for example, a long tail—they will *also* be selecting *indirectly* on particular mate choice genes, because they will be choosing mates whose *mothers* likely also had genes for preferring long tails.

The result is a strong, positive feedback loop in which mate choice becomes the selective agent in the evolution of mate preference itself. Fisher called this self-reinforcing sexual selection mechanism "a runaway process." Selection on specific display traits creates evolutionary change in mating preferences, and evo-

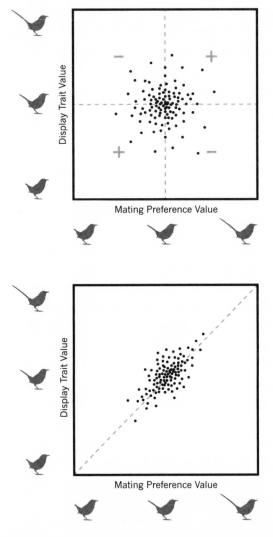

Evolution of genetic covariance between a display trait—for example, tail length—and a mating preference for it. (Top) A population begins with individuals (black dots) that have a random distribution of genetic variation for the display trait (vertical axis) and mating preference (horizontal axis). As a result of preference, many matings will occur among individuals in the upper right and lower left quandrants who have and prefer the same variations in tail length (+ signs). Few matings will occur in the other parts of the distribution where preferences and traits do not match (- signs). (Bottom) The result is the evolution of covariation between genes for the display and the preference (dotted line).

lutionary change in mating preferences will create further evolutionary change in display traits, and so on. The form of beauty, and the desire for it, shape each other through a coevolutionary process. In this way, Fisher provided an explicit genetic mechanism for how the display trait and the mating preference can "advance together," as Darwin first envisioned for the Argus Pheasant (see quotation on pages 25–26).

Fisher's coevolutionary mechanism also explains the potential evolutionary benefit of mating preference. If the female chooses a mate with a sexually attractive trait—again let's say a long tail—her male offspring will be more likely to inherit this sexually attractive trait. If other females in the population also prefer long tails, then the female will end up with a greater number of descendants, because her male offspring will be sexually attractive to them. This evolutionary advantage is the indirect, genetic benefit of mate choice alone. We call it indirect because it does not accrue *directly* to the chooser's own survival or fecundity (that is, her capacity to have and raise offspring), or even to the survival of her offspring. Rather, the benefit accrues through the reproductive success of her sexually attractive sons, which will result in a wider propagation of her genes (that is, more grandkids).

Fisher's runaway process works something like the Dutch tulip bulb craze of the 1630s, the speculative financial market bubble of the 1920s, or, to take something much more recent, the overvalued housing markets that led to the near collapse of the entire world banking system in 2008. All of these are examples of what happens when the value of something becomes unhinged from its "actual" worth and continues not only to be valued but to increase in value. What drives speculative market bubbles is desire itself. That is, something is desirable because it is desired, popular because it's popular. Thus, Fisherian mate choice is the genetic version of the "irrational exuberance" of a market bubble. (We will return to this economic analogy in chapter 2.)

Fisher asserted that mating preferences do not continue to evolve because the particular male that the female chooses is any better than any other male. In fact, sexually successful males could sometimes evolve to be *worse* at survival or *poorer* in health or condition. If a display trait becomes disconnected from any other, extrinsic measure of mate quality—that is, overall genetic quality, disease resistance, diet quality, or ability to make parental investments—then we say that that display trait is arbitrary. Arbitrary does not mean accidental, random, or unexplainable; it means only that the display trait communicates no other information than its presence. It simply exists to be observed and evaluated. Arbitrary traits are neither honest nor dishonest, because they do not encode any information that can be lied about. They are merely attractive, or *merely beautiful*.

This evolutionary mechanism is rather like high fashion. The difference between successful and unsuccessful clothes is determined not by variation in function or objective quality (really) but by evanescent ideas about what is subjectively appealing—the style of the season. Fisher's model of mate choice results in the evolution of traits that lack any functional advantages and may even be disadvantageous to the displayer—like stylish shoes that hurt one's feet, or garments so skimpy that they fail to protect the body from the elements. In a Fisherian world, animals are slaves to evolutionary fashion, evolving extravagant and arbitrary displays and tastes that are all "meaningless"; they do not involve anything other than perceivable qualities.

Fisher never presented an explicit mathematical model of his runaway process (something that later biologists did, as we shall soon see). Some have conjectured that he was such a skilled mathematician that he thought the results were obvious and needed no further explication. If so, then Fisher was sorely mistaken, because there were plenty of discoveries still to be made. Actually, I think Fisher probably knew there was more work to do. So why didn't he do it? I think Fisher did not pursue his runaway model any further because he realized that the implications of this evolutionary mechanism were completely antithetical to his personal support for the eugenics movement. Fisher's runaway model implied that adaptive mate choice—the kind of choice required to eugenically "improve" the species—was evolutionarily unstable and would almost inevitably be undermined by arbitrary mate choice, the irrational desire that beauty inspires. And he was right!

Around the centennial of Darwin's *Descent of Man,* the concept of sexual selection began to return to the evolutionary mainstream. Why did it take so long? Although it would require an extensive historical and sociological study to investigate my hunch, I don't think it was a coincidence that evolutionary biologists finally began to reconsider mate choice, particularly female mate choice, as a genuine evolutionary phenomenon at precisely that moment when women in the United States and Europe began to organize politically and to protest for equal rights, sexual freedom, and access to birth control. It would be nice to think that the insights from evolutionary biologists had an influence on these positive cultural developments, but unfortunately history shows that the opposite was true.

With the return of the scientific interest in mate choice, there came a renewed battle between the aesthetic Darwinian/Fisherian view and a rejuvenated version of neo-Wallacean adaptationism. In 1981 and 1982, more than fifty years after Fisher published his model of sexual selection, the mathematical biologists Russell Lande and Mark Kirkpatrick independently confirmed and expanded upon it. Inspired by Fisher's theory, Lande and Kirkpatrick applied different mathematical tools to explore the coevolutionary dynamics between mate choice and display traits and got very similar answers. They showed that traits and preferences can coevolve merely because of the advantage of sexually attractive offspring alone. Further, they demonstrated that the process of mate choice can create a covariance between the genes for a given display and the genes for the preference for that display.

The Lande-Kirkpatrick sexual selection models also confirmed mathematically that display traits evolve through a balance between natural selection and sexual selection. For example, a male may have the optimal tail length for survival (that is, favored by natural selection), but if he is not sexy enough to attract even a single mate (that is, disfavored by sexual selection), he will fail to pass on his genes to the next generation. Likewise, a male may have the perfect tail size for attracting mates (that is, favored by

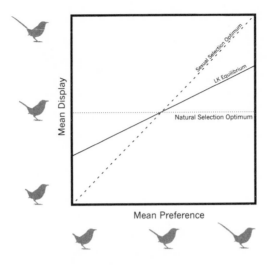

Lande-Kirkpatrick model for the evolution of a display trait—such as tail length—and a mating preference for it. The mean display trait in a population (vertical axis) will evolve toward an equilibrium (solid line) between the trait value favored by natural selection (horizontal line) and the trait value favored by sexual selection (broken line).

sexual selection), but if he is so sexually extravagant that he cannot survive long enough to attract a single mate (that is, disfavored by natural selection), he will also fail to pass on his genes. Lande and Kirkpatrick confirmed the intuition of Darwin and Fisher that natural and sexual selection on display traits will establish a balance between the two opposing forces. At this equilibrium, the male may still be quite far from the natural selection optimum, but that's the cost of doing business with sexually autonomous, choosy females.

However, Lande and Kirkpatrick went well beyond Fisher and Darwin in defining this equilibrium. Using different mathematical frameworks, they each discovered that this balance between natural and sexual selection is not restricted to a single point. Rather, there exists a line of equilibria—literally, an infinite number of possible stable points of balance between natural and sexual selection on a given display trait. Essentially, for *any* perceivable display trait, there is some conceivable combination of sexual selection and natural selection acting on that trait that could result in a stable equilibrium. That is the true meaning of an "arbitrary" trait; practically any perceivable feature could function as a sexual ornament. Of course, the further away a display trait is from the natural selection optimum, the stronger the sexual advantage must be for it to evolve.

How do sexual and natural selection on display traits reach a balance? In other words, how will populations evolve toward equilibrium? Here, too, Lande and Kirkpatrick provided a rich mathematical machinery to flesh out Fisher's verbal, nonmathematical model. In order to evolve to a stable equilibrium, *both* the mating display trait *and* the mating preference must *coevolve*. In other words, in order for females to get what they want (that is, evolve to an equilibrium), they must select on and change the male display trait. But because traits and preferences are genetically correlated, coevolution means that the females must also change what *they* want. By (a rather strained) analogy, this evolutionary process is a little bit like a marriage: spouses frequently attempt to change each other, and they frequently succeed. But the process of reaching a stable resolution usually requires a transformation *both* of one spouse's behavior *and* of the other spouse's opinion of that behavior.

In theory, aesthetic coevolution may sometimes occur so rapidly that display traits cannot evolve fast enough to satisfy the increasingly radical preferences of a population. Lande showed that if the genetic correlation between preference and traits is strong enough, it is theoretically possible for populations to evolve *away* from the line of equilibrium; that is, the line of equilibrium may become unstable. This process is considered the ultimate realization of Fisher's "runaway" process, in which mate choice ends up changing *itself* so rapidly that its ever-evolving preferences can *never* be met and desire can never be fully satisfied.

Last, Lande's and Kirkpatrick's mathematical models also explain how mate choice could drive the evolution of new species. When populations of a given species become isolated from one another (for example, as a new mountain range rises, or deserts form, or rivers are rerouted), these populations will be subject to different random influences. Each subpopulation will ultimately diverge in its own unique aesthetic direction to a distinct point on the equilibrium line, toward its own differentiated standard of beauty: longer tails or shorter tails; higher-pitched songs or lower-pitched songs; red bellies or yellow bellies; blue heads, bare heads, or even bare, blue heads. The possibilities are endless. If the isolated populations diverge far enough from each other, the process of aesthetic sexual selection may result in an entirely new species—a process called speciation. According to this theory, aesthetic evolution is like a spinning top. The action of mate choice creates an internal equilibrium that determines what is sexually beautiful within a population. But random perturbations of the top—either internal forces like mutation or external factors like population isolation by a geographic barrier—can cause the top to spin away toward a new equilibrium.

The overall result is that mate choice fosters the evolution of ever-escalating and ever-diversifying standards of beauty among populations and species. Practically anything is possible—an idea for which there is ample evidence in some of the birds that populate these pages. I call them aesthetic extremists for good reason.

Russ Lande and Mark Kirkpatrick were directly inspired by the nearly forgotten aesthetic mate choice mechanisms of Darwin and Fisher. However, the modern, adaptationist, neo-Wallacean mechanism of mate choice had to be reinvented from scratch because no one remembered Wallace's own honest advertisement theory. Yet the modern versions are strikingly similar to Wallace's in logic; that is, they share his fundamental insistence on the greater efficacy of natural selection. Natural selection *must* be true, and all sufficient, because it is such a powerful and rationally attractive idea.

In the 1970s and 1980s, the chief proponent of the neo-Wallacean view of adaptive mate choice was Amotz Zahavi, a charismatic and energetic Israeli ornithologist with a fierce independent streak. In 1975, Zahavi published his "handicap principle." A scientific megahit, this paper was a huge stimulus to the study of mate choice and has now been cited over twenty-five hundred times. Zahavi thought his ideas were entirely new.

According to him, "Wallace . . . dismissed altogether the theory of sexual selection by mate preference." However, the beautifully intuitive core idea of Zahavi's handicap principle is precisely neo-Wallacean: "I suggest that sexual selection is effective because it improves the ability of the selecting sex to detect quality in the selected sex."

Although Zahavi precisely restated Wallace's adaptive mate choice hypothesis, he abandoned Wallace's rhetoric by using the newly rehabilitated term "sexual selection," instead of "natural selection," to describe it. But Zahavi also added his own distinctive twist to Wallace's logic. To Zahavi, the entire point of any sexual display is that it is a costly *burden* to the signaler—literally, a handicap. By its very existence, the ornamental handicap demonstrates the superior quality of the signaler because the signaler has been able to survive it. He wrote, "Sexual selection is effective only by selecting for a character that lowers the survival of the organism . . . It is possible to consider a handicap as a kind of test."

The more elaborate the display trait, the greater the costs, the bigger the handicap, the more rigorous the test, and the better the mate. The individual who is attracted to a mate with such a costly trait is responding not to its subjective beauty, which is incidental to its costs, but to what it tells her about the male's ability to rise *above* its cost. This is the handicap principle.

In what way was the handicapped male better? To Zahavi, it was clear that he could be better in *any imaginable* way. However, those who followed Zahavi established that the adaptive benefits of honest signaling could be of two basic kinds—direct and indirect. The direct benefits of mate choice include any advantages to the health, survival, or fecundity of the choosers themselves. Such adaptive direct benefits could include choosing a mate who provides extra protection from predators, a better territory with more food or better nesting sites, no sexually transmitted diseases (STDs), a greater capacity to invest in the feeding and protection of offspring, or lower mate search costs. Alternatively, the adaptive indirect benefits are in the form of good genes that are inherited by the chooser's offspring and contribute to their survival and fecundity. Like the indirect Fisherian benefit of having sexy offspring, the good genes benefit doesn't help the chooser directly but results in a greater number of grandchildren. However, unlike the indirect Fisherian benefit, the chooser's offspring are not merely more attractive but actually *better* at surviving and reproducing, not merely at acquiring and fertilizing mates. Thus, good genes are *different* from the genes for the display trait itself, and theoretically they should provide heritable advantages to both male *and* female offspring.

Both direct benefits and good genes are adaptive benefits to mate choice; they can only occur when, as first proposed by Wallace, observable variation in the display trait among potential mates is correlated with some additional advantage that will contribute to the survival or fecundity of the choosers or their offspring. These correlations arise from an interaction between sexual selection on mating/fertilization success *and* natural selection on survival and fecundity. Zahavi's handicap principle was a new proposal about *how* the adaptive correlation between display and mate quality arises and how it can be maintained.

Zahavi promoted the handicap principle with a single-minded fervor. But his idea had one big flaw. If the sexual advantage of an ornament is directly proportional to its survival costs, then the two forces will cancel each other out, and neither the costly ornament nor a mating preference for it can evolve. In a 1986 paper boldly titled "The Handicap Mechanism of Sexual Selection Does Not Work," Mark Kirkpatrick provided a mathematical proof of this evolutionary trap.

To understand this problem, let's consider a corollary of Zahavi's handicap principle. I call it the "Smucker's principle." Smucker's jelly takes its name from its founder, Jerome Monroe Smucker, who opened a cider press in Orrville, Ohio, in 1897. Readers of a certain age may recall the company's catchy advertising slogan: "With a name like Smucker's, it has to be good!" The slogan claims that the Smucker's brand name is so unappealing, so off-putting, so *costly,* that the fact that the company has survived with this name *proves* that its jelly is of really high quality. The Smucker's slogan embodies the handicap principle.

But let's look a little more carefully at the implications of the Smucker's principle. What if Smucker's jelly were suddenly in competition with another jelly with an even worse, more costly name? Wouldn't an even *worse*, more off-putting name indicate a jelly of even *higher* quality? What limits the possibility of ever-worsening and more costly names indicative of ever-higher-quality jellies?

Luckily, this exact thought experiment has already been conducted in a parody of the Smucker's ad performed as a fake advertisement sketch on *Saturday Night Live* in the 1970s:

JANE CURTIN: And so, with a name like *Flucker's,* it's got to be good.

CHEVY CHASE: Hey, hold on a second, I have a jam here called *Nose Hair.* Now, with a name like *Nose Hair,* you can imagine how good it must be. MMM MMM!!

DAN AYKROYD: Hold it a minute, folks, but are you familiar with a jam called *Death Camp*? That's *Death Camp*! Just look for the barbed wire on the label. With a name like *Death Camp,* it must be so good it's incredible! Just amazingly good jam!

From there the names got worse and worse. John Belushi promoted a jelly called *Dog Vomit, Monkey Pus,* and then Chevy Chase returned with yet another new jelly named *Painful Rectal Itch.* The competition culminated with a jelly whose name was so repulsive it induced nausea and could not be spoken on the air. "So good, it's sick making!" Jane Curtin proclaimed, before signing off with "Ask for it by name!"

The "Smucker's principle" reveals the internal logical flaw of Zahavi's "handicap principle." As Kirkpatrick proved mathematically, if the sexual benefit of a signal is directly related to its costs, the signaler will never gain any advantage. Rather, handicaps will fail under their own costly burden. Fortunately, that means we can all rest easy that there will never be a jelly named Painful Rectal Itch.

The Smucker's principle further demonstrates that Zahavi's handicap principle is fundamentally incompatible with the aesthetic nature of sexual display. Sexual displays actually evolve because they are attractive, not disgustingly informative or repulsively honest. If the sole purpose of sexual display is to communicate the capacity to survive a great burden, then why are sexual traits ornamental? Why isn't acne sexually attractive? After all, acne is frequently an honest indicator of a surge of adolescent hormones and would therefore provide reliable information about youth and fertility. Why don't organisms evolve *genuine* handicaps like partially formed body parts? Why don't individual organisms gnaw off a limb to show how good they are at surviving without the missing appendage? Why not two limbs? That would really say something about how hardy they are! Or, why not poke out an eye? The reason, or course, is that the handicap principle is disconnected from the fundamentally aesthetic nature of mate choice and therefore nearly irrelevant to nature.

In 1990, Alan Grafen at Oxford came to the rescue of the failing handicap principle. The stakes were high. The entire neo-Wallacean mate choice paradigm was on the line. Of course, Grafen was forced to acknowledge Kirkpatrick's proof of the failure of the handicap principle as originally articulated by Zahavi. However, Grafen showed mathematically that a nonlinear relationship between display cost and mate quality could salvage the theory. In other words, if lower-quality males pay a proportionally higher cost to grow or display an attractive trait than do higher-quality males, then the handicap could evolve. If a handicap is like a test, then Grafen proposed that higher-quality individuals basically get an *easier* test. The only way to fix the handicap principle was to actually break it.

Having established a way to salvage handicaps, Grafen then asked how we should decide between two plausible evolutionary alternatives, the Zahavian handicap and the Fisherian runaway as elaborated by Lande and Kirkpatrick:

> According to the handicap principle, . . . there is a rhyme and reason in the incidence and form of sexual selection . . .

This is in contrast to the Fisher process, in which the form of the signal is more or less arbitrary and whether a species has undergone a bout of runaway selection is more or less a matter of chance.

In the Wallacean tradition, Grafen strongly endorsed the comforting "rhyme and reason" of adaptation over the unnerving arbitrariness of aesthetic Darwinism. Then Grafen went in for the kill: "To believe in the Fisher-Lande process as an explanation of sexual selection without abundant proof is methodologically wicked."

I do not know of any other contemporary scientific debate in which one side has actually been branded as *wicked*! Not even cold fusion! Clearly, this is not an everyday scientific debate. In a striking reprise of St. George Mivart's moralizing tone, Grafen's outsized response indicates the intellectual magnitude of what is at stake. Darwin's *really* dangerous idea—aesthetic evolution—is so threatening to adaptationism that it must be branded as wicked. Nearly one hundred years after Wallace advocated his pure form of Darwinism, Grafen deploys the same Wallacean insistence to try to win the debate again.

Grafen's reasoning struck a chord. Although personal comfort is not a scientifically justifiable criterion, many people, including scientists, *do* want to believe that the world is filled with "rhyme and reason." So, even though Grafen merely demonstrated that there were conditions under which the handicap principle *could* work, he so discredited the Fisherian theory that most evolutionary biologists concluded that the handicap principle not only could work but *would* work—all the time. If belief in the alternative hypothesis is "wicked," there's little choice to make. Adaptive mate choice has dominated the scientific discourse ever since.

In comparing the intellectual styles of Zahavi and Fisher, Grafen wrote that "Fisher's idea is too clever by half" but that "Zahavi's upward struggle from fact will triumph." This distinction between cleverness and fact also lent itself to a narrative in which the proponents of arbitrary Fisherian mate choice were cast as pointy-headed mathematicians with no appreciation of the natural world, while adaptationist advocates of the handicap principle were seen as salt-of-the-earth natural historians. Matt Ridley brought this distinction to vivid life in his 1993 book, *The Red Queen:*

> The split between Fisher and Good-genes began to emerge in the 1970s once the fact of female choice had been established to the satisfaction of most. Those of a theoretical or mathematical bent—the pale, eccentric types umbilically attached to their computers—became Fisherians. Field biologists and naturalists—bearded, besweatered, and booted— gradually found themselves to be Good-geners.

Ironically, I find that I have been written out of the historical narrative of my own discipline. I have spent cumulative years of my life in tropical forests on multiple continents studying avian courtship displays. I have been as "bearded, besweatered, and

The author—"bearded, besweatered, and booted"—in the field recording bird songs on a reel-to-reel tape recorder with a parabolic microphone at 2900 meters altitude near Laguna Puruhanta in the Ecuadorian Andes in 1987.

booted" as any field biologist. Yet I have also been an ardent and inquisitive "Fisherian" since the mid-1980s. According to the Grafen and Ridley narrative, I do not exist. Neither does Darwin, a naturalist who certainly put in his time in the field. Odder still, neither does Grafen, who is primarily a mathematician. Unfortunately, Ridley's scenario also eliminates from consideration all female field biologists and naturalists. (Sorry, Jane Goodall and Rosemary Grant!) Of course, the function of this kind of intellectual fable is to obscure the actual complexity of the issues, to use rhetoric to claim the higher ground by portraying adaptationists as romantic figures with deeper personal connections to nature and to knowledge.

The intellectual origins of aesthetic evolution are not in abstract mathematics but in Darwin's own, bold realization of the evolutionary consequences of the subjective aesthetic experiences of animals and the intellectual insufficiency of natural selection to explain the phenomenon of beauty in nature. Nearly 150 years later, the best path to appreciating how beauty has come into being is still to follow in Darwin's footsteps.

The Darwin versus Wallace, aesthetic versus adaptationist debate remains vital to science today. Whenever we study mate choice, we are using intellectual tools that were shaped by this debate, and we need to be aware of the history of our tools.

Among those tools is the language we use to define concepts in evolutionary biology. For example, let's examine the history of the word "fitness." To Darwin, fitness had the ordinary language meaning of physical fitness. Fitness meant fit to do a task. Darwinian fitness was the physical capacity to do the tasks necessary to ensure one's survival and capacity for reproduction. However, during the development of population genetics in the early twentieth century, fitness was redefined mathematically as the differential success of one's genes in subsequent generations. This broader and more general new definition combined all sources of differential genetic success—survival, fecundity, *and* mating/fertilization success—into a single variable under the common label of "adaptive natural selection." The redefinition of fitness was accomplished precisely during the period when sexual selection by mate choice had been entirely rejected as irrelevant to evolutionary biology. Thus, the effect of redefining fitness was to flatten and eliminate the original, subtle, Darwinian distinction between *natural selection* on traits that ensured survival and fecundity and *sexual selection* on traits that resulted in differential mating and fertilization success. Ever since, this mathematically convenient but intellectually muddled new concept of fitness has reshaped how people think evolution works and made it difficult to even articulate the possibility of a distinct, independent, nonadaptive sexual selection mechanism. *If it contributes to fitness, it must be adaptive, right?* The Darwinian/Fisherian concept of sexual selection by mate choice has been essentially written out of the language of biology. It has become linguistically impossible to be an authentic Darwinian.

The flattening of the intellectual complexity of aesthetic Darwinism was motivated, at least in part, by the belief that conceptual unification is a general scientific virtue, that the development of fewer more powerful, more broadly applicable, singular theories, laws, and frameworks is a fundamental goal of science itself. Sometimes unification in science works great, but it is doomed to fail when the distinctive, emergent properties of particular phenomena are reduced, eliminated, or ignored in the process. This loss of intellectual content is exactly what happens when something complex is *explained away* instead of being *explained* in its own right.

By claiming that evolution by mate choice was a special process with its own, distinctive internal logic, Darwin fought against the powerful scientific and intellectual bias toward simplicity and unification. Of course, many of Darwin's Victorian antagonists were recent converts from religious monotheism to materialist evolutionism. Their historic monotheism might have predisposed them to adopt a powerful new monoideism; they replaced a single omnipotent God with a single omnipotent idea—natural selection. Indeed, contemporary adaptationists should question why *they* feel it is necessary to explain all of nature with a single powerful theory or process. Is the desire for scientific unification simply the ghost of monotheism lurking within contemporary scientific explanation? This is another implication of Darwin's *really* dangerous idea.

If evolutionary biology is to adopt an authentically Darwinian view, it must recognize, as he did, that natural selection and sexual selection are *independent* evolutionary mechanisms. In this framework, adaptive mate choice is a process that occurs through the *interaction* of sexual selection and natural selection. I will use this language throughout this book.

To better understand the evolution of beauty and how to study it, we will now take a look at the sex lives of birds. There can be no better place to start than with Darwin's "eminently interesting" Great Argus pheasant.

CHAPTER 2

Beauty Happens

In the hilly rain forests of the Malay Peninsula, Sumatra, and Borneo lives one of the most aesthetically extreme animals on the planet—the Great Argus (*Argusianus grayi*), which Darwin described as affording "good evidence that the most refined beauty may serve as a sexual charm, and for no other purpose."

The female Great Argus is a large, robust pheasant with a complex, finely vermiculated, but dull camouflage pattern of chocolate-brown, reddish-brown, black, and tan swirls on her feathers. Her legs are bright red, and the feathers of her face are sparse, revealing bluish-gray skin beneath. At first view, the main thing that distinguishes the male Great Argus from the female is the great elongation of his tail and wing feathers. The feathers extend over a yard behind him. In total, the male Argus measures nearly six feet from the tip of his beak to the tip of his tail. But length aside, his plumage appears to be quite similar to that of the cryptic female, and he's not particularly impressive looking. His real charms remain hidden, not to be revealed until the peak of his courtship of the female, which very few people on earth have ever witnessed outside the confines of a zoo.

Seeing a Great Argus in the wild is very difficult. They are extraordinarily wary and disappear into the forest at the first sign of your approach. The early twentieth-century ornithologist and pheasant fanatic William Beebe was among the first scientists to see the display of Great Argus in the wild. Beebe was a curator at the New York Zoological Society who would later become world famous for exploring the depths of the oceans in a bathysphere—a primitive, deep-diving submarine. Beebe saw his first Great Argus—a male—descending a muddy bank in tropical Borneo to drink from a puddle of rainwater that had collected in a wild boar wallow. He describes this first sighting ecstatically in his 1922 *Monograph of the Pheasants,* expressing his feeling of triumph in the language of both a proud bird-watcher and the American, colonial-era adventurer that he was: "Brief as the glimpse had been, I felt a great superiority to my fellow white men the world over, who had not seen an Argus Pheasant in its native home."

As is typical of most avian aesthetic extremists, Great Argus are polygynous, which means that single males mate multiply with different females. However, the opportunity for multiple mating creates competition among males to attract mates. Some attractive males are highly successful, and others are not at all. The result is strong sexual selection for whatever display traits females prefer. After the female chooses a mate, the male's participation in the reproduction is complete, and he plays no further role in the life of his mate or their offspring. The female is entirely responsible for building a nest of leaves on the ground, incubating her clutch of two eggs, protecting her chicks, and feeding them and herself, which she does by foraging for fruits and insects on the forest floor. Both females and males are reluctant fliers. When threatened, they usually escape by running away on foot. However, at night they fly up to a low perch to roost—except when the female is incubating her eggs, when she remains on the nest.

The male Great Argus lives an entirely separate, bachelor life. To create a stage large and pristine enough to accommodate his extraordinary courtship display, he clears an area four to six yards wide right down to the bare dirt of the forest floor. Assiduously picking up all the leaves, roots, and sticks in the space he's chosen, often on a ridge or a hilltop within the forest, he carries them to

A male Great Argus maintaining his display courtyard.

the periphery of his court. Like a modern yardman (but without the ear protection), he also employs his huge wing feathers as a leaf blower by beating them rhythmically, sending all the remaining debris flying from his court until it is completely clear. He prunes any leafy vegetation or vines that grow into the court from above by snipping the branches with his beak. Once his court is ready for the business of mating, all he needs is a female visitor.

To attract an audience, the male Argus calls from his court in the early morning and evening and also on moonlit nights. The Great Argus call is a loud, haunting two-note yelp, *kwao-waao*, which is the source of the names for the species in several Southeast Asian languages—for example, *kuau* in Malay and *kuaow* in Sumatran. The call is loud and piercing enough to be heard from great distances. Because the bird is so elusive, that's usually all that a human visitor is likely to experience of the Great Argus in the wild.

A few years ago, I spent five days at a research station in the

Danum Valley Conservation Area in northern Borneo within the range of the Great Argus. Late one afternoon, we wandered along a heavily wooded trail near the river, and I heard the loud *kwao-waao* of the male Great Argus, exactly as Beebe described. The call was so loud that I thought the bird must be just around the next bend in the trail, and I froze with excitement. However, I soon realized that he was calling from a considerable distance away on the *other side* of the river. Even if the male had kept calling, it would have taken us more time to reach him than there was sunlight left in the day. And even if we had been lucky enough to track him down at his court, he would almost certainly have fallen silent as we approached, only to melt away into the surrounding forest undetected. With nothing but the tantalizing echo of his call to confirm his existence, I could only imagine what it must have been like for Beebe to see this amazing bird.

When we returned to the research station that evening, having birded the leech-infested forests since well before dawn, we met the French artist boyfriend of a researcher at the camp. He was there to "paint the forest," he told us. He then casually asked us to identify an unusual bird he had come across while taking a stroll in the late morning near camp. With complete nonchalance, he proceeded to describe a large fowl nearly two yards long that had walked across the dirt access road only three hundred yards from the main compound. After tromping through the forests for days without so much as a glimpse of the bird he had managed to see without even trying—or appreciating—I could barely conceal my envy at his great, unearned fortune. As I scratched my leech bites, I experienced the opposite of Beebe's feeling of "great superiority" and could only mutter private curses to the Gods of Birding.

If catching even a glimpse of the Great Argus in the wild is a great challenge, to see what the male Argus actually does with his enormous wing and tail feathers during his courting of the female requires elaborate preparations and can turn into quite a protracted ordeal. William Beebe tried watching Great Argus from a pup tent set up by a court and from a blind suspended in

The strutting display of the male Great Argus.

a tree above a court, but both efforts were unsuccessful. Finally, he had his assistants dig a large foxhole in the ground behind a buttress root of the tree that was next to a male's court. Seated in this foxhole and hidden by branches, he waited daily for most of a week until at last he observed the male enact a full-on courtship performance for a visiting female. Little did he know it, but Beebe had it easy! Fifty years later, the ornithologist G. W. H. Davison spent 191 days over a three-year period observing male Argus Pheasants in Malaysia. During his seven hundred hours of observations, Davison saw only *one female visit*. That is the equivalent of working forty-hour weeks for more than half a year. Needless to say, few people have ever had enough patience to do this, and most observations of Argus behavior come from birds in captivity.

Here's what happens when the female Argus arrives at a male's court. The male first performs several preliminary displays, which include a ritualized pecking at the ground and elaborate, stylized strutting on his bright red legs. Eventually, he rushes around her in wide circles with his wings hunched up at an angle that exposes their upper surfaces. Then, without warning, when

he is just a foot or two away from the female, the male transforms himself instantly into an entirely different shape, revealing
unimaginably intricate color patterns on his four-foot-long wing
feathers. In what biologists have come to refer to, with inexplicable reserve, as the "frontal movement," the male bows down to
the female, unfurling the elaborate feathers of his open wings into
a huge hemispherical disk that extends forward, over his head,
and partly surrounds the female from one side. In 1926, the pioneering Dutch animal behaviorist Johan Bierens de Haan compared this cone to the shape of an inverted umbrella blown out by
a gust of wind.

In this extraordinary posture, the male tucks his head *under*
one of his wings and peeks out at the female from behind the gap

The "frontal movement" display of the male Great Argus.

in his feathers formed at the "wrist" of his wing to gauge her reaction to his display. The deep blue of the facial skin that surrounds the male's tiny black eye will be just visible to the female through the gap in his flexing wings. To support this extraordinary posture, the male perches athletically with one set of talons in front of the other like a sprinter in starting blocks. While bowing before the female, he raises his rear, cocks his long tail feathers, and pumps them rhythmically up and down so that the female can get sporadic glimpses of them over the top of the inverted cone of his wing feathers, or in the gap that sometimes opens up between the left and the right wings. The tips of the cone of wing feathers wave over the female's head like a mini portable amphitheater. After repeated, throbbing shakes of the inverted feathery cone, lasting a total of two to fifteen seconds, the male transforms back into a "normal" bird shape and resumes his ritualized pecking of the ground for a few seconds before repeating the display.

So far, this description of the male's theatrical display postures, dramatic as it is, has ignored what is really *most* remarkable about the "frontal movement" of the Great Argus—the over-the-top patterns on his wing feathers. When he assumes this blown-out-umbrella posture, the male reveals the upper surfaces of the wing feathers, which are largely hidden when his wings are folded and closed. The transformation is unimaginably stunning. Although the hues of his wing feathers are in a subdued palette of black, deep brown, red brown, golden brown, tan, white, and gray, the ornateness and complexity of the pattern in which they are arranged is perhaps the most highly elaborated of any creature on earth. From the tiniest submillimeter-sized dots on individual feathers to the overall pattern of the fully extended four-foot-wide feathery cone, the forty wing feathers of the Argus Pheasant combine to create a paisley effect of such staggering complexity that it simply blows the peacock's tail away (color plate 3). Nothing else I know in nature can rival the fantastic intricacy of this design.

Each individual feather encompasses all the pattern complexity of a zebra, a leopard, a tropical reef butterfly fish, a flock of butterflies, and a bunch of orchids. The overall appearance is as richly worked as the design of a Persian carpet. Each wing feather is so densely packed with varied zones of dotted, striped, and swirling waves of color that it could rightly merit its own monograph.

The shorter, primary wing feathers, which are attached to the bones of the "fingers" and "hand" at the tip of the bird wing, form the bottom half of the cone. These feathers have dark shafts, light gray tips, and various zones of tan with intricately spaced brown dots or reddish brown with tiny white speckles. But the most celebrated color patterns are found on the secondary wing feathers, which are attached to the trailing bones of the forewing, or ulnas; they create the top half of the feathery cone. Each secondary feather is over three feet long and nearly six inches wide at its tip. The central shaft, or rachis, of each feather is bright white and divides the feather into two halves that are adorned with entirely distinct color patterns. The inner vanes are an array of blackish dots on a gradient of gray. On the outer vane of each secondary feather, the twisted bars of deep brown and light tan (which camouflage the bird so well when the wings are folded at rest) grade into wavy, striped patterns of tan and black. Nearest the rachis on the outer vane is a series of remarkable golden yellowish-brown spheres outlined heavily in black (color plate 4). It is these spheres—often called ocelli or eyespots—that give the species its name. In 1766, Carl Linnaeus named this pheasant after the all-seeing, hundred-eyed giant of Greek mythology, Argus Panoptes. However, the Great Argus has three times as many "eyes" as his namesake!

Twelve to twenty of these lovely golden spheres radiate in a line from the base to the tip of each secondary feather. I refer to these round golden patches as "spheres" because they are exquisitely and subtly counter-shaded, as if by the skillful brush of a painter, to create a stunningly realistic optical illusion of three-dimensional depth. The golden tan at the center of the sphere is outlined from below with a dark, mascara-like smudge, creating the impression of a shadow being cast. On the opposite side of the circle, the golden yellow blends subtly into a bright white crescent that looks like a "specular" highlight—like the shine from the

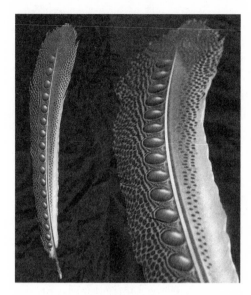

(Left) The "golden spheres" on the male Great Argus secondary feathers gradually increase in size toward the tip of the feather. (Right) A forced perspective illusion makes the spheres appear to be nearly uniform in size when viewed at an angle, similar to the view of the female during the display. *Photos by Michael Doolittle.*

surface of a glossy round apple. As Darwin noted, the color shading on each sphere is precisely oriented so that when the secondary feathers are suspended above and around the female in the giant cone, they produce the startling impression that the golden spheres are three-dimensional objects suspended in space and illuminated *from above* as if by a shaft of light piercing through the forest canopy. The three-dimensional illusion is further enhanced by the fact that when the male holds these secondary feathers up in the air during the display, ambient light will be transmitted *through* these unpigmented white highlights, giving them an extra brilliant and luminous quality.

An additional optical illusion is created by the fact that the golden spheres at the bottom of each secondary wing feather are about half an inch wide at the base and gradually increase in size to over an inch wide at the tip. Because the spots become physically larger the farther they are from the female's eye, they appear to create a *forced perspective* illusion in which the spheres appear *uniform* in size from her point of view.

Taken together, the elements of the male display add up to a sensory experience of mind-boggling complexity—a throbbing, shimmering hemisphere of three hundred vertically illuminated golden spheres that instantaneously appear suspended in the air against a feathery background tapestry of speckles, dots, and swirls. The golden balls emanate outward from the center of the display, where the male's black eye and blue face can be glimpsed peeking out. The whole effect is magnificent.

How do all these marvelous ornaments impress the female Argus? Observers are unanimous in describing the female's response as completely underwhelming, or even undetectable. William Beebe wrote, "There is no question in my mind that the wonderful colouring, the elaborate ball-and-socket illusion of the ocelli, the rhythmical shivering of the feathers which makes these balls revolve—all are lost, as aesthetic phenomena, upon the nonchalant little hen."

In rejecting the possibility that the female Argus is having any aesthetic experience, Beebe exercised an odd kind of reverse anthropomorphism. If we humans find the male's display to be awe inspiring, shouldn't the "little hen" exhibit a stronger, visible response to it? Shouldn't she be acting more like how we feel? Maybe because Beebe had spent months in the jungle trying to observe this display and many weeks huddled in his various hideouts, he expected the female Argus to evince at least some of the excitement that he himself experienced when he finally saw the display from his muddy foxhole. His conclusion that she did not share his excitement led him to be skeptical of the possibility that the male's display had any aesthetic impact on her at all. However, sexual selection theory holds that every elaborate ornament is the result of an equally elaborate, coevolved capacity for aesthetic discernment. Extreme aesthetic expression is always a consequence of extreme rates of aesthetic failure—that is, rejection by potential mates. Male Argus have such extreme ornaments precisely because most males are *not* chosen as mates. Thus, a calm, under-impressed female Argus is actually acting as we should expect—more like an experienced, well-educated connoisseur evaluating one of the many extraordinary works available to her scrutiny than an excited naturalist having a once-in-a-lifetime encounter. And from what I've seen of videos of these courtship performances, that's exactly how I would describe her—rigid with highly focused attention as she casts her discerning eye over the displaying male. The female Argus may appear dispassionate as she watches the male's efforts, but it's her coolheaded mating decisions over the course of millions of years that have provided the coevolutionary engine that has culminated in the male Argus's display of hundreds of golden balls shimmering and gyrating in the air.

The magnificent feathers and elaborate displays of the Great Argus have long been a prime piece of evidence in our struggle to understand the origin of beauty in nature, but this evidence has led thinkers to diametrically opposite conclusions. In his 1867 antievolution tract, *The Reign of Law,* the Duke of Argyll cited the "ball and socket" designs of the Great Argus wing feathers as a sign of God's hand in creation. Darwin countered that the Great Argus is evidence of the evolution of beauty by mate choice, concluding that "it is undoubtedly a marvelous fact that the female [Great Argus] should possess this almost human degree of taste."

During the century-long intellectual eclipse of mate choice theory, biologists were hard-pressed to explain the reason for aesthetic extremities like those of the Great Argus. William Beebe described Darwin's theory as intellectually tempting—"Darwin's ideas are those which we human beings would prefer to accept"— but ultimately unpersuasive. Given his low opinion of the cognitive and aesthetic capacities of female pheasants, Beebe simply could not accept the idea of sexual selection: "It seems impossible to conceive, much as we would like to believe in it, and personally, I should be willing to strain a point here and there to admit this pleasant psychologically aesthetic possibility; but I cannot."

Then how did Beebe explain the evolution of the male Great Argus? He could not. He concluded, "It is one of those cases where we should be brave enough to say, 'I do not know.'" Ironically, a man who spent years of his life tracking down the displays of this fabulously beautiful creature, and many other pheasants, found Darwin's explanation for its beauty "impossible." This is a real measure of the intellectual loss that followed in the wake of Wallace's rout of Darwin's theory of mate choice.

Today, however, all biologists embrace the fundamental concept of mate choice. Thus there is complete consensus that the ornamental plumage and behavior of the Great Argus have evolved through the agency of female sexual preferences and desire—that is, sexual choice. We now agree that ornament evolves because individuals have the capacity, and the freedom, to choose their mates, and they choose the mates whose ornaments they prefer. In the process of choosing what they like, choosers evolutionarily transform *both* the objects of their desires *and* the form of their own desires. It is a true coevolutionary dance between beauty and desire.

What biologists *don't* agree on is whether mating preferences evolve for those ornaments that provide consistently honest, practical information—about good genes or direct benefits like health, vigor, cognitive ability, or other attributes that would help the chooser—or whether they are merely meaningless, arbitrary (albeit fabulous) results of coevolutionary fashion. Actually, most biologists *are* in agreement with the former hypothesis. I am not. More precisely, I think that adaptive mate choice *can* occur but it is probably rather rare, whereas the mechanisms of mate choice envisioned by Darwin and Fisher, and modeled by Lande and Kirkpatrick, are likely to be nearly ubiquitous.

But it nonetheless remains true that since Darwin's *Descent of Man* the beauty-as-utility argument has been rampantly successful. The purpose of this chapter is to show how this flawed consensus persists. It persists in large part because it has been propped up by an unscientific faith in the ultimate validity of its own conclusions.

In 1997, I submitted a manuscript to the *American Naturalist,* a first-class science journal in ecology and evolutionary biology. The paper discussed both the arbitrary and the honest advertisement mechanisms of mate choice to try to determine which were operative in the evolution of certain avian courtship displays I had observed. In one section of the manuscript, I discussed a specific sequence of display behavior within a group of birds called manakins (which I discuss further in chapters 3, 4, and 7). Through a comparative examination of the display behaviors of multiple species within the group, I described how the males of one of the species, the White-throated Manakins, evolved a novel bill-pointing posture that replaced an ancestral tail-pointing posture that had been a routine part of the standard display repertoire. It was as though evolution had edited out the old posture with a cookie-cutter and pasted in the new one in the same exact position within the behavioral sequence. I proposed that this change was unlikely to have evolved because it provided better information about mate quality—if it did, then *all* of the manakin species would have evolved it—and more likely to have evolved in response to arbitrary, coevolved aesthetic mate preferences.

In science, journal editors send your work out to anonymous peer reviewers—other scientists who often include your intellectual competitors. The reviewers' comments on the work are used by the editor to help decide whether the work should be published and to guide the author on improvements to the work. In this case, the anonymous reviewers hated this section of the paper. They argued that I could not state that this new posture had evolved through arbitrary mate choice because I had not specifically rejected each of the many adaptive hypotheses that they could imagine. For example, I had not tested whether the bill-pointing White-throated Manakin males were revealing their superior vigor or disease resistance. I responded that standing motionless in one posture as opposed to another was unlikely to be able to communicate any additional information about vigor or genetic quality, unless we were to hypothesize that the tail-pointing posture in the ancestral birds had evolved in order to reveal whether they were infested with butt mites, and the bill-pointing posture must have evolved in order to reveal the possibility of some more recent problem in evolutionary history, such as infestations of throat mites. This seemed unlikely to me, but the reviewers insisted that the burden of proof was on me to demonstrate that the display traits were arbitrary. Of course, this made it impossible to "prove" my point, and I ultimately cut this section out of the manuscript in order to publish the paper.

This exchange continued to bother me long after the paper appeared in 1997. How many of these adaptive hypotheses, I wondered, would I have to test before I could conclude that any given display trait was arbitrary—that is, that it lacked information about any quality other than its attractiveness? When would I ever be done with this task? Even if I were able to test every adaptive explanation they could think of, pleasing one set of reviewers would only be the first of my hurdles. Their reasoning implied that I would have to test other hypotheses in order to satisfy other skeptical reviewers, and then others, ad infinitum. Because there would be no end to the creative imaginations of the reviewers, there would be no end to the process of trying to demonstrate that any specific trait is arbitrary. I was trapped. The prevailing standard of evidence meant it would be impossible for me to ever conclude that any trait had evolved to be arbitrarily beautiful. It had actually become impossible to be a contemporary Darwinian.

I realized that it was Alan Grafen's standard of evidence that had put me in this bind: "To believe in the Fisher-Lande process as an explanation of sexual selection without abundant proof is methodologically wicked."

Of course, Grafen was not the first to deploy the "abundant proof" standard, which has a long, respected history in science. In the 1970s, in regard to paranormal psychology, Carl Sagan claimed, "Extraordinary claims require extraordinary evidence." This famous "Sagan Standard" can actually be traced back to the French mathematician Pierre-Simon Laplace, who wrote, "The weight of evidence for an extraordinary claim must be proportioned to its strangeness."

Thus, whether Grafen's abundant proof standard should be invoked depends on our perceptions of the *strangeness* of the Darwin-Fisher theory of mate choice. But what dictates the *strangeness* of a hypothesis? Should we allow our gut feelings about the way the world *should* work to dictate our scientific investigation of the way it *does*? Grafen argued that the comforting "rhyme and reason" of Zahavi's handicap principle should compel us to reject the terrible strangeness of arbitrary mate choice.

Of course, it's human nature to want to believe in a universe that is rational and orderly. No less a scientist than Albert Einstein backed away from quantum mechanics—for which he had laid much of the intellectual groundwork himself—because it brought uncertainty and unpredictability into the world of physics. In rejecting quantum mechanics, Einstein famously wrote, "God does not play dice." But eventually quantum mechanics triumphed despite its enduring strangeness, because the predictive power of the theory was too great to ignore. Our understanding of the physical laws of the universe has progressed immeasurably since then. Physics was forced to embrace a stranger universe.

Unfortunately, it has been difficult to dislodge the taste for "rhyme and reason" in evolutionary biology. In mate choice, the longing for rhyme and reason has left us with a tired, worn-out science that consistently fails to account for the evolution of beauty in the natural world. The current adaptationist "consensus" rests on surprisingly weak foundations. To get to the heart of what is wrong with it, we have to explore the basics of the scientific process.

When we test a scientific hypothesis, we must compare a conjecture—say, that a specific mechanism is responsible for producing the observations that we have made of the world—with a more general conjecture that nothing special is happening; that is, no specific, or special, explanation is required to account for the observations we have made. In science and statistics, this "nothing special is happening" hypothesis is known as the null hypothesis, or null model. In an incredibly pleasing and serendipitous coincidence that has no influence whatsoever on the validity of my argument, the concept of the null hypothesis was actually invented in 1935 by none other than Ronald A. "Runaway" Fisher, who coined the term and described it this way: "We may speak of this hypothesis as the 'null hypothesis,' and it should be noted that the null hypothesis is never proved or established, but is possibly disproved, in the course of experimentation."

Thus, before we can assert that some specific process or mechanism of interest is happening, we must first reject the null hypothesis that *nothing special* is happening. The rejection of the null results in an affirmative conclusion that *something* distinctive is, indeed, going on. But, as Fisher observed, the null hypothesis is intellectually asymmetrical. One can find evidence to reject the null hypothesis, but one can *never* really prove it. In other words, given the logical structure of scientific inference, it is possible to provide enough evidence to establish that something special *is* happening but impossible to definitively establish that *nothing special* is happening.

Of course, the null hypothesis is more than just a temporary intellectual tool that we deploy to get a scientific job done. Sometimes, it is actually an accurate description of reality. Sometimes, "nothing special" really *is* happening! And when the null is an accurate description of the world, its function is to prevent science from going off on unsupportable flights of fancy. Null hypotheses actually protect science from its own crazy conjectures and faith-based fantasies.

Unfortunately, there are fundamental reasons why humans, including professional scientists, are biased toward thinking that *something special* must be happening. The human brain gets lots of rewards for detecting hard-to-see patterns in the flow of sensory information and cognitive details. Being able to figure out what's going on when it's not obvious is perhaps the most fundamental advantage of intelligence. Think, "I see the fresh tracks of the water buffalo in the mud. I've noticed that they come here to drink every morning. If I come early tomorrow morning and hide behind that bush, I can kill one to eat!" But the cognitive capacity to interpret the world as filled with meaning and governed by rational cause and effect can also guide us to mistaken conclusions, convincing us that something special *must be* going on when in fact nothing specific is happening. Ghost stories, miracles, magic, astrology, conspiracy theories, hot streaks in sports, lucky dice, or team curses are all examples of the boundless human desire for explanatory rhyme and reason where none is required.

Lots of people indulge in their irrational desires for *meaningful* explanations of our chaotic world, often in ways that are so mainstream that it never occurs to us to wonder about their validity. For example, an entire industry of business news provides continuous explanations of what's going on in economic marketplaces when in all likelihood there is absolutely nothing special going on most of the time. Business news channels broadcast an endless stream of financial reports about "events" in global financial markets. They confidently explain that the Hang Seng Index is up, or the London FTSE is down, or Dow futures are unchanged because of the latest unemployment report, negotiated sovereign debt settlement, or quarterly profit reports. Of course, the null hypothesis is that market activities are the result of the aggregate effects of millions of independent decisions by individuals who are each trying, as John Maynard Keynes memorably stated, "to guess better than the crowd how the crowd will behave." But the null model that market fluctuations lack a common or generalizable external cause is *never* entertained on the business news. This may be because business news is, after all, a business itself. Honest reporting of the null hypothesis would be very bad for their bottom line. Audiences are unlikely to tune in following a null news promo: "Random stuff happened on Wall Street today! *Details at twenty past the hour!*" The business news reporters assume that everything is the result of some rhyme and reason and that their job is to report that as true, even if it has to be invented.

Null hypotheses are essential to science, even when they are horribly wrong, because it's only in the attempt to find evidence to reject them that a better understanding emerges. For example, "cigarettes do not cause lung cancer" is a null hypothesis. According to this null, lung cancer has many diverse causes, and smoking has no generalized effects on lung cancer risks. Many people do smoke cigarettes, and many smokers do get lung cancer, but according to the null there is no causal association here.

Interestingly, in the 1950s, Ronald A. Fisher was an enthusiastic and energetic public advocate of this particular, dismally incorrect null hypothesis, which has since been definitively disproven. Another, more contemporary null hypothesis is "global warming is not caused by the human production of atmospheric greenhouse gases." The job for the scientist in such instances is to prove the null hypothesis wrong by gathering the requisite evidence to reject it. In other words, the scientific burden of proof always lies with those who want to show that something specific is happening, not on those who think that it is not.

After years of struggling against Grafen's abundant proof standard, I came to realize that the field of evolutionary biology had become like the financial market news reports. Evolutionary biologists have become convinced that a special kind of rhyme and reason—adaptive mate choice—*must* be happening everywhere and all the time. Why are they so convinced? When you examine it, it is mostly just a belief that the world *must* be that way. Remember, in rejecting Darwinian mate choice, Wallace asserted as a matter of principle that "natural selection acts perpetually and on an enormous scale." The intellectual justification remains largely unchanged.

Despite its enduring strangeness to many, the Lande-Kirkpatrick sexual selection mechanism is not merely an alternative hypothesis to adaptive mate choice; it is the appropriate *null model* for the evolution of sexual display traits and mating preferences. It describes how evolution by mate choice works when nothing special is happening—that is, when mates are choosing what they prefer, period. Because evolution requires genetic variation to occur, the Lande-Kirkpatrick model assumes genetic variation in trait and preference. But it does not assume that mates vary in quality, that any display traits are correlated with that quality, or that mating preferences are under natural selection to prefer those traits. That is why it is the null model.

If the Lande-Kirkpatrick mechanism is the appropriate null model for evolution of traits and preferences, then it cannot be proven. Thus, Grafen's demand for "abundant proof" of the Fisher-Lande process was so rhetorically effective precisely because it demanded *the impossible*. Checkmate! This was the trap I experienced when I realized that I could never satisfy my reviewers. And this is why, nearly 150 years after *The Descent of Man* and 25 years after Grafen's 1990 paper, there are still no generally accepted, textbook examples of arbitrary mate choice. Period. Grafen's gambit triumphed.

The contemporary science of mate choice is a case study in the intellectual pitfalls that can befall a science that does not incorporate any null hypothesis or model. In the absence of a null model, adaptive mate choice is unscientifically protected from falsification. It becomes the preordained answer to every question about the evolution and function of an aesthetic trait. When a trait can be shown to be correlated with good genes or direct benefits, the adaptive model is declared to be correct. When no such correlation is found, the result is interpreted merely as a failure to try hard enough to establish how the adaptive model is correct. In this framework, the ultimate research goal for every young scientist or graduate student is to demonstrate what everyone already knows to be true in some delightfully unexpected, new way that no one has ever imagined before. Because it has been embraced for the comforting rhyme and reason it provides, the entire adaptive mate choice enterprise has devolved into a faith-based empirical program to generate evidence to *confirm* a generally agreed-upon truth. The function of null models is to prevent this kind of faith-based confirmationism from taking over science.

"**S**tuff happens." The phrase may sound ridiculous or even flippant, but in its simplicity it actually captures the essence of the null model. Within the context of evolution through mate choice, we can restate this null as "Beauty Happens." (Remember, we mean beauty as the animal perceives it.) As the null model for the origins of aesthetic traits in nature, Beauty Happens provides an invigorating new perspective on the evolution of sexual beauty. It's a slogan that I think Darwin would have both understood and embraced.

At this point, it is important to emphasize again that a fully aesthetic theory of mate choice includes the possibilities of *both* the arbitrary null model (Beauty Happens) *and* the adaptive mate choice model (honest indicators of good genes and direct benefits). After all, a Maserati or a Rolex can be aesthetically pleasing while also performing utilitarian functions like driving at race car speeds or keeping accurate time. Thus, the aesthetic perspective is inclusive of other possible explanations for the evolution of specific display traits. The adaptive view, by contrast, does not allow for the possibility that arbitrary Fisherian mate choice occurs. It is the very opposite of inclusive.

How should the science of mate choice proceed from here? When looking at a given sexual ornament or display behavior, we must ask this basic question: Has the trait evolved because it provides honest information about good genes or direct benefits or because it is merely sexually attractive? Only by first disproving the null model that Beauty Happens can this scientific research program make progress.

The science of mate choice needs a null model revolution. Although researchers who joined the field in order to pursue their interests in adaptation will not find this message comforting, we have good evidence from other fields of evolutionary biology that null model revolutions are both successful *and* intellectually productive, even for adaptationists. In molecular evolution, a null model revolution in the 1970s and 1980s led to the universal adoption of the neutral theory of DNA sequence evolution. Now, before one can claim that certain DNA substitutions are adaptations, one must reject the null hypothesis that such changes are merely neutral variations that evolved by random drift in the population. In community ecology, a null model revolution in the 1980s and 1990s led to the universal adoption of null models of community structure. Now, before one can claim that an ecological community has been structured by competition, one must first reject a random, null model of community composition. In both fields, even the most ardent natural selectionists have ultimately embraced null and neutral models, because they advance their ability to test and support hypotheses of adaptation. It is critical that the science of evolution embrace a null model of sexual selection.

Opponents of adopting null and neutral models in evolutionary biology sometimes complain that the proposed null models are too "complex" to be an appropriate null model. To them, null models should be simpler and more parsimonious. But this view misconstrues the intellectual function of the null model. For example, if cigarettes cause lung cancer, then the causal explanation of most lung cancers is actually quite simple—cigarettes. If the null hypothesis that cigarettes do not cause cancer were true, then the actual causes of lung cancers would be much more variable, individualized, *and* complex. So null models are not necessarily simpler explanations. Rather, the null model is the hypothesis that the proposed, generalized causal mechanism is absent. In evolution, that critical causal mechanism is natural selection, which is why the Beauty Happens hypothesis is the appropriate null.

With an understanding of what is at stake if we forgo the null model, we can return to a consideration of the male Great Argus. First, we need to grapple with the *full* breadth of the aesthetic complexity that requires evolutionary explanation. The totality of the sexual ornaments in the Great Argus includes the male territory and court-clearing behavior, court attendance, vocalizations, the diverse display repertoire including each of his movements, the facial skin color, and the size, shape, patterning, and pigmentation of each feather. The full display behavior of the Great Argus is like an opera or a Broadway musical. It consists of music, dancing, elaborate costumes, lighting, and even trompe l'oeil effects, albeit on an intimate stage with a solo cast.

One way to try to think about this aesthetic complexity is to conceive of each and every detail as an evolutionary design "decision." How many total decisions would be required to describe the "Full Monty" of the Great Argus? Starting at the tip of one primary wing feather, we see that the broad tip of the feather is gray, not brown, with large dots that are reddish brown, not white, tan, or black. Toward the base of that same feather, the background color changes to tan, but the dots stay the same color, become smaller, get closer together, and then converge into a true honeycomb pattern. Each and every one of these details could be different. Indeed, every one of these details *is* different in every other species of bird in the world. Evolutionary biologists who believe that natural selection dictates the form of various display traits are not only required to describe the mere existence of ornament; they are charged with explaining the origin and maintenance of each and every *specific* detail of its form. In the case of the Great Argus, the number of independent aesthetic dimensions adds up to hundreds or even thousands—a practically unfathomable degree of complexity.

The adaptive mate choice paradigm asserts that each and every one of these features has *specifically* evolved as an honest indicator of good genes or direct benefits. In other words, each detail evolved as it did because it was *better* at providing quality information than all other available variations. Most mate choice researchers see their job as demonstrating *how* this is true, not testing *whether* it is true. Without a null model that allows one to reject the adaptationist account, they cannot do otherwise. In any given study, researchers will measure multiple aspects of male ornament and try to correlate them to the health and genetic information they are presumably providing, but at best only one or a few of the many aesthetic features of the full display repertoire will show *any* sign of a correlation with mate quality. Biologists then use this very limited subsample of their data to draw general conclusions about the role of honest signaling in the process of sexual selection as a whole. The vast majority of the data inevitably fail to confirm the adaptive theory of mate choice. As a result, the vast majority of the ornamental details remain unexplained even as the adaptive explanation of mate choice triumphs.

We will never establish a satisfactory explanation of evolution by studying only those data that turn out to "work" the way the researcher hopes. Because those investigations that are not able to confirm the adaptive value of any ornamental features are considered failures—failures to work hard enough to find the data to demonstrate how adaptive mate choice is true—such studies don't get published. In this way, the current paradigm prevents us from ever seeing these data, which are actually a legitimate description of the way the world is, and how it got that way. Indeed, they are exactly consistent with the Beauty Happens model. In this way, the adaptationist worldview can make us blind to the true nature of reality. And this blindness certainly affects our ability to "see" the Great Argus.

Unfortunately, studying mate choice in the Great Argus in the wild would be extremely difficult. Recall that G. W. H. Davison observed males for seven hundred hours over three years and only managed to witness one female visit. He saw no copulations. Perhaps if one could find dozens of Argus nests, one could use DNA analyses of the chicks to identify all their fathers. However, one would also have to place arrays of hidden cameras at multiple male courts to record the patterns of female visits and the variations in display behavior among successful and unsuccessful males. And one would need to capture these males and record information about their health, condition, and genetic variation. It would be a *huge* and expensive undertaking.

Setting aside the difficulty in obtaining these data from the wild, let's consider whether female pheasants might be gaining either of the two kinds of adaptive benefits from their mate choices. The most fundamental benefit is good genes—heritable genetic variations that would endow the female's offspring, both male and female, with survival and fecundity advantages.

Although the good genes hypothesis has had a good run in intellectual history and remains popular, empirically it has fallen on hard times. Many studies have failed to find any evidence of a correlation between good genes and female sexual preferences. For example, a recent "meta-analysis"—that is, a big statistical study of multiple data sets from many independent investigations of different species—*did* find significant evidence in support of arbitrary Fisherian mate choice while *failing* to find support for the idea that males who are preferred provide any good genes. These results were based on the scientific literature, which is likely to have a publication bias toward the publication of "positive" results—that is, results that support good genes. As discussed, "negative" results are more frequently considered scientific failures and consigned to the rubbish heap. Thus, the failure of meta-analysis to find support for good genes is probably just the tip of the data iceberg. The vast volume of data remains unseen, lurking below the surface, and this giant bolus of unpublished, privately held data is likely to be overwhelmingly negative. It's becoming more and more apparent that good genes is an intriguing idea that is failing to find much support in nature.

The other adaptive benefit that Great Argus males may provide to females that choose them as mates is in the form of direct benefits, which accrue to the survival and fecundity of the female herself. In monogamous birds that form social pairs to raise their young, these direct benefits may include defending a shared territory rich in high-quality resources, helping with parental care, defending against predators, and making other contributions to a successful family life. But the male Great Argus provides no parental care or reproductive investment whatsoever. He merely provides sperm. Because females mate and leave immediately to incubate their eggs and raise their young on their own, their interactions with males are limited to the visits they make to various males in order to choose their mates and the brief moment of copulation that ensues once they've made their choice. Thus, they have only two possible ways of obtaining any direct benefits whatsoever from male Great Argus. First, preferred males could be those with display signals that make female mate choice more efficient, minimizing the investment of time and the risk of predation incurred during the female's visits to the males. However, there is nothing remotely efficient about what's involved in assessing the Great Argus display. The female must travel widely (probably miles) to visit different males, and she must observe each one at a really intimate proximity in order to properly observe his display. The other possibility is that male displays could be providing honest information about their lack of infection by sexually transmitted diseases. However, this seems highly unlikely as well. Selection to avoid sexually transmitted disease would result in strong natural selection against the polygynous breeding system, which would greatly foster STD transmission, and not to selection for extreme coevolved aesthetic traits and preferences.

In conclusion, even without further data from the wild, there are excellent reasons to think that the Great Argus is an evolutionary example of the Beauty Happens mechanism.

Another intellectual hurdle for adaptive mate choice is the sheer complexity of the Great Argus display. According to the handicap principle, the honesty of any display is ensured by the costs it imposes on the individual. These costs include both the developmental costs of making it and the survival costs of having it. But the costs of signal honesty create another burden to an adaptive explanation of the many multiple ornaments in the Great Argus display repertoire. According to the theory, each of these ornamental dimensions must provide an *independent channel* of quality information in order to sustain the *additional costs* that ensure its honesty. If some costly ornamental detail within a repertoire does not provide some independent information about quality, then it would either never evolve or be eliminated by natural selection as redundant and superfluous. Thus, the handicap principle establishes real constraints on the evolution of aesthetically complex repertoires of multiple display traits. Yet aesthetic complexity is present not just in the Great Argus but throughout nature.

Of course, multi-trait repertoires with many independent ornamental dimensions pose no challenge at all to the Beauty Happens evolutionary mechanism. Indeed, the model predicts them. Given free rein, mate choice is likely to produce evolutionary runaways in the complexity of the *repertoire* of ornaments as well as in the complexity of any individual ornaments.

Some honest advertisement theorists have proposed that complex ornamental repertoires could function as adaptive *multimodal* displays. In this view, the Great Argus aesthetic repertoire is like a Swiss Army knife; each aspect of the display is a different adaptively optimized blade for a distinct communication task within the general mission of honest and efficient mate attraction.

Each display communicates a distinct channel of quality information through a specific sensory modality. The concept of "multimodal" display is an attempt to flatten aesthetic complexity into a manageable set of individualized, rational utilities. But it doesn't avoid the problem of multiple redundant costs.

Before we go further, however, we should ask, "Is this even possible?" How many independent channels of mate quality information are there for the female to evaluate? It is hard to know because no one, as far as I know, has ever asked this question before. However, I think there are a few relevant ways to think about it. If you wanted to accurately evaluate the health and genetic quality of a human being, how would you go about it? This is, in part, what doctors try to do during regular checkups. How much can you tell about a person's future health from the results of an annual physical examination? Well, the American Academy of Family Physicians has recently determined that beyond routine weight and blood pressure monitoring, there is *no evidence* of the medical effectiveness of regular physical examinations. Except for the assessment of body weight and blood pressure, a doctor's observations do not detect enough information relevant to future health outcomes with sufficient frequency to make annual checkups cost-effective. Of course, a doctor's exam involves asking a lot of specific questions and the use of many invasive procedures—like blood tests—that are not available to female Great Argus as they evaluate potential mates. Female pheasants do not have sphygmomanometers, stethoscopes, or EKG machines. Yet even with all our equipment and our advanced medical knowledge, regular detailed inspection of the human body and verbal interviews are *not* able to provide sufficiently useful information about human health outcomes to make them worth doing.

The truth is that it is very difficult to accurately assess the genetic quality of an animal and predict its future health even with advanced knowledge and scientific tools. Can we expect the female Great Argus to be able to make better assessments of the health of their potential mates than human physicians?

But let's go further than your typical family physician and imagine that we can sequence the entire genome of every individual patient. What can we learn from information about potential health risks to those individuals from their genomes? Well, we can learn about the possibility of developing rare diseases that are caused by single genes like cystic fibrosis and Tay-Sachs. But we would learn surprisingly little about the risks of any of the complex diseases that cause most deaths—like heart disease, stroke, cancer, Alzheimer's, mental illness, or drug addiction. Indeed, since the early years of the twenty-first century, the juggernaut of initiatives in genomic medicine has been hampered by the *failure* of the genomic data to provide much predictive information about any complex diseases. For example, it is easy to find dozens of genetic variations that are significantly associated with heart disease. But, except for a few rare genetic variations that are particular to certain ethnic groups, when the effects of all these genes are added together, they explain less than 10 percent of the heritable risk of heart disease. So, even with *complete genomic* information, predicting genetic quality and future health outcomes is fundamentally challenging. This fact is why the Food and Drug Administration, in 2013, prohibited personal genomic companies like 23andMe from marketing information to their customers about their genetic risks of disease without specific approval. Most of the statistical associations between single genes and disease are currently so vague and tenuous that reporting such information to customers was considered fundamentally misleading.

So again we must ask, is it likely that a female Great Argus could draw any more valid conclusions about the genetic suitability of a potential mate than a scientist armed with complete genomic information? Of course, it's theoretically possible that she might be able to do this, but this is an empirical issue that should actually be investigated, not accepted on blind faith. The failure of human genomic medicine to find reliable tools to predict most complex health outcomes is highly relevant to the good genes hypothesis, providing even more reason to be skeptical about the prospect of assessing adaptive value of a mate from every ornament.

The intellectual collapse of one infamous honest signaling mechanism provides amusing insights into the social phenomenon of mate choice science. In papers published in 1990 and 1992, the Danish evolutionary biologist Anders Møller proposed that body symmetry reveals an individual's genetic quality and that bilaterally symmetrical displays evolve through adaptive mate choice for higher-genetic-quality mates. Møller's data indicated that female Barn Swallows (*Hirundo rustica*) prefer males with the longest and most symmetrical outer tail feathers. Soon, there was a burgeoning cottage industry supporting mate choice based on symmetry in a wide variety of organisms.

Ironically, like an irrational Fisherian runaway, the idea of symmetry as an honest indicator of genetic quality got ever more popular, merely because it was so popular. One scientist who was excited by the idea and attempted to replicate its findings in his own research was distressed to find that he could not do so. "Unfortunately, I couldn't find the effect," he was quoted as saying in a *New Yorker* article published in 2010. "But the worst part was that when I submitted these null results I had difficulty getting them published. The journals only wanted confirming data. It was too exciting an idea to disprove, at least back then." The adaptationist confirmation bias at work once again.

But in the late 1990s, support for the idea that symmetry indicates genetic quality suddenly began to wane. A few critical papers came out, and then a few more. By 1999, meta-analyses of multiple data sets showed that support for the idea had simply evaporated.

Of course, scientists are loath to admit that they are slaves to fashion like everyone else. So, contemporary reviews of mate choice in the animal kingdom rarely even mention this embarrassing episode. Yet the enthusiasm for honest symmetry is such a prime example of bandwagon science that it was prominently featured in the *New Yorker* article, mentioned above, about the sociology of failure in science. Unfortunately, it still lives on in adaptive theories of human sexual attraction, neurobiology, and cognitive science. You would think that, decades on, news of its collapse and discredit would eventually reach the evolutionary psychology researchers who continue to preach it. But the "honesty of symmetry" has become a zombie idea—an idea so attractive that it lives on and on despite being repeatedly falsified.

In any case, the symmetry hypothesis could never have provided more than a very partial explanation of the evolution of complex ornaments like the patterns on the Great Argus's wing and tail feathers. Even if it did exist, natural selection for perfectly symmetrical signals would fail to explain any of the myriad other specific and complex details within the Great Argus's plumage and display.

A newly emerging adaptive mate choice hypothesis takes a page right out of Wallace's critiques of Darwin. It has recently been proposed that elaborate courtship displays evolve in order to indicate male vigor, energy, and performance skill to their prospective mates. Accordingly, females prefer such displays *because* they raise the male's heart rate, exhaust his energy reserves, or push him to the limits of his physiological capacity. The best dances indicate strong, fit fellows. Unfortunately, this popular idea fails in several ways to explain *specific* details of complex display repertoires like that of the Great Argus. There are many imaginable displays that would create far greater physiological challenges to the male than his relatively low-energy performance. So why haven't more extreme tests of his physiology evolved instead?

Of course, I acknowledge that the males of many species do engage in displays that are physiologically demanding. But the fact that physiological costs are incurred does not mean that those costs are honest indicators of quality. Display traits evolve to a balance between natural and sexual selection advantages, and this equilibrium may be far from the optimum for either health or survival. When Beauty Happens, costs will happen too.

The question is whether the physiological challenges are incidental consequences of extreme aesthetic performance or the entire point of the display. By analogy, do people like the extraordinary leaps, pirouettes, and so on of ballet dancers because such performances push the performers to the limits of their physiological and anatomical capacities? Or do performers encounter these physiological challenges in the process of producing art that audiences enjoy? Do we value these feats of physical skill because of their aesthetic effect on us? Or because the effort of achieving them requires that many ballet dancers will experience painful and debilitating foot and leg injuries?

There is no reason to believe that the love of ballet, or of any other human art form, is based on how much pain and effort they cost to the performers. Likewise, there is no reason to believe that the female of the Great Argus or any other species chooses a mate because of how much he endures in the course of his courting performance. It is always the artfulness of the performance that matters; the physiological demands of producing it are secondary. To believe otherwise is to confuse evolutionary cause and effect. Last, just as in the Great Argus, there are many *more* costly performances that we can imagine that are not preferred. By analogy, atonal twentieth-century concert music, from Berg to Boulez, is incredibly difficult for performers to play well, but that doesn't make audiences like it.

An interesting way to understand the Darwin/Wallace debate about mate choice is to compare the value of beauty to the value of money. Under the old "gold standard," the value of a dollar existed because each dollar could be redeemed for a tiny piece of gold. The value of a dollar was *extrinsic;* dollars had value because they stood for something else of value—that is, gold. By the mid-twentieth century, however, economists and governments realized that the value of money is merely a "social contrivance." Today, the value of a dollar is *intrinsic;* dollars have value because people in general agree that dollars have value. There is no gold behind them.

The adaptationist view of beauty works like the gold standard. Accordingly, beauty has no value in and of itself; its value only arises because beauty stands for other *extrinsic* values, either good genes or direct benefits. In contrast, the Darwinian/Fisherian view of beauty works like all modern currencies. Beauty has value only because animals have evolved to agree that it has value. Its value is intrinsic, and it can evolve for its own sake. Beauty, like money, is a "social contrivance," and the Lande-Kirkpatrick null model is the mathematical description of that process.

Hard-core advocates of a return to the gold standard, called goldbugs, still believe that the abandonment of the gold standard was a reckless and immoral flight from reason. Like evolutionary goldbugs, neo-Wallaceans are certain that behind every sexual ornament there must be an evolutionary pot of gold, filled with good genes or direct benefits to mate choice, and they defend this view as simple rhyme and reason. Like goldbugs, neo-Wallaceans are quick to label other views as "wicked."

This analogy also provides insights into why Beauty Happens is the appropriate null model of evolution by sexual selection. Imagine that the next time you see a beautiful rainbow, a small, green-suited leprechaun suddenly appears and promises you that there is a pot of gold at the end. Ask yourself, "What is the null hypothesis?" Obviously, the null hypothesis is that the value of the rainbow is intrinsic and that there is no gold at its end. And until you find that pot of gold at the end of the rainbow and can reject the null hypothesis, you have to stick with it. Likewise, adaptive mate choice posits that behind each and every sexual ornament is a pot of evolutionary gold laden with good genes and direct benefits. What's the null hypothesis? Obviously, the null hypothesis is that there are no good genes or direct benefits until you can prove that there are. The burden of proof lies with those who believe in adaptive mate choice. Some of those ornaments will indeed be found to be signals of quality. Others (most, in my opinion) will not. We should no more place our trust in evolutionary leprechauns than we do in little green-suited ones!

There are other similarities between the science of mate choice and the "dismal science" of economics. Both disciplines have active debates about the nature and importance of "market bubbles." The last decades of the twentieth century saw the development of a new, American-style capitalism characterized by increasingly complex mathematical models of investment and risk management and the systematic dismantling of the regulatory controls that had curbed some of the riskier behaviors of financial institutions. The result was supposed to be an unprecedented new era of global growth and prosperity. What happened instead was the global financial crisis in 2008. Obviously, something went fundamentally wrong with the economic model that was expected to prevent such instability. How did economists get this so wrong?

At the core of this failure was the a priori belief in a powerfully rational idea, the efficient market hypothesis, which states that, given open access to accurate information, free markets will always establish the true, correct value of an asset. According to the efficient market hypothesis, economic bubbles are impossible. Sound familiar? As the economist Paul Krugman concluded, "The belief in the efficient market hypothesis blinded many if not most economists to the emergence of the biggest financial bubble in history."

I think that most evolutionary biologists are equivalently blind to the reality of arbitrary mate choice.

To explore the parallels between the science of mate choice and the business cycle, I had lunch one day with my Yale colleague and neighbor Robert Shiller, the Nobel Prize–winning economist. A well-known expert on housing markets and an advocate of behavioral economics, Shiller was dubbed "Mr. Bubble" in a 2005 *New York Times* story in which he presciently warned that real estate prices could drop by 40 percent over the next generation. It took only three years for his predictions to be realized.

In his now-classic 2000 book, *Irrational Exuberance,* Shiller presented the case for the role played by human psychology in the volatility of many economic markets. A speculative financial market bubble, he wrote, occurs when price increases spur investor confidence and lead to increased expectations of future gains. The result is a positive feedback loop in which each increase in asset prices begets greater confidence, increased expectations, increased investment, and higher prices. These economic feedback loops involve some of the same basic dynamics as the Beauty

Happens mechanism. Both sexual displays and asset prices can be driven by popularity alone, decoupled from extrinsic sources of value.

I asked Bob what he thought about the idea that there might be similarities between the intellectual frameworks of macroeconomics and evolutionary biology. He was particularly struck by how closely the arguments waged by efficient market theorists and adaptationist evolutionary biologists resembled each other. What he said made perfect sense to me:

> To many economists, the mere existence of an asset at a given price indicates that its price must accurately reflect its value. That's very similar to arguing that the existence of a given tree or bird in a certain environment demonstrates that it must have achieved an optimal solution to the challenge of survival because it has not yet been displaced by some other ecological competitor. Both use their views to interpret the world in a way that reinforces those views.

Such logic results in empirical intellectual disciplines that are more dedicated to confirming their own worldviews than to establishing an accurate understanding of the world.

For the title of their 2009 book about behavioral economics, Bob and his co-author, George Akerlof, revived the term "animal spirits," which John Maynard Keynes coined to refer to the psychological motivations that influence people's economic decisions. In the book, they document that research on "animal spirits" has been discouraged in economics precisely because these irrational influences are viewed as inherently unscientific and beneath consideration of a quantitative, scientific discipline. Ironically, I think there has been a parallel intellectual movement in evolutionary biology to banish consideration of the "animal spirits" of animals! Adaptive mate choice proposes that sexual desire always remains under strict control of the ultimately rational need for extrinsically better mates. In a curious anthropomorphic inversion of nature, animal passions are now seen as being more rational than our own.

A few weeks after my lunch with Bob, a team of economists published the results of a randomized, controlled experiment on the dynamics of Internet popularity. By randomly introducing thumbs-up or thumbs-down ratings into the comments section of stories on a major news website, the researchers demonstrated that popularity can be driven *merely* by popularity itself—what the authors called a positive herding effect—completely independent of variations in actual content quality. In other words, going viral on the web is often just a matter of stuff happening. When I next ran into Bob, I mentioned this new study as a vivid, experimental demonstration of the role of feedback loops in driving arbitrary popularity bubbles. "Are you going to write about that in your book?" he asked. "Because I was thinking about writing about that study in my book too!" Who would have imagined that an ornithologist and an economist would be in competition to report on the same research?

The Great Argus and the many other birds we will meet in these pages provide aesthetically extreme challenges to conventional, adaptive evolutionary theory. Neo-Wallacean adaptive mate choice may be more popular at the moment, but without Darwin's broadly aesthetic perspective, we can never account for all the complexity, diversity, and evolutionary radiation of intersexual beauty in nature. Only the Beauty Happens hypothesis allows for a genuine engagement with the full, explosive diversity of sexual ornament.

I do not doubt, however, that meaningful, honest, and efficient signals of mate quality *can* evolve. There are circumstances in which mating preferences do indeed come under natural selection. Further, there may be circumstances in which signal honesty evolves to be so robust that it cannot be eroded away by the irrational exuberance of aesthetic desire. But we will never arrive at a genuine understanding of the diversity of nature by assuming that this is always true. We must use a nonadaptive null model to maintain the falsifiability of adaptive mate choice. Otherwise, it ceases to be science.

Although I am skeptical of adaptive mate choice, I do not claim that the "Emperor wears no clothes." Actually, I believe that the "Emperor wears a loincloth." In other words, I predict that the vast majority of intersexual signals can only be explained as the arbitrary evolutionary consequences of Beauty Happening, while the adaptive mate choice paradigm likely explains about the same proportion of the total "corpus" of intersexual signals in nature as is covered by that humble garment. How will we ever know if this prediction is correct? The only way for evolutionary biologists to proceed is to embrace the Beauty Happens mechanism as the null model of evolution by mate choice and see where the science leads.

CHAPTER 3

Manakin Dances

How, and why, has beauty changed within and among bird species over the course of millions of years? What determines what any given species finds beautiful? What, in short, is the evolutionary history of avian beauty?

These questions might seem impossible to answer, but we actually have many of the scientific tools we need to address them productively. One of the challenges to understanding the evolution of beauty is the complexity of animal displays and mating preferences. Fortunately, we do not need to invent a trendy new brand of "systems science" in order to investigate these complex aesthetic repertoires, because the science of natural history—the observation and description of the lives of organisms in their natural environments—provides us with exactly the tools we need. Natural history was a critical component of Darwin's scientific method and remains a bedrock foundation of much of evolutionary biology today.

Once we have gathered information about individual species, we need other scientific methods to compare and analyze them and to uncover their complicated, often hierarchical evolutionary histories. The scientific discipline that enables us to do that is called phylogenetics. Phylogeny is the history of evolutionary relationships among organisms—what Darwin called the "great Tree of Life."

Darwin proposed that discovery of the Tree of Life should become a major branch of evolutionary biology. Unfortunately, research interest in phylogeny was largely abandoned by evolutionary biology during most of the twentieth century. However, powerful new methods for reconstructing and analyzing phylogenies have been developed in recent decades, which has led to a revival of interest. So, now that the two critical intellectual tools necessary to study the evolution of beauty—natural history and phylogenetics—are available, there has never been a better time to be asking questions about how beauty, and the taste for it, evolve.

Doing so will help us to understand the process of evolutionary radiation—diversification *among* species—in a new way. In evolutionary biology, adaptive radiation is the process by which a single common ancestor evolves through natural selection into a diversity of species that have a great variety of ecologies or anatomical structures. The amazing diversity of Darwin's Finches (Geospizinae) on the Galápagos Islands is a canonical example of adaptive radiation. In this chapter, however, we will investigate another group of birds—the neotropical manakins—in order to understand a different kind of evolutionary process: *aesthetic radiation*. Aesthetic radiation is the process of diversification and elaboration from a single common ancestor through some mechanism of aesthetic selection—especially mate choice. Aesthetic radiation does not preclude the occurrence of adaptive mate choice, but also includes arbitrary mate choice for sexual beauty alone, with all of its often dramatic coevolutionary consequences.

The science of beauty requires that we get out of the laboratory and the museum and into the field. Fortunately, my bird-watching youth was great basic training for doing natural history research on birds in the field. I discovered the second critical element of this branch of beauty studies—phylogenetics—as an undergraduate at Harvard University. My immersion in formal ornithological studies began in the fall of 1979 with a freshman seminar, the Biogeography of South American Birds taught by Dr. Raymond A. Paynter Jr., the curator of birds at the Museum of Comparative Zoology (MCZ). Dr. Paynter introduced me to the intellectual magic of natural history museums. Up on the fifth floor of the huge and ancient brick building that housed the Bird Department was a series of rooms where hundreds of thousands of scientific bird specimens were curated. During my undergraduate years, the MCZ was my intellectual home. I hung out a lot in the bird collections doing bibliographic work and curatorial tasks for Paynter and generally smelling like mothballs.

Dr. Paynter himself was far too intellectually conservative and cautious to be interested in the revolutionary new field of phylogenetics. But I soon discovered that the latest concepts and methods in this field were being hotly debated downstairs in the Romer Library in the weekly meetings of the Biogeography and Systematics Discussion Group. In retrospect, this time at Harvard was a golden era for phylogenetics. From the meetings of this "revolutionary cell" in the Romer Library, multiple graduate students went out into the world and made fundamental contributions to the field, helping to bring phylogeny back into the mainstream of evolutionary biology.

My own work was profoundly shaped by those weekly discussions in the early 1980s. I became fascinated by phylogenetic methods and eager to reconstruct avian family trees. For my senior honors project, I worked on the phylogeny and biogeography of toucans and barbets. Working at a desk I made for myself on a big table beneath the towering skeleton of an extinct moa in room 507 of the bird collection, I was excited to make observations of toucan plumage and skeletal characters and to construct my first phylogenies. I am happy to say that I have been continuously associated with world-class scientific collections of birds ever since. Only, I don't smell like mothballs anymore.

As graduation approached, I was casting about for what to do next, searching for a research program that would combine my bird-watching skills and passion with my new obsession with avian phylogeny. Before going on to graduate school, I was desperate to get to South America and to see more of the birds I had met in the drawers at the MCZ. (There were very few tropical bird field guides in those days, so browsing through a museum collection was actually the best way to learn about the birds before actually seeing them in real life.) Intrigued by the Harvard graduate student Jonathan Coddington's research using the phylogeny of spiders to test hypotheses about the evolution of orb-web-weaving behavior, I wanted to make a similar use of phylogeny to study the evolution of bird behavior.

At about that time, I met Kurt Fristrup, a Harvard graduate student, who had worked on the behavior of the flamboyantly orange Guianan Cock-of-the-Rock (*Rupicola rupicola*, Cotingidae) (color plate 5), one of the planet's most amazing birds. Kurt suggested, "Why don't you go to Suriname to map manakin leks?" In retrospect, this was one of the most consequential pieces of professional advice I ever received.

On a thin branch twenty-five feet high in the sun-dappled understory of a tropical rain forest in Suriname perches a tiny glossy black bird with a brilliantly golden yellow head, bright white eyes, and ruby-red thighs—a male Golden-headed Manakin (*Ceratopipra erythrocephala*)(color plate 6). He weighs about a third of an ounce (ten grams), or a bit less than two U.S. quarters. He has a short neck and short tail, giving him a compact body, but he has a nervous energy that belies his almost dumpy appearance. He sings a high, soft, descending whistled *puuu* and peers intently around, hyperaware of his surroundings. In moments, a second male whistles back from his perch in an adjacent tree, and then a third nearby. The male answers immediately. His social environment is obviously the focus of his keen attention. In all, there are five males clustered together in the forest. They are obscured from one another by foliage, but they are all within earshot of each other.

In response to the neighboring calls, the first male draws himself up into a statuesque upright posture with his light-colored bill pointing upward. After singing an energetic, syncopated, and raspy *puu-prrrrr-pt!* call, he suddenly flies from his perch to another branch twenty-five yards away. After a few seconds, he flies rapidly back to his main perch singing an accelerating crescendo of seven or more *kew* calls in flight. His flight path traces a subtle S-curve trajectory, first down below the level of the perch and then up above it. He lands on the perch from above while uttering a sharp buzzy *szzzkkkt!* Immediately upon landing, the male lowers his head, holds his body horizontal to the branch, and raises his rear up with his legs extended, revealing bright red thighs against his black belly, like a provocatively colored pair of breeches. He then slides *backward* along the perch in the tiny rapid steps of an elegant "moonwalk," as if on roller skates. In the middle of the moonwalk, he flicks his rounded black wings open vertically above his back for a moment. After sliding backward for twelve inches along the branch, the male suddenly lowers and fans his tail, flicks his wings vertically again, and resumes his normal posture.

Moments later, the second male Golden-headed Manakin flies in and perches on another branch about five yards away. The first male immediately flies to join him, and they sit quietly side by side—but facing away from each other—in the dramatic upright posture. Intense, competitive, but mutually tolerant, the two males are deeply engaged with each other.

This scene is just a few moments in the bizarre social world of a Golden-headed Manakin lek. A lek is an aggregation of male display territories. Lekking males defend territories, but these territories lack any resources that females might need for reproduction other than sperm: no significant food, nest sites, nest materials, or other material assistance to the female. Golden-headed Manakins defend individual territories between five and ten yards wide, with two to five such territories grouped together. Leks are essentially sites where males put themselves on display in order to lure females to mate with them. Over the breeding season, individual females visit one or more leks, observe male displays, evaluate these displays, and then choose one of those males as their mate.

Lek breeding is a form of polygyny (one male with many potential mates) that results from female mate choice. In a lek-

The backward slide display of the male Golden-headed Manakin.

breeding system, females can select any mate they want, and they are often nearly unanimous in preferring a small fraction of the available males. So a relatively few males get to mate with a relatively large number of females. The skew in mating success is rather like the contemporary skew in income distribution. The most sexually successful males are very successful and account for half or more of all the matings, while other males will never have any opportunity to mate in a given year. Some males go their whole lives without mating.

After mating, female manakins build nests, lay clutches of two eggs, incubate them, and care for the developing young entirely on their own without any help from the males, whose contributions to reproduction end with their sperm donations. Because females do all the work, they don't depend on the males for anything, and their independence allows them almost total sexual autonomy. This freedom of mate choice has allowed extreme preferences to evolve; females only choose the few males whose behavioral and morphological features meet their very high standards. The rest will be losers in the mating game. Thus the aesthetic extremity of male manakins is an evolutionary consequence of extreme aesthetic *failure*, which results from strong sexual selection by mate choice.

Female manakins have been choosing their mates in leks for about fifteen million years. Over the course of time, the features they have preferred have evolved into an extraordinary diversity of traits and behaviors among the approximately fifty-four species of manakins distributed from southern Mexico to northern Argentina. Manakin leks are among nature's most creative and extreme laboratories of aesthetic evolution. For me, they proved the perfect place to study Beauty Happening.

Inspired by Coddington's revolutionary spider research and Fristrup's helpful suggestion, I headed off in the fall of 1982 to the nation of Suriname, a small, culturally Caribbean, former Dutch colony in northeastern South America, for what turned out to be a five-month sojourn in search of manakins. In Suriname, I worked at the Brownsberg National Park, a fifteen-hundred-foot-high, table-topped mountain covered in tropical rain forest, which is just a few hours south of the capital city of Paramaribo, down red dirt roads. Within a couple days of observing my first Golden-headed Manakins, I also found the White-bearded Manakin (*Manacus manacus*). One morning while walking through the young secondary forest along the main road through the park, I heard a sharp *snap* within a shrubby thicket, which sounded like a tiny popgun or a toy firecracker. In the thick shrubs along the road edge, I spied a boldly plumaged White-bearded Manakin (color plate 7) . The male of the species has a black crown, back, wings, and tail and bright white underparts that extend in a collar around his nape. Perched only a yard above the ground, this male gave a loud *chee-poo* call, which was quickly answered by another male a few yards away.

Unlike the Golden-headed, the White-bearded Manakin displays on and near the forest floor, and the males cluster closely together in tiny display territories within a few yards of each other. After I waited patiently for a few minutes, a flurry of displays suddenly broke out. The first male flew down to a small court—that is, a patch of bare dirt on the forest floor about a yard wide—and began to bounce rapidly back and forth between small saplings around the edges of the court. Each flight was punctuated by a sharp *Snap!* that is made by the wing feathers. When perched, his body was transformed. The previously smooth white feathers of his throat were now fluffed out and forward to form a puffy white beard that extended *beyond* the tip of his bill. Soon several males were all snapping and calling simultaneously. When perching, the males would occasionally make a sudden, explosive, and rapid series of *snaps* so quickly that they blurred together in a flatulent Bronx cheer. As suddenly as the excitement started, the wave of displays ended, and the lek quieted down to a few *chee-pooos,* with long waits in between.

Unlike the elegant flight and perch displays of the Golden-headed Manakins, the White-bearded Manakin displays are rowdy and rambunctious. The males are packed together, hopping and popping vigorously. White-bearded Manakin males are like buff gymnasts, executing short flights and rebounds with muscular precision.

Comparing the radically different display repertoires of just these two manakin species introduces the central dilemma of their aesthetic evolution. How did they evolve to be so different from each other? The true magnitude of this mystery emerges when we realize that every one of the approximately fifty-four species of manakins has evolved its own distinct repertoire of plumage ornaments, display behaviors, and acoustic signals; that is fifty-four distinctive "ideals" of beauty. Because nearly all species of the family are lekking, we can be confident that all manakins evolved from a single lekking common ancestor, which, we can infer from time-calibrated molecular phylogenies, lived about fifteen million years ago. So, why did the females of each manakin species evolve such highly diverse mating preferences—their own Darwinian standards of beauty? And how did this aesthetic

A male White-bearded Manakin landing on a sapling on his display court with his throat feathers erected.

radiation occur? Learning the answer requires that we explore the history of beauty through the Tree of Life.

There is a reason manakins are such a good example of the evolution of beauty, and it has to do with family life. Over 95 percent of the world's more than ten thousand bird species are raised by two attentive, hardworking parents. But not manakins. The British ornithologist and pioneering manakin man David Snow first proposed an evolutionary explanation for their distinctive breeding system in his enchanting 1976 book, *The Web of Adaptation*. The book is an evocative account of his and his wife's adventures studying lekking manakins and cotingas in Trinidad, Guyana, and Costa Rica. (I read the book with great excitement when I was in high school, and my still vivid memory of it was one reason why I responded so positively to Kurt Fristrup's suggestion to go study manakins in Suriname.) Snow hypothesized that eating a diet consisting largely of fruits, as manakins do, can rearrange an animal's family life and unleash a cascade of effects on its social evolution.

Imagine that you eat insects for a living. You are probably thinking that this would not be an easy life, and you would be right. Insects make themselves difficult to find, prickly, hard to handle, distasteful, and sometimes even toxic. Living on a diet of insects is hard work quite simply because insects do not want to be eaten. That's why raising a family on insects is almost always a two-bird job.

By comparison, feeding mostly on fruit is like a dream—a land of milk and honey—because fruit *wants* to be eaten. Fruits are highly caloric, nutritional bribes created by a plant to entice animals to swallow, transport, and deposit their seeds far away from the parent plant. Fruit is the plant's way of seducing mobile organisms to do its bidding and disperse its young. As a result, fruit advertises itself, is easy to find, often easy to handle, and abundantly available. Fruit-eating animals, like manakins, oblige the plant by regurgitating and defecating the seeds from the fruits they eat as they move through the forest.

If the living is so easy for fruit eaters, why don't they just use both parents to raise *lots* more kids? The problem, Snow proposed, is predation at the nest. Lots of chicks means lots of activity to attract predators and therefore lots more risk of losing the whole brood. Snow argued that limiting the clutch size—that is, the number of eggs laid in each bout of breeding—to two allows a single female to raise the family safely and successfully all on her own. By feeding mainly on abundant fruit, a single female manakin can build her own nest, lay the eggs, incubate the clutch, feed the young until fledging entirely by herself, and reduce predation at the nest.

Snow hypothesized that lek display in manakins evolved when an evolutionary shift in diet to fruit meant males were "emancipated from parental care." Females used their capacity for mate choice to select among available mates, and the result was tremendous aesthetic elaboration and diversification of male display. Of course, Snow's scenario for how this would happen was incomplete because he did not yet have an understanding of sexual selection. We now know that unconstrained opportunities for mate choice will lead to the evolution of selective mate preferences—that is, pickiness.

Lekking birds feature so prominently in this book because lek-breeding systems create the strongest sexual selection forces in nature and give rise to the most aesthetically extreme—and often enchanting—forms of sexual communication.

I was excited by my sightings of Golden-headed and White-bearded Manakin leks at the Brownsberg, and I did start to try to map out the male territories within the leks, as Kurt Fristrup had suggested. However, I was much more intrigued by the actual dances the males did than by the spatial relationships of their territories. Besides, David Snow and Alan Lill had already published extensively about these two common and broadly distributed species. I wanted to focus on manakins that hadn't been as well studied.

My real intellectual goal was to find the virtually unknown White-throated Manakin (*Corapipo gutturalis*) and the White-fronted Manakin (*Lepidothrix serena*), which were both reported to occur at the Brownsberg. The male White-throated Manakin is a deep, glossy iridescent blue-black color with an elegant snowy-white throat patch that extends down the breast in a pointed V-shape (color plate 8). The species was so poorly known that it had been left out of François Haverschmidt's *Birds of Surinam*, published in 1968, but birders had recently reported it from the Brownsberg. In contrast, the male White-fronted Manakin is a velvety black with a royal-blue rump, a snowy-white forehead, a banana-yellow belly, and an orange-yellow spot on its black breast (color plate 9). Very little was known about the species in the wild.

Finding a specific bird species among the hundreds of species in a tropical rain forest is a real challenge. At the time, the songs of the White-fronted and White-throated Manakins had not been described for science, and no recordings were available. The only way to find these birds was to persistently bird-watch my way through the entire avifauna until I found them. This method consisted of going out every day, listening for new bird songs, tracking them down, identifying them, learning them, and adding them to my growing mental catalog of bird sounds that were *not* the manakins I was looking for. Of course, this was still spectacularly exciting, because virtually *all* the birds were new to me. Along the way, I would find legendary neotropical birds like the Ornate Hawk-Eagle (*Spizaetus ornatus*), the Crimson Topaz (*Topaza pella*) hummingbird, the Variegated Antpitta (*Grallaria varia*), the Sharpbill (*Oxyruncus cristatus*), the White-throated Peewee (*Contopus albogularis*), the Red-and-black Grosbeak (*Periporphyrus erythromelas*), and the Blue-backed Tanager (*Cyanicterus cyanicterus*). But the checklist of the birds of Brownsberg listed over three hundred species. So, if I wanted to find the two manakins that were my focus, I had my work cut out for me.

At the end of the first week, I found my first territorial male White-fronted Manakin just off a trail on the flattop of the Brownsberg. The advertisement song of this species turned out to be one of the least impressive of all the manakins. It is a single, simple *whreeep* note with the casual, rolling, froggy richness of a brief toot on a police whistle. In my notes from that first day of discovery, I described the song as a "short, sporadic farty trill." The display repertoire of the White-fronted Manakin turned out to be relatively simple, too—on the vanilla end of the diversity in manakin aesthetics. The main male display consists of a series of to-and-fro flights about two feet above the ground, which take him back and forth between thin, vertical saplings that surround a central "court" about a yard wide.

These display flights were of two types. Some were direct "beeline" flights between saplings, with the bird flipping around in midair so that when he landed he would be facing inward toward the court for his return flight. The series of beeline flights would continue for up to twenty seconds. During these displays, the male sometimes perched momentarily on a sapling with his azure-blue rump and white fore crown showing boldly. In the alternative "bumblebee" flight displays, the male flew back and forth between two saplings, springing off the branches as soon as he touched them and hovering in the air with his body held nearly vertical, his wings beating in a rapid whirr. This gave the rather eerie visual impression of a multicolored ball hovering between the saplings at knee height above the ground.

In many days of observation, I saw two probable female visits. I say "probable," because all young male manakins have green plumage like the females. In neither case was I able to observe a copulation, which would have confirmed the sex of the visitor. Marc Théry made later observations of the same species in French Guiana. He observed that females follow the male around the court during several to-and-fro flights and then alight on a small horizontal perch on the court edge. The male then flies up and mounts the female in copulation.

After starting my observations of the White-fronted Manakin, I alternated mornings of watching them at their leks with the search for other manakin species elsewhere in the park. I soon found the male White-crowned Manakin (*Dixiphia pipra*), which is coal black with a bright white crown and bright red eyes, and I observed it for several days. It took a little longer to find the Tiny-tyrant Manakin (*Tyranneutes virescens*), a truly diminutive and amazingly nondescript olive-green bird with an oft-hidden, tiny central yellow crown stripe that weighs in at only seven grams—or about as much as one and two-thirds teaspoons of salt. The male sings a soft, hiccuping little trill from a thin branch about three to five yards high. The first time I found a male singing, he was so motionless and inconspicuous that it took me ten minutes to spot the bird, even though he was perched in plain sight.

I enjoyed my sightings of these birds, but because the display behaviors of both the White-crowned and the Tiny-tyrant Manakin had already been described by David Snow in the early 1960s, I was still determined to find the mysterious White-throated Manakin.

The courtship of the White-throated Manakin was only

known from a brief note published in the British ornithological journal the *Ibis* in 1949, which described a single anecdotal
observation by T. A. W. Davis. One morning in nearby British
Guiana, Davis saw a group of males and "females" consorting
together. (Davis did not consider whether any of these green
"females" could actually have been young males.) He observed
some remarkable male displays and even saw a pair copulating
on a mossy fallen log on the forest floor. The displays included a
posture with the bill pointed upward, revealing the white throat,
and another with the wings held open and the male moving across
the log in a "slow undulating crawl." No one had ever reported a
display like this in any other manakin species, and I was desperate
to see it for myself.

One day in mid-October, I descended the slopes of the mountain to lower-altitude forests along the Irene Val Trail, named
for the lovely Irene Waterfall. It was an active morning in a very
birdy tropical forest. At one point, I heard a whooshing sound
immediately by my head. At first, I thought I might have been
dive-bombed by a hummingbird, but when I looked up, I was
surprised to see a male White-throated Manakin perched on a
branch immediately above the trail. I then realized that I had
just stepped over a large log that was lying in the middle of the
trail. Intrigued by the possibility that I had interrupted him in
mid-display, I backed off the trail to use the forest foliage as a
temporary blind. Immediately, the male flew back down to the
log in the trail with a rapid flurry of whirring wings, bounding
leaps, popping noises, and squeaky calls. The first male was soon
joined by two other adult males and two immature males—which
were identifiable by their mostly green, female-like plumage and
black, Zorro-like face masks. Within the space of a few minutes, I
saw more White-throated Manakin displays than T. A. W. Davis
had in 1949, and I knew that I had a great scientific opportunity
ahead. In the months that followed, I would spend dozens of days
observing the White-throated Manakins and, in the process, get
totally hooked on studying lek behavior.

Although manakin display repertoires are usually dramatic,
the displays of the male White-throated Manakin had a degree

of complexity that was completely new to me, comprising an extraordinarily rich array of behavioral elements. His advertisement song is a high, thin, whistled *seeu-seee-ee-ee-ee*, sometimes shortened to *seeu-seee*. He sings this call quite calmly, only a few times a minute at most, from a perch two to six yards high. The astounding acoustic and acrobatic tour de force in his display repertoire is the log-approach flight display. Starting from a perch five to ten yards away, the male flies toward the log, giving a crescendo of three to five insistent *seee* notes as he goes. In the air, about a foot above the log, the male suddenly stalls in mid-flight with a prominent flap of his wings, producing a sharp *pop,* and drops to the log. Immediately upon landing, he rebounds into the air, turns around in mid-flight, gives a squeaky, raspy, cranky-sounding *tickee-yeah* call, and lands about a foot and a half down the log. He lands instantly frozen in a crouching, bill-pointing posture with his beak held straight aloft and his pointy, V-shaped snowy-white throat patch exposed. I also observed an alternative log approach—the "mothlike flight" in which a male fluttered slowly, undulatingly down to the log with a series of labored, exaggerated wing flaps, all the while holding his body in a vertical position.

Once on the log, the glossy blue-black male performs additional displays. Sometimes, he crouches and lowers his beak to the log, holding the wrists of the wings slightly shrugged above his back, while running back and forth across the log. In the "wing-shiver" display, he holds his body horizontal and opens and closes each wing in rapid, alternating succession, flashing the brilliant white patches that are concealed when his wings are closed. With each alternating wing opening, the male shuffles the foot on the same side to creep backward along the log. This is Davis's "slow undulating crawl."

Each male displays at a few logs within a territory about twenty yards wide. The display excitement within a male's territory is occasionally enhanced by the arrival of a rowdy, traveling band of two to six males of mixed age that display together and with the territory holder. The groups include both adult males that may have their own territories but have temporarily joined

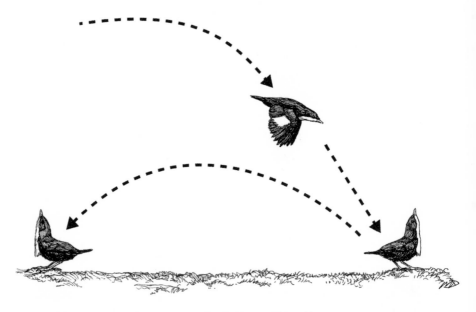

The log-approach display of the male White-throated Manakin.

the wandering, group display and young males in various stages of preadult plumage, who apparently do not hold territories. These group displays are *not* coordinated but more like a highly competitive form of rabble-rousing. Males vie for access to the same display log, performing a rapid flurry of log-approach displays one after another and frequently displacing each other from the log. During the competition for control of the log, males "strafe" each other by flying low over the log and producing only a mechanical *pop* just over the male on the log, right at the nadir in flight. The result can be an exciting flurry of *pops* and log-approach calls in rapid succession by different males: *POP-tickee-yeah—POP—POP-tickee-yeah—POP!*

During months of observations at White-throated Manakin logs, I saw only two female visits. One or two green-plumaged individuals perched on a log and intently observed a displaying male while he performed a series of log-approach displays or wing-shiver displays. Interestingly, when performing the wing-shiver display for a visiting female, the male turned his back and crawled *backward* toward the female. Even during the bill-pointing pos-

The bill-pointing (left) and wing-shiver (right) displays of the male White-throated Manakin.

ture, which displays his bright white throat, he turned his back on the female. With his beak held high, he often peered nervously over his shoulder to monitor how the visiting female was responding to his display. I myself saw no copulations. But both T. A. W. Davis in British Guiana back in the 1940s and Marc Théry in French Guiana many years later documented that copulation takes place on the log after a series of these displays, with the male mounting the female directly on the rebound from a log-approach display.

In November 1982, an unusual, and unusually talented, birder arrived at the Brownsberg. Tom Davis was a lanky, six feet eight, foulmouthed telephone company engineer and legendary New York birder from Woodhaven, Queens, with great identification skills and an audiophile's obsession for recording bird song in the field. Through a series of birding vacations, Tom had become an outstanding expert on the birds of Suriname. When Tom arrived, he told me that during the previous year, while sitting on a bench overlooking the forested valley where he had been birding for so many years, he had discovered a spectacular *above-the-canopy* flight display by White-throated Manakins.

In our very first day together in the field, Tom took me to a viewing point from which he was able to show me this novel flight display, which took place more than fifty to a hundred feet *above* the tallest trees in the forest. After waiting for about thirty minutes, I saw a male ascending skyward while vocaliz-

ing an emphatic series of *SEEEE . . . SEEEEE . . . SEEEEE* notes
that were even louder, more intense, and more emphatic than the
similar notes I'd heard at the logs during log-approach displays.
The ascending male flew in a bizarre fluffed-out posture looking
rather like a black-and-white cotton ball. After the male reaches
the apex of his flight, he suddenly plummets back down into the
forest. In the previous year, Tom had made a tantalizing observation; some of the above-the-canopy flight displays end with a
loud, mechanical *Pop!* note after the male disappears back into
the forest.

In the weeks that followed, I was able to piece together the
entire display sequence. One day during observations at a display
log, I heard the especially intense version of the *SEEEE* calls that
the male makes during his above-the-canopy flight from overhead
and suddenly saw the male come careening downward through a
hole in the forest canopy toward the log and perform a full log-
approach display. Only then did I realize that I should have been
looking up! Within a few days, I made multiple observations of
males plummeting down through the forest canopy to the log
after their above-the-canopy flights.

I am sure that I would never have discovered these flight displays by myself, given that I was spending all my time inside the
forest at the display logs themselves. So, Tom Davis's fantastic
observations were essential to the story. The specific function of
this especially extravagant behavior—advertising to females over
many acres of forest?—remains enigmatic.

My ornithological *Wanderjahr* in Suriname was a transformative personal and intellectual experience. I had made it out of
the university to a distant and exotic corner of the world, and I
had thrived. During my five months there, I had used my birding
skills to observe hundreds of species of birds. I came away with
unique scientific observations of previously unknown lek behaviors, which were significant enough to constitute my first scientific
papers, published a few years later in the canonical ornithological

journals the *Auk* and the *Ibis*. I had also made good progress on devising a doctoral project on the evolution of manakin behavior.

The next year, I had the opportunity to return to South America to work as a field assistant to a Princeton graduate student, Nina Pierpont, who was studying woodcreeper ecology at Cocha Cashu—a remote, Amazonian field station in southeastern Peru. My research at Cocha Cashu proved to be critical to my future life, for it was there that I met Ann Johnson, a Bowdoin College student who was working as an assistant for a Princeton undergraduate student, Jenny Price, on the social behavior of White-winged Trumpeters (*Psophia leucoptera*). Ann and I became sweethearts that summer, and we have been partners ever since. Ann is a producer and cinematographer of nature and science documentaries for television. We have three sons.

In the fall of 1984, I started graduate school in evolutionary biology at the University of Michigan. Inspired by the diversity and complexity of manakin displays from Suriname, I proposed for my dissertation a grand, comparative analysis of the evolution of manakin behavior across the entire family. I wanted to use manakin phylogeny—their family tree—to study the evolution of manakin lek display behavior. This emerging scientific field combined phylogeny with the study of animal behavior, called ethology, into a vibrant new discipline—phylogenetic ethology. The goal was to investigate the evolution of behavior comparatively through its history. Although I didn't realize it at the time, this was my first step into the study of aesthetic radiation.

During my first year in graduate school, my office mate, Rebecca Irwin, introduced me to the classic work of Ronald A. Fisher and to the revolutionary new papers on mate choice by Russell Lande and Mark Kirkpatrick. This was my first exposure to the science of mate choice and to the deep intellectual conflicts between the aesthetic/Darwinian and the adaptationist worldviews. But even then I could sense that the open-ended and arbitrary qualities of the Fisher hypothesis looked a lot more like how nature worked than the honest signaling theories did.

I was desperate to get back to South America and continue

with my manakin fieldwork. I did not know where to go, but I was particularly intrigued by the idea of going to the Andes, which would provide so many great birding experiences. So, for my first summer in graduate school in 1985, I proposed that Ann and I would conduct field research in the Ecuadorean Andes to discover the unknown lek display behavior of the nearly mythical Golden-winged Manakin (*Masius chrysopterus*). I had no better justification for the research than the fact that the bird was entirely unknown. I certainly did not tell my advisers or the grant agencies that I had chosen this bird in particular because it was beautiful and happened to live in the Andes, where hunting for it would be so birdy, fun, and rewarding. But thanks in part to my new track record of published manakin display descriptions, I managed to get a few small grants to fund this high-risk project. Even the local camping outfitter, Bivouac in Ann Arbor, agreed to subsidize the purchase of the camping equipment that we would need for the fieldwork, which helped make my few dollars go further.

By any measure, the Golden-winged Manakin is a strikingly gorgeous bird (color plate 10). The male's plumage is mostly velvety black with a brilliant, plush yellow crown that extends slightly forward in a brushy crest over the beak, like a 1950s greaser hairstyle. The hind crown is brilliant red in the populations located on the east slope of the Andes and reddish brown in populations on the west slopes. On either side of the crown, the male sports two tiny, black, feathery horns. However, the truly stunning features of the male's plumage are usually discreetly hidden. The wing and tail appear completely black when the bird is perched. But once in flight, the inner vane of each wing feather is revealed to be a vivid golden yellow, the same color as his crown. As we would discover, the sudden golden flash of his wings in flight is a major feature of the male's courtship display, producing a visual effect that is as breathtaking as it is unexpected.

When Ann and I arrived in Ecuador, all that we knew about this bird came from what we had learned from fifty-year-old museum specimens. In 1985, there were no recordings of the Golden-winged Manakin in the collections of the Cornell Lab of Ornithology or the British Library of Wildlife Sounds, so we didn't know what the bird sounded like. We also knew nothing about its breeding season, because this was among the many things completely unknown about the species.

We started our search in Mindo, a little town on the western slopes of the Andes at sixteen hundred meters in altitude, to the west of the capital of Quito. Mindo has since become a bustling ecotourism destination, but in 1985 it was a sleepy village with only a few dozen houses lining its mud streets. The forests around Mindo, however, were filled with diverse birdlife. We were thrilled to find Golden-winged Manakins foraging for fruit among flocks of brilliant *Tangara* tanagers. But we were unable to find any territorial males or any evidence of song or display activity. When asked by the curious locals if we had found the bird we were looking for, we had to explain, "La epoca no está buena." It's not the right season. Of course, we had no idea what the right season *was*.

After a month in Mindo without success, we got a great tip from an expatriate American ornithologist and bird artist, Paul Greenfield, who would later co-author the excellent *Birds of Ecuador* with Robert Ridgely. Paul had recently been birding along a mini railway line that ran parallel to the Colombian border from the north Andean town of Ibarra down to San Lorenzo on the Pacific coast. In the cloud forest around the tiny settlement of El Placer, he had seen plenty of Golden-winged Manakins. Perhaps, he suggested, if we went to a new locality with different geography, altitude, and weather conditions, it would be breeding season there, and we would be able to find the displaying males we were looking for.

We decamped to El Placer—literally "Pleasure"—via a train that consisted of a single car, like a city bus with small-gauge railroad wheels. This one car made a single round-trip to the coast and back each day. The "town" of El Placer was really just a collection of about ten rough-hewn, tin-roofed plank houses for the families of the workers who maintained that stretch of the railroad track. Besides the houses there was nothing in El Placer except an empty school, a railroad company office that doubled as a small store, and a few muddy footpaths into the surrounding forest.

El Placer must surely rank among the rainiest places on earth. It rained or drizzled continuously throughout the six weeks we were there. Even at the quite low altitude of five hundred to six hundred meters, the forest was very cool and mossy. The forest was second-growth cloud forest that had regenerated since the construction of the railroad decades before. We found a beautiful community of birds there, including Golden-winged Manakins, on the very first morning.

The first Golden-winged Manakin we saw was perched calmly on a branch about six feet above the ground, inside the dense mossy forest. In these very low light conditions, his velvety black plumage was like a light sponge, but his golden crown was brilliantly visible. He uttered a brief, low, raspy, frog-like *nurrt* call about three times a minute, a vocalization that was so underwhelming that we could easily have passed it off as the occasional call of a frog or insect. Between displays, male manakins often look like idle workers waiting out a long shift at a rather boring job. So, this male's quite sedentary and indolent attitude was an excellent indication that he was on territory. My hunch was soon confirmed when we heard and located a second calling male about twenty meters away across the trail. This was clear evidence of a lek with multiple males and a great find after our weeks of fruitless observations in Mindo.

Given the unpredictability of wild bird behavior, you can never really know if the first moments of observation will be your last or the start of months of subsequent study. So you must always proceed as if the first sightings are the only opportunities you will ever have. We immediately deployed tape recorders and notebooks to record the behavior and songs of the two Golden-winged males, noting the qualities of the song, the rate of countersinging between them, and the positions of their song perches.

After an hour or so, I heard a remarkably familiar sound coming from the area of the first singing male. It started with a high, thin descending whistle and ended with an accelerated and syn-copated riff—like *seeeeeeeeeeeeeeeeee-tseet-tseee-nurrrt!* I was immediately reminded of the log-approach display flight song of the White-throated Manakin from Suriname. The similari-ties were so strong that I became confused. We were thousands of miles away from the range of the White-throated Manakin in northeast South America; how could this be? The unexpected, even unimaginable solution to this conundrum would soon become clear, but in my mind there was still a lot of resistance to realizing it.

I returned to watch the first male Golden-winged Manakin in his territory, and what I observed over the next few minutes was profoundly surprising. Indeed, it was a scientific revelation. The male continued counter-singing, trading *nurrt* calls with the neighboring male, but he then flew off his habitual perch and into the dark forest. In a few moments, however, I heard the long, thin, high-pitched, continuous, descending *seeeeeeeeeeeee* note approaching through the air. I then saw the male Golden-winged Manakin drop rapidly in flight to land on a large, exposed but-tress root of a tree right in front of me. As he landed, he immedi-ately rebounded into the air, turning around in mid-flight, vividly flashing his brilliant golden wing patches, and landed back down on the root facing back in the direction of his first landing posi-tion. As he landed, he froze in an elongate tail-pointing posture with his beak held down against the surface of the root, his body plumage sleek, and his tail held up at a forty-five- to sixty-degree angle in the air.

As rapidly as the brain converts an optical illusion from one image into an entirely new picture that was previously impercepti-ble, a rich and highly detailed set of scientific conclusions became immediately clear to me. The calls that were surprisingly similar to the White-throated Manakin's were the log-approach display call of the Golden-winged Manakin. The host of remarkable sim-ilarities between the display behaviors of these two species were behavioral homologies—similar behaviors that they had both inherited from an ancient, shared ancestor, a common ancestor that no one had ever even conjectured might exist. Because the

The log-approach display of the male Golden-winged Manakin.

males of these two species look completely different from each other and are in two different genera, no one had ever before hypothesized that they were closely related to each other. However, after I saw their displays, it was immediately and vividly clear to me that the White-throated Manakin (*Corapipo gutturalis*) and the other *Corapipo* manakins were the closest relatives of the Golden-winged Manakin.

It is hard to express how astounded I was by this discovery. It was a true epiphany, the culmination of weeks of futile searching, nine months of planning for the trip to the Andes, five months of previous fieldwork in Suriname, years of academic studies in ornithology and the sciences, and a parallel life of birding. All these influences had coalesced in an instant to reveal a heretofore entirely unsuspected connection. Never once during all my planning for this Andean expedition for the Golden-winged Manakin had I imagined such a possibility, that I could rewrite the phylogeny of the manakin family. Nor could I have, in my wildest dreams.

Of course, the stunning result of the expedition was personal proof that it really pays to listen carefully to the voice of one's private ornithological muse. It pays to be lucky, too, for obviously I could never have come to this moment without my previous observations of the White-throated Manakin, which I was among the very few people on earth to have seen. My observations of White-throated Manakins in Suriname proved to be a unique and essential preparation for understanding the evolutionary implications of what I had witnessed in El Placer. What's more, this newly revealed evolutionary pattern also implied something fundamental about the process of sexual selection by mate choice and the consequences for the assembly of complex repertoires of ornamental traits and seductive signals. Thirty years later, these discoveries still resonate in my work.

In the coming weeks, Ann and I would spend over 150 hours watching, tape-recording, and filming the display behaviors of the Golden-winged Manakin. It would take me much more analysis in order to establish the exact details of the host of behavioral homologies shared between these species since their common ancestor. It was obvious that long ago a common ancestral species had evolved a unique display repertoire, elements of which the Golden-winged and White-throated Manakins still exhibit in the present day.

But it was also clear that over time parts of that repertoire had diverged and transformed, with each species evolving its own unique display elements. I discovered many such differences between them. For example, once on the log, the Golden-winged males do not perform the bill-pointing posture and the to-and-fro display like those of the White-throated Manakin. Nor do Golden-wings perform anything like the wing-shiver display, even though they have a glorious golden wing patch to show off during such a display. Male Golden-winged Manakins do, however, have a unique display of their own. Once on the log, the male performs an elaborate "side-to-side bowing display," in which he fluffs out his body plumage, cocks his tail slightly, and erects the tiny black hornlets on either side of his golden crown. Then, with the mechanical rhythm of a windup toy in a davening trance, he bows forward, nearly touching his bill to the log, rises up, takes a few steps to the side and rotates a bit, bows again, takes a few steps back in the original direction, and bows, and so on. The males we observed continued this display for ten to sixty seconds without interruption. Nothing remotely like this occurs among the White-throated Manakins or any other manakin species.

These exciting discoveries helped establish that the aesthetic repertoires of manakins are hierarchically complex. The visual, acoustic, and acrobatic displays of manakins are composed of some behavioral elements that have been handed down from their ancient common ancestors, and others that have subsequently evolved in unique ways in each of those species. The beauty of manakins cannot be understood solely in terms of the current environment or population context, but is contingent upon phylogenetic history. The full evolutionary history of beauty can only be understood in the context of phylogeny. The history of beauty is a tree.

Fleshing out the details of what behaviors had changed evolutionarily on what branching of the tree required my finding a third manakin species to which I could compare the Golden-winged and White-throated Manakins. In the same sense that it takes more than two data points to describe a statistical trend, it is difficult to make conclusions about the details of evolutionary

The tail-pointing (left) and side-to-side bow (right) displays of the male Golden-winged Manakin.

history from a comparison of just two species. For example, spider monkeys have tails, but humans do not. Clearly, some evolution in tails has happened since these two species had a common ancestor, but which way did it go? Did the spider monkey evolve a tail? Or did the humans lose one? Only by looking at a third, more distantly related species—say, a lemur, tree shrew, or dog—can we infer that the evolutionary event was the *loss* of the tail in an ancestor of humans after shared ancestry with the spider monkey.

So, what third species could I use to reconstruct the evolutionary history of the Golden-winged and White-throated Manakins? It would have to be closely related enough to Golden-winged and White-throated Manakins to be useful. (In the example above, comparing primates with sea urchins, worms, or jellyfish would not have helped me infer the evolutionary history of their tails.) Luckily, in the fall of 1985, soon after my return from Ecuador, Barbara and David Snow published a beautiful description of the poorly known courtship display behavior of the Pin-tailed Manakin (*Ilicura militaris*) from the lower montane forests of southeastern Brazil. The male Pin-tailed Manakin has the bright, crisp, bold plumage color patterns of a toy soldier, as the scientific name of the species suggests (color plate 11). The male is gray below, black on the back and tail, green on the wings, with a red rump and a bright red plush fore crown. The central feathers of the male's black tail are narrowly pointed and twice the length of the other tail feathers. The female is olive green above, and dull greenish-gray below, with somewhat elongate central tail feathers.

Because the male Pin-tailed Manakins look *entirely* different from the male Golden-winged and White-throated Manakins, these three species had never been hypothesized to be closely related. However, as I read the Snows' descriptions of the display repertoire of the Pin-tailed Manakin, I could see that many of its elements resembled the behaviors of Golden-winged and White-throated Manakins, and I was certain that the Pin-tailed Manakin was closely related to the Golden-winged and White-throated Manakins. By including the Pin-tailed Manakin in my analysis, I was able to resolve many outstanding questions about the evolution of the behavioral repertoires of the Golden-winged and White-throated Manakins. By comparing all three species, I could identify which display behaviors had evolved in the common ancestor of all three species, which behavior novelties had evolved in the exclusive ancestor of the Golden-winged and White-throated Manakins, and which behavioral elements had evolved uniquely in each of the three species.

For example, I first considered the evolution of the male display sites. Most manakins display on thin tree branches. Golden-winged and White-throated Manakins are unique in the family in displaying on mossy fallen logs on the forest floor. Pin-tailed Manakins, on the other hand, display on upper surfaces of thick horizontal branches of trees, which are basically like living logs up in the trees. So, it appears that displaying on thick branches evolved in the common ancestor of all three species from the thin perches of ancestral manakins. Then displaying on fallen logs or buttress roots evolved subsequently in the exclusive common ancestor of the Golden-winged and White-throated Manakins.

Another trait I examined was the tail-pointing posture. On their thick display branches, Pin-tailed Manakins perform a tail-pointing posture that is homologous with the Golden-winged Manakin's but doesn't resemble anything the White-throated male does. Thus, I concluded that the tail-pointing posture had evolved in the common ancestor of all three species but was lost

The tail-pointing display of the male Pin-tailed Manakin.

in the White-throated Manakin lineage and replaced by the novel bill-pointing posture.

By thoroughly comparing the behaviors of all three species, I developed a comprehensive hypothesis of the history of behavioral diversification in the group. The display repertoires of each species had included physical, vocal, and display elements and had evolved in many creative ways: by insertion of entirely novel elements into the repertoire; by the elaboration of current elements in new ways; and by the combinations of elements and the loss of ancestral elements. I was able to propose an entirely new hierarchical view of the coevolutionary history of manakin beauty.

For my doctoral dissertation, I went a step further, using new information about manakin anatomy to produce a reasonably complete and well-resolved phylogeny of the entire manakin family. This research involved hundreds of dissections of the syrinx— the unique little gizmo the birds sing with—of all manakin species. I then used this evolutionary tree to test my hypotheses about behavioral homology. For example, I found common features of syringeal structure that confirmed my hypothesis that the Pin-tailed, Golden-winged, and White-throated Manakin genera had an exclusive common ancestor. And, as I had proposed based on

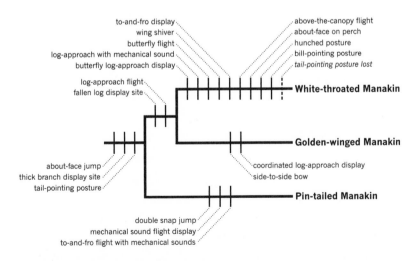

Phylogeny of the White-throated, Golden-winged, and Pin-tailed Manakins depicting the evolutionary origins and losses of the behavioral elements within the display repertoires of each species and their shared ancestors. *Based on Prum (1997).*

their display behavior, these features also pointed to the Golden-winged and White-throated Manakins' being more closely related to each other than either was to the Pin-tailed Manakin.

Today, what we know of the aesthetic radiation of manakins provides many evolutionary lessons about how Beauty Happens over the Tree of Life. We've learned that manakin aesthetic repertoires include many elements that are older than the individual species themselves. We can see that each species' display repertoire is contingent upon both the evolutionary legacy of that species—what it inherited from its various ancestors—and any new display elements—aesthetic elaborations, innovations, or losses—that have evolved in that species alone.

How the elements of a given display repertoire come into being over the course of time shows us the inherently serendipitous and unpredictable nature of aesthetic evolutionary process. From a common history, sister species evolve in many different and unpredictable aesthetic directions. Through each aesthetic change, mate choice also creates new aesthetic opportunities, which can unleash an evolutionary cascade of effects. These include the evolution of further aesthetic extremity and complexity. As Beauty Happens, different species evolve off in ever more different, arbitrary directions from their shared ancestral repertoires. Especially when sexual selection is strong, as in manakins and other lekking birds, Beauty Happening over the course of long evolutionary timescales results in explosive aesthetic radiations.

My fieldwork in Suriname in 1982 launched me on a path of exploration that I have continued to follow down to the present day (though with diminished capacity in recent decades due to major hearing loss). In the intervening decades, I have conducted ornithological research in twelve neotropical countries and have had the good fortune to observe nearly forty species of manakins in the wild. (I am still working eagerly to see the rest.) For some of those species, I spent hours, days, or even months of observation getting to know their habits, observing their daily rhythms, describing their courtship songs and dances, and mapping out their social relationships. This helped me to build a rich database of natural history knowledge about manakin behavioral complexity and aesthetic diversity.

But my ever-expanding knowledge of manakin diversity also taught me to ask bigger, more fundamental questions about the evolutionary workings of the natural world. Early on, I had thought of manakins as colorful birds with delightfully bizarre display and social behaviors. Later, I conceptualized the manakins as a great example of how the complex mechanisms of mate choice affect behavioral evolution among species. Most recently, I have come to think of manakins as one of the world's premier examples of aesthetic radiation. And as we'll see in a later discussion of manakins (see chapter 7), the female manakins haven't only transformed male display repertoires; they've changed the very nature of male social relations. It's an astonishing story of the transformative power of female mate choice.

Manakins are just one small piece of a vast tapestry of avian beauty. There are over ten thousand species of birds in the world, ranging from the plainest of sparrows to the most exquisite of manakins. Because every single bird species exhibits some specific sexual ornaments that are employed in courtship communication and mate choice, it is clear that the capacity for mate choice in birds originated in an ancestor common to *all* birds, perhaps even in a lineage of feathered theropod dinosaurs dating all the way back to the Jurassic. From this single common ancestor, the repertoire of aesthetic traits and mating preferences has continued to coevolve and radiate into the many thousands of distinct forms of avian beauty that exist today. On different phylogenetic branches at different times, the pace of coevolutionary change has slowed or increased as new ecologies have contributed to variations in breeding systems and parental care arrangements, which in turn have given rise to tremendous variation in the nature and strength of sexual selection by mate choice. Along the way, mate preferences have continued to evolve in various avian lineages, sometimes occurring in both sexes, sometimes in females only, or, much less often, in males only, and the aesthetic repertoires of the sexes have coevolved accordingly. Each lineage and species has evolved along its own distinctive and unpredictable aesthetic trajectory. The result has been the flowering of more than ten thousand distinctive aesthetic worlds comprising over ten thousand coevolved repertoires of displays and desires.

Something comparable has occurred on myriad different branches across the entire Tree of Life. From poison dart frogs and chameleons to peacock spiders and balloon flies, whenever the social opportunity and sensory/cognitive capacity for mate choice has arisen, an aesthetic evolutionary process has taken hold. This aesthetic evolutionary process has arisen hundreds or thousands of times during the history of life, even in plants that have evolved ornamental flowers of distinct shapes, sizes, colors, and fragrances to seduce animal pollinators into dispersing their gametes (in the form of pollen) to other flowers waiting to be fertilized.

Throughout the living world whenever the opportunity has arisen, the subjective experiences and cognitive choices of animals have aesthetically shaped the evolution of biodiversity. The history of beauty in nature is a vast and never-ending story.

CHAPTER 4

Aesthetic Innovation and Decadence

In the understory of a mossy cloud forest in the western Andes of Ecuador, a small cocoa-brown bird with a red fore crown sings from a slim perch. *Bip-Bip-WANNGG!* The tonal sound rings like feedback from an elfin electric guitar. Three other males within earshot call back in rapid response with increasing excitement. These are territorial male Club-winged Manakins (*Machaeropterus deliciosus*) at a lek displaying to attract mates. The strange acoustic quality of their songs is associated with an even stranger movement. Instead of opening their beaks to make their electronic-sounding songs, the male Club-wings flick their wings open at their sides to make the initial *Bips* and then snap their wings up over their backs to set their swollen and twisted inner wing feathers into rapid sideways oscillation to produce the extraordinary *WANNGG* sound (color plate 12). These male Club-winged Manakins are *singing with their wings*.

We have seen that many other manakins make *pop* and *snap* sounds with their wing feathers during courtship display. White-throated Manakins make a loud *pop* as they stall in flight over their display logs. White-bearded Manakins make their explosive *snaps* as they leap between the display court and the surrounding saplings, and they produce a loud flatulent roll—a rapid series of *snaps*—while perched above their courts. The many variations on *snap, crackle,* and *pop* in the manakins are all feather sounds.

The existence of these nonvocal communication sounds is evolutionarily baffling, because manakins all have perfectly good vocal songs that remain an important part of their aesthetic repertoires. Why would any species—let alone many separate species—evolve an entirely new way to sing when the traditional avian vocal songs had been working fine, even gloriously, for over seventy million years?

Like eyes, limbs, and feathers, the mechanical sounds of manakins are examples of evolutionary innovations—entirely novel biological features that are not homologous with any ancestral, or antecedent, feature. Evolutionary innovations are intellectually exciting because they require more than simple, incremental, quantitative change—more than mere evolutionary tinkering, if you will. Innovations involve the evolution of genuinely new phenomena and features, or qualitative evolutionary novelties.

The evolution of limbs, eyes, and feathers is an important subject in evolutionary biology. Indeed, I have worked a lot myself on the evolutionary origin of feathers. But the mechanical sounds of manakins are distinct from all of these evolutionary novelties

The male White-bearded Manakin produces the roll-snap wing sound by clapping its wings together rapidly over its back.

because they are *aesthetic* innovations that have evolved by mate choice. Aesthetic innovations provide us with a unique opportunity to investigate both how sexual coevolution works and how evolutionary innovations happen. In recent years, biologists have discovered that adaptation provides at best an incomplete account of the process of evolutionary innovation. I hope that by exploring aesthetic innovation here, we will see that adaptive mate choice provides an insufficient explanation of the origin and diversification of ornament as well.

So, how did the innovative mechanical sounds of manakins evolve? The best hypothesis is that manakin display movements produced incidental noises—the whirrs or shuffles or other sounds of moving feathers—in the same way that running and dancing produce incidental noises as feet touch the ground. However, through aesthetic coevolution, these incidental sounds became subject to female preferences along with the rest of the display. Consequently, distinct preferences for such sounds evolved and diversified, until the sounds themselves became a distinct part of the aesthetic repertoire of the species, much as tap dancing became its own genre of dance. Mating preferences for mechanical wing songs probably evolved from earlier acoustic preferences for vocal advertisement songs and became distinct, new preferences over evolutionary time.

The Club-winged Manakin has gone in for innovation in a big way. Most manakins, like tap dancers, are satisfied making percussive *pops, snaps,* and *riffles,* but the male Club-winged Manakin really *sings.* Sings, perhaps, even better than he flies. As we'll see, the Club-winged Manakin is not only an example of aesthetic innovation; it also shows us how adaptation and aesthetic selection can be at odds with each other and how decadent beauty can win.

I first heard the wing songs of the Club-winged Manakin in 1985 on our first morning at El Placer, where Ann and I discovered the lovely and unexpected log dances of the Golden-winged Manakin. Among all the sounds in the busy morning chorus coming from the mossy forest that day, I thought at first that these odd electronic notes might be the musical musings of a parrot—a brief, half-heard snippet of the highly variable, quiet, warbling chatter that parrots sometimes sing to one another while perched in close-knit groups. Later that day, I was stunned to discover that this sound came from inside the forest understory and was made by the legendary, and poorly known, Club-winged Manakin. In the coming weeks, during our searches for additional Golden-winged Manakin territories, we found a few leks of Club-wings in the same forests, and I gorged myself on watching them and tape-recording their contorted musical performances. The wing songs are a major component of the lek display of the species. Indeed, unlike other manakins, male Club-winged Manakins have a greatly reduced vocal repertoire and no vocal advertise-ment song. One very simple vocalization—a series of sharp *keah* notes—is produced during its crouching display.

At El Placer, we caught Club-winged Manakins in the same mist nets we used to capture the Golden-winged Manakins for color banding. The wing feathers of female Club-wings were nor-mal in every way, but the inner secondary flight feathers of the adult males—the feathers that attach to the trailing upper fore-wing bone called the ulna—were truly bizarre. Indeed, they had been illustrated in 1860 by the British ornithologist Philip Lutley Sclater in his description of the species. Sclater's illustrations were reproduced by Darwin in the section of his *Descent of Man* on the instrumental music of birds, in which Darwin hypothesized that the mechanical sounds of manakins and other birds evolved by mate choice. Specifically, the male Club-winged Manakin's fifth, sixth, and seventh secondary feathers (counting inward from the wrist) have greatly thickened, swollen central shafts, or rachises. At the tip, sixth and seventh secondaries form twisted knobs, like the handles on the tops of tiny shillelaghs, or the tips of mis-shapen soft-serve ice cream cones. In contrast, the fifth secondary feather has a sharp forty-five-degree bend near its tip that creates a smooth blade pointing inward toward the body.

When I first saw these songs being produced, I struggled to

The secondary wing feathers of the male Club-winged Manakin. (Left) The open wing viewed from below. (Below left) The crooked, blade-like tip of fifth secondary feather. (Below right) The swollen tip of sixth secondary with a row of prominent bumps. *From Bostwick and Prum (2005).*

imagine how feathers could make such a sound—even the stiffened and twisted flight feathers of male Club-winged Manakins. It would take another twenty years to figure it out. This long delay had a few sources. The first problem was technological. We had to wait until high-speed video technology was invented and became rugged enough for use in a cloud forest. The second was personnel. Eventually, in the late 1990s, I was lucky enough to recruit an enterprising and ambitious graduate student, Kimberly Bostwick, to my lab after her undergraduate work at Cornell University just as the first generation of field-worthy, high-speed video cameras became available. As always, perhaps the biggest barrier of all was *intellectual*. The sound production mechanism the birds actually use turned out to be a mechanism that I had considered and rejected immediately in El Placer in 1985 as ridiculously out-landish. Luckily, Kim's perseverance led both to discovering the answer and to convincing me that I was totally wrong.

Kim Bostwick started her pioneering doctoral research on the functional morphology of feather sound production with the "easy" manakins. For example, using high-speed video cameras, Bostwick showed that White-bearded (*Manacus manacus*) and White-Collared Manakins (*Manacus candei*) make their *snaps* by slapping the upper surfaces of the wings together *over* their backs. Likewise, their Bronx cheer "roll snaps" are made with an incredibly rapid series of the same wing-slapping movements.

The *Manacus* wing sounds are certainly behavioral innova-tions, but the sound production mechanism is quite simple. Wing *snaps, pops,* and *clicks* are made by feather percussion, and these sounds are as acoustically sharp and abrupt as the movements that make them. However, the ringing, musical wing song of the Club-winged Manakin is unique. It has a genuine frequency, pitch, or tone like a violin or the dial tone of a phone, and the longest note rings for more than one-third of a second.

In 2002, Kim conducted weeks of fieldwork in northwest-ern Ecuador and ultimately captured beautiful high-speed video sequences of Club-winged Manakin males singing their wing songs. At five hundred or a thousand frames per second, the videos revealed that during the production of the sustained *WANNGG* sound the wing feathers oscillate from side to side in a nearly vertical plane over the bird's back and that these oscillations are driven by tiny, rapid, side-to-side movements of the wrists. The flight feathers of the left and right wings swing outward and then inward, in sync with each other. At the end of the inward swing, the swollen flight feathers of the left and right wings collide in the center over the male's back and rebound outward again. The feathery oscillations continue at the blistering rate of nearly one hundred cycles per second for one-third of a second. The tiny pumping movements of the wrists are among the fastest verte-brate muscle movements ever observed.

Bostwick's beautiful videos answered lots of questions but also posed new problems. The frequency of the wing oscillations

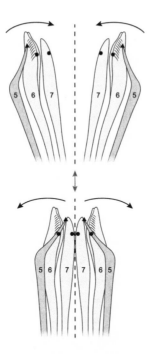

Graphic model of sound production by the secondaries of the male Club-winged Manakin. As the secondary feathers oscillate rapidly inward (top) and outward (bottom) over the back of the bird at one hundred cycles per second, the blade on the tip of the fifth secondary rubs against the bumps on the swollen sixth secondary, inducing it to vibrate at the frequency of the sound (1500 cyles per second). *Based on Bostwick and Prum (2005).*

is near a hundred cycles per second, but the frequency of the wing song is around fifteen hundred cycles per second. That is a pitch between high F-sharp and high G, or about one musical fifth above high C (keys 70 or 71 on the piano). In other words, the frequency of the sound was about fifteen times faster than the frequency of the wing feather oscillations. How is the frequency of the movement multiplied to produce the frequency of the sound? How could this work?

Bostwick realized (and then convinced me!) that interactions *among* the feathers were critical to producing the sound. With each oscillation, the sharp blade on the bent end of the fifth secondary rubs up and down the swollen knob on the sixth secondary. And the surface of the thickened sixth secondary has a series of tiny ridges exactly on the surface that contacts the blade of the fifth. Like bowing a violin or strumming your fingers back and forth over the tines of a comb, the blade of the fifth secondary applies a series of mechanical impulses into the sixth secondary, which drives the sixth and seventh secondaries to resonate loudly at the frequency of a high F-sharp/G.

This mechanism of sound production, called stridulation, is the same way that crickets, katydids, and cicadas make their chirps and whines. Stridulation was the ridiculous hypothesis that I had rejected completely as impossible from the very start while watching these birds twenty years before. So much for scientific intuition.

Just as the pitch of a violin string is determined by its length, mass, and tension, the frequency of sound produced by any resonator is determined by its physical properties. In 1985, I just could not imagine a feather—even a thick Club-winged secondary feather—as an effective resonator. However, just as our analysis of the high-speed video predicted, Bostwick and other collaborators later showed that the fifth, sixth, and seventh secondary feathers of male Club-winged Manakins have extraordinary resonance properties at fifteen hundred cycles per second, which other, normal manakin feathers lack. Furthermore, the coupled oscillations among the secondary feathers function to further amplify the volume of the sound. It is the acoustic collaboration among the multiple feathers attached to the male's ulnas that gives the sound its distinctive harmonic structure and decidedly musical, ringing, violin-like quality. Bostwick's analyses showed that avian beauty can be both innovative and almost ridiculously complex.

The Club-winged Manakin's aesthetic innovations pose enormous challenges to adaptive mate choice. It is possible that manakin wing songs could be correlated with variation in male quality, but the vocal songs of birds are also supposedly correlated with quality. If vocal songs are already robust indicators of quality, then why would any species abandon one highly evolved, honest indicator in favor of an entirely new and yet unproven sound production technique? Adaptive mate choice explanations often seem like Rudyard Kipling's *Just So Stories,* in which the extraordinary features of animals—the giraffe's neck, the elephant's trunk, and the leopard's spots—are explained with a set of outlandish events. However, in the case of the Club-winged Manakin, the *Just So Stories* for vocal songs and wing sounds conflict. They cannot both be entirely true.

Alternatively, it could be that Beauty Happens—that arbitrary mating displays and preferences coevolve in the absence of natural selection on preference for quality information or mating efficiency. According to this hypothesis, the Club-winged Manakins' stridulating wing songs are merely another delightful and unexpected event in the marvelous aesthetic radiation of the manakins.

If Beauty Happens, then sexual display traits do not always improve survival, and can instead evolve to be highly costly to the individuals that have them. Each display trait is predicted to evolve to an equilibrium between its sexual advantage and its survival costs, and this equilibrium may be *far* from the optimum preferred by natural selection for male survival and fecundity alone. The sexual advantages of attracting mates can outweigh the survival advantages of being well adapted. In other words, a handsome, reckless, die-young James Dean–type may leave more offspring than a quiet librarian who lives to be an octogenarian.

How far will beauty, and the preference for it, go to ensure sexual advantage? Pretty far. In subsequent research on the Club-winged Manakin, Kim Bostwick has provided a definitive scientific answer to an immortal question. She documented that beauty is *not* only skin deep, and her discovery provides profound insights into how aesthetic evolution works.

Making those unusual wing songs requires more than just unusual feathers and movements. It requires major evolutionary changes in the shape and composition of the wing bones and the sizes and attachments of the wing muscles. Wing bones and muscles are surprisingly invariant among birds. Bird flight places such precise functional requirements on the structure of the wing that the birds of the world have evolved relatively minor changes to the basic design. Birds have only tinkered with the highly functional

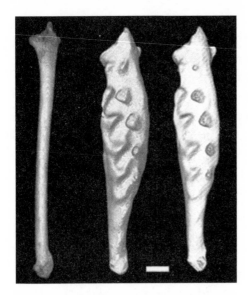

X-ray tomography images of the ulnas of (left) a male White-crowned Manakin (*Dixiphia pipra*), (center) a male Club-winged Manakin, and (right) a female Club-winged Manakin. Scale bar equals 2 mm.

design that was perfected over 135 million years ago, when Mesozoic birds first evolved the modern flight stroke.

By comparison to other birds, the major changes in wing anatomy that Bostwick discovered in Club-winged Manakins are truly startling. The ulnas of other manakins are simple hollow, columnar tubes. But the ulnas of male Club-winged Manakins are so wildly different they are nearly unidentifiable as the same bone. They are four times wider and three times larger in volume than those of other manakins, despite actually being shorter in length. The upper surface of the ulna of male Club-wings also features a prominent, wide shelf with deep sculpted grooves and peaks for ligamentous attachments to the oscillating secondary feathers. There is nothing else like it in any other bird in the world. Even more surprisingly, however, the ulnas of male Club-winged Manakins are *solid bone,* and the calcium in these bones is two or three times denser than in the wing bones of other manakins. In contrast, more than half the volume of other manakin ulnas is occupied by a hollow internal space. In fact, *every* other species of bird on the planet has hollow ulnas. Even theropod dinosaurs like *Tyrannosaurus rex* and *Velociraptor* have hollow ulnas! Thus, in order to sing their wing feather songs, male Club-winged Manakins have dramatically transformed anatomical features of their wings that have been consistently present for over 150 million years. Sexual selection for these innovative wing songs has forced male Club-winged Manakins to abandon a forelimb bone design that even predates bird flight itself.

Kim Bostwick hypothesized that the broader, solid ulna and its complex surface for attachment of feather ligaments function in two ways: to enhance stridulatory sound production by providing a more substantial, fixed anchor for the base of the feathers; and to enhance the resonance and coupling among secondary feathers within the wing.

Clearly, the wings of male Club-winged Manakins have evolved to serve two completely distinct functions—flight and tonal song production. Apparently, their wing bones cannot do both jobs equally well with the traditional anatomical design shared by all other flying birds (and even some of their nonflying ancestors). Some anatomical compromise is necessary. However, compromise in the design of wing morphology to accommodate song production is highly likely to create new survival and energetic costs to males. In the field, it is easy to see that male Club-wings fly awkwardly. There are no data yet on how the bizarre ulna morphology of male Club-wings affects their flight mechanics and energetics. But it is nearly impossible to imagine that the multiple anatomical changes to flight feathers, wing bones, and muscles necessary for singing these wing songs do not diminish male flight capacity, maneuverability, flight performance, and energetic efficiency.

The overwhelming uniformity of wing anatomy of flying birds is powerful evidence that this morphology has been maintained in all these species by natural selection and that male Club-winged Manakins have evolved far from the natural selection optimum for flight efficiency. If these derived anatomical features do not impose any function or survival costs on male Club-wings, then we would expect that many other bird species should also have evolved similar variations in wing morphology. But they haven't.

The Club-winged Manakin's wing song provides a likely stark example of evolutionary decadence—an evolved *decrease* in the overall survival capacity and fecundity of a population through mate choice. It is this discomfiting prospect of evolutionary decadence that was so threatening to adaptationism that entertaining arbitrary sexual selection without abundant proof was branded as "methodologically wicked." According to the adaptive mate choice theory, these costly wing bones are evidence that attractive males are good enough to survive these extra physiological and functional challenges. However, recall from chapter 1 that Zahavi's original handicap (that is, Smucker's) principle does not actually work; if the costs of the ornament are directly related to the benefits, there can be no payoff. The only way to fix the handicap principle is to break it, by assuming that better males get to cheat by paying relatively *lower* costs for each incremental advantage in quality. There is no evidence of such costs in any organism, and certainly not Club-winged Manakins. I think the aesthetically transformed wing anatomy of male Club-winged Manakins is compelling and excellent evidence that sexual decadence evolves in nature, but without physiological evidence about the costs the case could still be considered inconclusive. To resolve this deadlock, we will have to look even deeper.

Recently, I began to look for evidence of maladaptive and decadent evolutionary consequences of mating preferences in *female* Club-winged Manakins. The extraordinarily bizarre changes to the wing bones of Club-winged Manakins are very likely to be detrimental to male flight function. But what has happened to the wing bones of *female* Club-wings? These birds are so rare in natural history museums that there are no skeletal specimens of this species in any museum in the world. However, from X-rays and micro-CT scans of museum study skins, I have discovered that ulnas of *female* Club-wings have the *same* greatly distorted and highly derived size and shape as males do. However, unlike the males', the ulnas of the females are not solid bone but hollow in the center.

How could this have happened? Apparently, in selecting on male wing song production capacity through mate choice, female

Club-winged Manakins have evolutionarily transformed both the male's wing morphology *and* their own. Again, we do not yet have physiological evidence that these morphological changes affect the female's flight capacity or energetics. However, the best explanation of why these wing bones are so invariant across *all* of the rest of birds is that natural selection has maintained their highly functional, tubular, columnar design to achieve optimal flight function and capacity. In other words, the morphological consistency in wing bone design among birds is strong evidence that other variations in wing bone shape are functionally inferior and costly to survival and fecundity. Although female Club-wings will never use their wings to sing a song, they appear to incur at least some of the functional costs of the extraordinary wing bone changes necessary for males to make these attractive songs. By not completely ossifying these bones, as males do, and maintaining a hollow space in the center, female Club-wings appear to avoid at least some of the costs of growing extreme ulnas that males incur.

The observation that male Club-wings are likely made worse by the action of female mate choice—less functional, capable, and efficient—could still be rationalized as providing honest information about mate quality. But the observation that *female* Club-wings have also likely made *themselves* less functional, capable, and efficient at flight as a consequence of their mating preferences for exotic male wing songs can only be described as decadent.

Interestingly, females will not be harming their *own* survival and fecundity by preferring males that make attractive songs with extreme wing bones. Rather, females with preferences for males with maladaptive wing bones will only pay an indirect, genetic cost for their preferences, because their daughters may inherit more awkward wing bones, which will interfere with *their* daughters' survival and fecundity. However, this indirect genetic cost to mate choice can be outweighed by a simultaneous indirect, genetic benefit of having sexually attractive male offspring. Because the maladaptive costs of aesthetically extreme mate choices are deferred by each generation of choosers, the whole population can ease further and further into decadence and dysfunction generation by generation. The population will not be saved from decadence by natural selection, because the maladaptive functional costs are indirect and will be more than balanced by the advantages of having beautiful, sexually attractive offspring. Nevertheless, the entire population becomes increasingly maladapted because the fit between the organisms and the environment gets worse and worse over time. The survival and fecundity of all individuals—both males *and* females—suffers.

In Club-winged Manakins, the evolution of decadent wing bones has apparently been facilitated by a quirk of avian biology. In all birds, the wing bones begin to develop very early in the life of an embryo, around six days after incubation begins, which is *before* the embryo has begun sexual differentiation. In other words, the six-day avian embryo does not yet have a sex. So, selection for evolutionary changes in the shape and size of *male* wing bones will affect *female* wing bones, too. As a result, female mating preferences that aesthetically transform males will decadently transform the whole species. As soon as the embryo becomes sexually differentiated, however, there is an opportunity for the sexes to diverge developmentally. Events that happen later in development—like the full ossification of the wing bones—can be sexually differentiated. This is why female Club-winged Manakins do not have completely ossified, solid wing bones like the males.

The stridulating wing songs of the Club-winged Manakin are more than just a bizarre, innovative way for one bird to sing. They demonstrate again that natural selection is not a universally strong and deterministic force in evolution. Some of the evolutionary consequences of sexual desire and choice in nature are *not* adaptive. Some outcomes are truly decadent. Natural selection is not the only source of organic design in nature.

How far can decadence go? New theoretical models being developed in my lab show that decadence can indeed evolve through the indirect costs of mating preferences. Mathematical genetic models of a similar evolutionary process further imply that the costs of decadent display traits can lead to the *extinction* of whole populations or species. This means that in addition to recognizing the role of sexual selection in fostering the evolution of new species, we should recognize that sexual selection may facilitate species decline and extinction. Is it any wonder that many of the world's most exquisitely beautiful and aesthetically extreme creatures are so rare? I don't think so.

Once we clearly conceive of the possibility, we see that the phenomenon of evolutionary decadence may not be rare or even unusual. There are many other examples in which female mate choice has resulted in female versions of male display ornaments that are useless to them. This phenomenon ignited a big debate between Charles Darwin and Alfred Russel Wallace on the nature of sexual difference in bird plumages. In retrospect, their heated discussions of this topic were unproductive because neither one had any clear notions about the mechanisms of genetics or inheritance. But the intensity of their debate demonstrates that the topic is still central to the issue of whether evolution by mate choice is necessarily an adaptive process, as Wallace insisted.

The existence of useless ornament in females challenges the logic and likelihood of honest advertisement. If male sexual ornaments are costly to make, to maintain, or to survive with, and these costs are critical to ensure ornament honesty, then how can females bear these costs when they cannot benefit from them? On the contrary, if the ornaments are not costly to females to make or survive with, then how can these traits be robust and honest indicators of mate quality in males? It is a big conundrum for adaptive mate choice, and evidence of this problem is abundant and largely ignored.

Like the decadent wing bones of the Club-winged Manakin, some of the most conspicuous examples of this phenomenon come from other traits that originate early in development. For example, the male Wilson's Bird of Paradise (*Cicinnurus respublica*) from western New Guinea has a bare, brilliantly light blue crown that is crisscrossed by a pattern of narrow stripes made of very short, very black feathers (color plate 13). His bizarre blue tonsure is one of nearly a dozen colorful plumage ornaments that are featured in his bizarre courtship displays, which females observe from a very close distance. The male Wilson's Bird of Paradise displays on the trunk of a small sapling in a bare dirt court on the forest floor. When the female approaches the male from above, he spreads his deep glittering green breast shield, cocks his bright red tail with its twin green curlicue feathers, and draws his head in, displaying his brilliant blue crown skin. Although they completely lack any use for it, female Wilson's Birds of Paradise *also* have the same bare crown patches as the male, but in a slightly deeper shade of blue.

Likewise, the lek-breeding Capuchinbird (*Perissocephalus tricolor*)—a fruitcrow (Cotingidae) from South America, closely related to manakins—exhibits bald, ornamental bluish crowns in both males and females, even though females will never use them in display.

Like the wing bones of manakins, the evolution of truly featherless, ornamental, bare skin in birds requires evolutionary changes to the distribution of feather follicles on the skin, which develop early in the life of the embryo before it has begun sexual differentiation. The bare crowns of Wilson's Bird of Paradise and Capuchinbird require the suppression of feather follicle development in these patches of embryonic crown skin. Thus, female mate choice for males with sexy bald tonsures will result in the correlated evolution of useless, or even decadent, female baldness.

Are blue crowns detrimental to the survival of female Wilson's Birds of Paradise or Capuchinbirds? Certainly, having a bright blue crown will not help a female avoid predation as she solitarily incubates her eggs on her open nest. So, there are very likely to be both survival and fecundity costs to the females' useless blue crowns. Regardless, they certainly cannot be called adaptations, because they do not enhance the functional fit between the female and her environment in any way.

The same phenomenon is evident in the brilliantly orange "Mohawk" crest of the male Guianan Cock-of-the-Rock (color plate 14). Normally, feathers on the crowns of birds grow out of their follicles toward the tail so that they lie flat on the surface of the skull and create a smooth plumage outline. However, in the curious crest of the male Guianan Cock-of-the-Rock, the feathers on each side of the crown grow *toward* the midline of the crown so that they stand up to create the elegant "Mohawk" effect. These feathers do not bend toward the center. Rather, the orientation of the individual feather follicles are rotated ninety degrees *clockwise* on the right side of the crown, and ninety degrees *counterclockwise* on the left side of the crown, so that the crown feathers grow inward. This is fancy stuff! And, like wing bones and bald heads, the critical *orientation* of the feather follicles is established with the origin of the feather follicles themselves around day 7 or 8 of development when the embryo has no sex yet. Again, as we would predict, a close look at the drab brown female Guianan Cock-of-the-Rock reveals that her dainty, little brown crown feathers are *also* reoriented ninety degrees on either side of the midline, creating a subtle, discreet, little tufted pleat on the top of her crown. Of course, the female has no use for even this modest crown tuft.

The examples go on and on. Among polygynous bird species with extraordinary ornaments, useless non-ornaments in females are very common. Together, all these traits constitute more evidence of the decadent consequences of Beauty Happening.

If you were educated to think that evolution is synonymous with adaptation by natural selection and the persistent improvement of the species, then the evolution of aesthetic decadence may seem troubling. Yet a simple consideration of our own human capacity for irrational and impractical desires should help us reconsider that simplistic view. Why should animals be *more* rational than we are?

As the American Jazz Age poet Edna St. Vincent Millay wrote in her poem "First Fig,"

> *My candle burns at both ends;*
> *It will not last the night;*
> *But ah, my foes, and oh, my friends—*
> *It gives a lovely light!*

As Darwin and Millay understood, survival is not the only priority in life when sexual success is determined by mate choice. Sexiness can trade off with survival and fecundity—natural selection with sexual selection—and the frequent result is evolutionary decadence, the degradation of the adaptive fit between the organism and its environment. In many species like the Club-winged Manakin, the costs of sexual success may be very high indeed. Even females can be made adaptively worse off—that is, lower in survival and fecundity—through the evolution of their own aesthetically extreme sexual desires. Yet the escape from adaptive constraint that makes evolutionary decadence possible also facilitates aesthetic innovation and inspires the deep creativity of avian beauty.

One day in 2007, the Yale paleontology professor Derek Briggs and his graduate student Jakob Vinther walked into my office in New Haven. They wanted to show me a picture that Jakob had taken—a scanning electron microscope image of a feather at twenty thousand times magnification. The grayscale image showed dozens of tiny, sausage-shaped objects lying roughly parallel to one another. "What do these look like to you?" they asked. "Those look like melanosomes," I responded. "I *told* you so!" Jakob exclaimed triumphantly to Derek. Apparently, something important was at stake here.

Melanosomes are the microscopic packages of melanin pigments that give feathers their black, gray, or brown coloration. What Jakob and Derek hadn't told me at first was that the electron microscope image was taken from the feather of a fossil bird from the Early Eocene Fur Formation in Denmark. If these were melanosomes, they were about *fifty-five million years old*.

The melanin pigments in bird feathers are synthesized by special melanin-producing pigment cells and packed into tiny membrane-bound organelles, which are called melanosomes. Similar to the pigmentation of human hair, in birds the melanin pigment cells transfer completed melanosomes into individual feather cells during feather development. As the feather cells mature, the melanosomes are walled into the hard beta-keratin protein of the feather to produce the color of the mature feather. Melanins are ancient pigments and are produced by almost all animals. Melanins are also diverse in chemical structure. For example, the plumage colors of a black American Crow (*Corvus brachyrhynchos*) and the color of human black hair are made by eumelanin molecules. The rufous brown plumage color of a Wood Thrush (*Hylocichla mustelina*) and the color of human red hair are made by the distinctive molecule pheomelanin.

Paleontologists had been examining fossil feathers with scanning electronic microscopes since the early 1980s. They had observed these cylindrical objects and even confirmed that they were made of carbon-containing organic molecules unlike the surrounding rock. However, paleontologists are mostly bone people, and they have not traditionally thought a lot about cell biology. So, based on the shape and size of these objects, they concluded that these structures were *fossil bacteria* that had consumed the feather during its fossilization. Because paleontologists are keenly interested in the specific mechanisms by which different fossils are preserved, this was treated like an important discovery. However, the hypothesis never made a lot of sense. For example, why were bacteria more commonly preserved while eating the dry, nearly indigestible feathers and never found consuming all the juicy, appetizing bits of the decomposing body? In any case, the bacterial hypothesis became an accepted fact in paleontology. Jakob's discovery presented an exciting opportunity to challenge this dogma.

To test whether these microscopic fossil structures were indeed bacteria or melanosomes, we needed an indisputable example of a fossil feather with a melanin pigment pattern preserved. Luckily, Derek Briggs has an encyclopedic knowledge of extraordinarily well-preserved fossils from the museums of the world, and he remembered a gorgeous horizontally *striped* fossil feather from the Crato Formation of Brazil, approximately 108 million years old, in the geology museum of the University of Leicester. The fossil preserved amazing details of feather structure, including the finest filaments of the feather barbules. Furthermore, the striped color pattern on the feather exhibited distinct characteristics of the natural pigment patterns of feathers and could not be confused with fossil bacteria.

With an electron microscope, we confirmed that the black stripes on the feather contained abundant tiny "sausages" a few microns long and about one hundred to two hundred nanometers wide, which strongly resemble the eumelanosomes from the feathers of living birds. In contrast, the white stripes on the fossil feather were entirely *devoid* of any such structures at all. Clearly, the best explanation is that the microscopic structures are preserved melanosomes from the original feather itself. Somehow, under the right conditions, melanosomes fossilize beautifully and

Melanin pigmentation in fossil and living bird feathers. (a) A fossil feather from the Crato Formation, Early Cretaceous, Brazil, showing black and light bands. (b) Dark bands reveal melanosomes. (c) Light areas reveal only the rock matrix. (d) Melanosomes from the feather of a modern Red-winged Blackbird (*Agelaius phoeniceus*) are nearly identical in form to those preserved in the fossil. Scale bars: (a) 3 mm, insert 1 mm; (b) 1 µm; (c) 10 µm; (d) 1 µm. *From Vinther et al. (2008).*

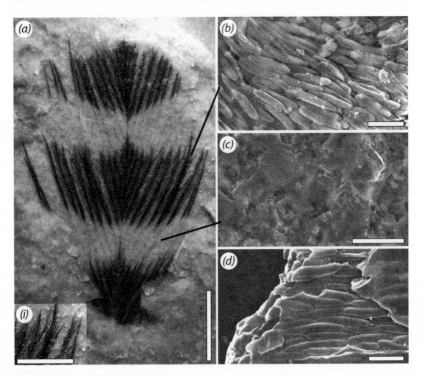

can endure for hundreds of millions of years, preserving aspects of the original color pattern of these ancient animals.

Our discovery of fossilized melanosomes has inspired an entirely new generation of research on the coloration of fossil vertebrates, including their feathers, hair, skin, scales, nails, and even retinas. Of course, the most exciting *possible* question in this new field of color paleontology was, what colors were the *dinosaurs*? Following our discovery, this question was no longer the stuff of science fiction but actually an entirely plausible question to investigate. Feathers first evolved in a lineage of meat-eating, bipedal theropod dinosaurs before the origin of birds or the origin of flight. We had shown, in principle, that we could reconstruct the melanin coloration of the plumages of the non-avian dinosaurs. Indeed, the striped Brazilian feather fossil is old enough that it might have been from the plumage of a non-avian dinosaur! All we would need were tiny samples of dinosaur feather fossils for electron microscopy. The feathered dinosaurs, which come mostly from early Cretaceous and late Jurassic deposits in Liaoning, northeastern China, are among the most exciting and revolutionary paleontological discoveries of the last century. But reconstructing their plumage coloration would take the excitement to a whole new level!

With an expanded team of collaborators, in the following year we started research on a late Jurassic specimen of the raptor-like dinosaur *Anchiornis huxleyi* from the Liaoning Formation in northeastern China at the Beijing Museum of Natural History. *Anchiornis* was a small bipedal theropod with a long bony tail, tiny teeth, and long, winglike feathers on both its forelimbs *and* its hind limbs. *Anchiornis* is one of those enigmatic "four-winged" dinosaurs, which are closely related to the raptor dinosaurs (like *Velociraptor,* who chased the kids around the kitchen in the movie *Jurassic Park*), to *Archaeopteryx lithographica,* the earliest bird fossil, and to the ancestor of all living birds.

Although the Liaoning Formation is known for its exquisite preservation, this particular specimen of *Anchiornis* did not look very promising. Actually, it looked like Jurassic roadkill—all

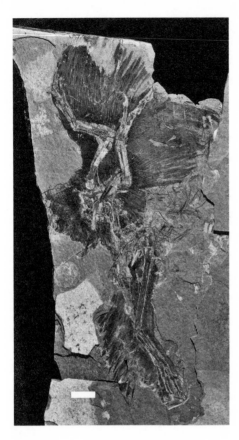

Specimen of the theropod dinosaur *Anchiornis huxleyi* from the Beijing Museum of Natural History (BMNHC PH828). Scale bar is 2 cm.

mangled, head removed and preserved on another slab, and limbs splayed out in all directions—but it did have a thick mat of dark feathers surrounding its bones. We ended up taking very tiny, mustard-seed-sized samples from three dozen locations around the body for electron microscopy. Given the poor appearance of the specimen, we were just hoping to find any melanosomes at all.

Back in New Haven, electron microscopy of the different samples revealed that some had well-preserved melanosomes, others preserved impressions of melanosomes, and some areas had no preserved melanosomes at all. Our next innovation was to compare the size, shape, and density of the melanosomes from the *Anchiornis* fossil with those of living birds. It turns out that eumelanosomes from black and gray feathers tend to be long and sausage shaped, whereas pheomelanosomes from rufous or red-brown feathers are more rounded and jelly bean shaped. By comparing measurements from *Anchiornis* melanosomes with those of living birds, we could diagnose the color of the fossil feathers. Because we had sampled many places from all across the specimen, we could reconstruct the color of nearly its *entire plumage.*

One of the most exciting moments in my scientific career was watching the plumage of *Anchiornis* come to life as I mapped the newly diagnosed colors—black, gray, rufous brown, and plain white—from the sample numbers back onto their anatomical positions in the animal's plumage. The resulting picture was more stunning than we could ever have imagined!

Describing the plumage coloration of *Anchiornis huxleyi* was like writing the very first entry in the *Field Guide to Jurassic Dinosaurs*. As a child, I had been inspired by field guides to go out into the world and study birds. Now, as a scientist, I had the opportunity to reimagine them in an entirely new way.

What did *Anchiornis huxleyi* look like? Its body plumage was largely dark gray with black on the forewings (color plate 15). The long crest feathers on the top of the head were rufous brown. Most striking of all, the long feathers on both its forelimbs and its hind limbs were white with black tips, or spangles—like the modern breed of Spangled Hamburg chicken. The effect of these black spangled limb feathers was to boldly highlight the trailing edge of the feather and to produce a series of black bars on the wings.

Interestingly, the long limb feathers on *Anchiornis* were not asymmetrical in shape, like modern avian flight feathers. So, it is not clear that this creature used its limbs as gliding "wings" at all. Furthermore, *Anchiornis* was heavily feathered all the way down to its toes and lacked the scaly legs and toes of most living birds.

Discovering the color of a dinosaur is more than just fun; it raises a host of fundamentally new questions about dinosaur biology and about the origins of what we think of as bird biology. The bold and complex plumage pigment patterns of *Anchiornis* were obviously used as sexual or social signals. Thus, the evolution of aesthetic plumage ornaments originated not within birds but way back in terrestrial theropod dinosaurs. The dinosaurs coevolved to be beautiful—beautiful to dinosaurs themselves—long before one exceptional lineage of dinosaurs evolved to become flying birds. The rich aesthetic history of the birds goes all the way back to their theropod roots in the Jurassic age.

Even more important, is it possible that the evolution of beauty contributed to the evolution of feathers themselves? Since the late 1990s, another, previously unrelated area of my research has focused on the evolutionary origin and diversification of feathers. Specifically, in 1999, I proposed a model of the stages of feather evolution based on the details of how feathers grow. This general area of research is called developmental evolution, or "evo-devo" for short. Since then, the evo-devo theory of feather evolution has been strongly supported by both paleontological data from fossil theropod dinosaur feathers and experimental tests of the molecular mechanisms of feather development.

Very briefly, my evo-devo theory of the origin of feathers proposed that feathers originated as simple tubes—imagine hollow ziti pasta growing out of the skin. In the next evolutionary stage, the tube was subdivided to produce a downy tuft. Only then, in the subsequent stages, did feathers evolve the capacity to create the planar surface—called the feather vane—that birds ultimately evolved to use to create the physical forces for flight.

The evo-devo theory of feather evolution implies that feathers originated and diversified in nearly all their morphological complexity *prior* to the origin of birds and *prior* to the origin of flight. Thus, planar feathers evolved in theropod dinosaurs for some other function and were later *co-opted* to function in flight by the lineage of dinosaurs that gave rise to the modern birds. In this way, the evo-devo theory of the origin of feathers and the new paleontological finds of feathered dinosaurs overturned the century-old hypothesis that feathers evolved through natural selection for aerodynamic capacity—that is, gliding and flight. Saying that feathers evolved for flight is like saying that digits evolved to play the piano. In truth, only the most advanced structures could function in such a complex capacity.

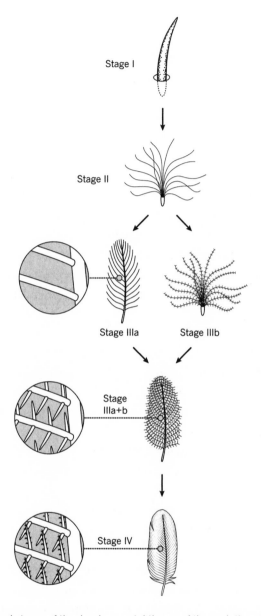

Stage I

Stage II

Stage IIIa   Stage IIIb

Stage IIIa+b

Stage IV

The hypothesized stages of the developmental theory of the evolution of feathers (Prum 1999). Feathers evolved through a series of developmental innovations starting with a hollow tube (Stage I) to a downy tuft (Stage II) to greater and greater complexity. The coherent planar vane (Stage IV) may have first evolved to provide a surface for the presentation of complex within-feather pigment patterns that functioned in aesthetic social and sexual signaling.

The aerodynamic theory of the origin of feathers was an example of an adaptationist approach to the origin of novelty. However, this big twentieth-century intellectual project failed. During the hundred years in which everyone was certain that feathers had evolved through natural selection on scales for flight, we learned nothing about how feathers actually evolved. We only made progress on the evolution of feather innovations by shelving the questions of the selective function of each innovation and searching for evidence and predictions concerning feather evolution within the details of feather development. The advantage of this evo-devo approach is that we can figure out *what* happened in feather evolution before we try to investigate *why* it happened.

Once we understand the progress of feather evolution, we can return to questions about the selective advantages of different stages in feather evolution. Early tubular and tufty stages have been convincingly hypothesized to have evolved for thermoregulation and water repellency. However, there is still no accepted hypothesis for why tufty down feathers (stage II) evolved into vaned feathers (Stage IIIa to Stage IV). What evolutionary advantage could the planar feather vane have provided prior to the evolution of flight? It is clear that a plumage made up of downy feathers, like a modern chick, would be warm enough or water repellent enough to provide any thermoregulatory requirements. After all, baby ducklings manage to stay very warm and very dry with downy plumages.

Is it possible that the original selective advantage of the planar vane was actually *aesthetic*? Obviously, down is fuzzy. Although downy chicks are cute, the plumage coloration pattern complexity that can be produced with fuzzy down feathers is aesthetically quite limited. Just like hair, you can make different down feathers different colors, and you have a limited ability to make the tips and bases of down feathers different colors. But that's it. The innovative planar feather vane, however, creates a well-defined, two-dimensional surface on which it is possible to create a whole new world of complex color patterns within every feather. In aggregate, many planar feathers can create complex plumage patches and a crisp, smooth, new outline to the entire body plumage.

In other words, the planar vane of the feather might have evolved through aesthetic selection to create a two-dimensional *canvas* upon which to depict complex pigment patterns—including stripes, spots, dots, and spangles. The key innovation of the planar feather vane might have evolved because it provided a whole new way to be beautiful.

This is a *really* big deal, because birds later evolved to use these same planar, vaned feathers to create aerodynamic forces required for flight. Feathers did not evolve for flight; rather, flight evolved *from* feathers. And among the best hypotheses for the key innovation that allowed birds to launch into the air is the desire for beauty.

The elaborate aesthetic capacities of birds are more than a vivid characteristic of living species. The coevolved desire for beauty might have made the evolution of birds possible in the first place.

As spectacular as that realization may be, there is more! About sixty-six million years ago, an enormous meteor hit the earth, leaving a crater 110 miles wide near what is now Chicxulub, Yucatán, Mexico. The cascade of environmental and ecological changes that followed this impact led to a mass extinction of terrestrial and aquatic life on earth including, most famously, the dinosaurs. Of course, we now know that the dinosaurs did *not* go extinct. Rather, *three* dinosaur lineages survived the mass extinction at the end of the Cretaceous; they were the flying ancestors of the three main lineages of living birds. These three lineages would later thrive, diversify (one explosively), and evolve into more than ten thousand species of birds that inhabit the planet today.

Why did birds survive the Cretaceous-Paleogene extinction event, and other dinosaurs did not? This is a tough problem, but we can be certain that merely having feathers was not enough, because there were many other lineages of feathered theropods that did not survive the Cretaceous-Paleogene boundary—including the fully plumaged raptor dinosaurs, like *Velociraptor,* the ornithomimids, and the troodontids. In fact, the only lineages of dinosaurs to survive the Cretaceous-Paleogene extinction were species that could *fly* with their feathers. Perhaps the capacity to fly allowed these birds to escape or avoid the worst ecological consequences of the Chicxulub impact or to disperse rapidly and find ephemeral refuges in the ecological chaos that followed. We don't know for sure. However, were it not for their ability to fly, the ancestors of the modern birds would likely have gone extinct along with all the other dinosaurs. Thus, the potentially *aesthetic* innovation of planar feathers facilitated the evolution of flight *and* the avian dinosaur survival of the Cretaceous-Paleogene extinction. It's harder to imagine a bigger possible impact for the role of beauty and desire in the history of life.

Throughout this book, I have argued that most of the abundant beauty in nature is likely to be meaningless and arbitrary and presents nothing to choosers other than the opportunity to be admired and preferred. But the investigation of the evolution of aesthetic complexity, innovation, and decadence demonstrates that this perspective is not a bleak, frivolous, or nihilistic view of the role of beauty in the natural world. In fact, the more we investigate the history of life from an aesthetic perspective, the more we will discover that aesthetic coevolution has had a powerful, innovative, and decisive impact on the quantity and form of biological diversity. When mating preferences are unconstrained by the narrow task of providing adaptive advantages, beauty and desire are free to explore and to innovate, and thereby transform the natural world. Thankfully, as a result, today we have the birds.

CHAPTER 5

Make Way for Duck Sex

A few years ago my wife, Ann, and I attended a lovely dinner party in our New Haven neighborhood with four other couples. We dined by candlelight on a delicious meal at a table set with beautiful linens, crystal wineglasses, and hefty heirloom silverware, while a passel of young children ate in front of an animated cartoon in another room. Many of us were meeting for the first time, so we engaged in the usual polite introductions and chitchat.

A short way into our meal, the mother of a few of the spaghetti-eating children in the other room spoke to me from down the table. "Oh, you're an ornithologist! You're just the person I need to ask." I expected to field another of the innumerable identification questions that arise from people's personal encounters with birds, but her question proved to be much more thought provoking. "The other day I was reading *Make Way for Ducklings* to my kids." I nodded in recognition of the classic story by Robert McCloskey, a book that had been read to me as a child and that I in turn had read to my three boys—so many times that I had nearly memorized it. "So, you know when the pair of Mallards settles down and builds their nest, and she lays her eggs? It seems that they're just getting started with a nice family together, but then he just *takes off*! What's with *that*?"

Before I could even inhale, from the other side of the table Ann gave me the anxious look we refer to in our house as the "hairy eyeball." She murmured the verbal warning shot "Don't go there!" Soon, all attention was on us, and everyone wanted to know exactly where it was I was not supposed to go. As if to warn all involved, Ann asked the curious mom, "You didn't just ask my husband about *duck sex*, did you?"

From this casual inquiry into the family life of ducks, our conversation veered into territory I knew in far greater depth than anyone might have expected. Thanks to Dr. Patricia Brennan, who spent from 2005 to 2010 as a remarkably enterprising postdoc in my lab at Yale, my research in those years had taken an unexpected detour into the study of the sexual behavior and genital anatomy of waterfowl. So, just as my wife feared, discussion of the kinky qualities of duck sex came to dominate the conversation that evening.

Duck sex can be elaborately aesthetic or shockingly violent and deeply troubling, but it is a fascinating topic. It may not be the best subject for dinner table conversation among new acquaintances—perhaps that's why we've never again been in the company of the woman who asked the question—but after all the disturbing details have been examined and understood, the story of duck sex actually concludes with a rather redeeming insight into the relationship between the sexes, the nature of desire, female sexual autonomy, and the evolution of beauty in the natural world.

The drama of duck sex brings to mind the ancient Greek myth of Leda and the Swan, in which Zeus took sexual possession of the lovely young Leda after assuming the physical form of a swan. This mythic scene has attracted the interest of artists ranging from the Greeks to Leonardo da Vinci to William Butler Yeats. Although often referred to as "the Rape of Leda," it has usually been depicted with a note of sexual ambiguity, there being an element of mutual desire mixed in with the suddenness of the act. Perhaps the Greeks intuited that something about waterfowl sex is intriguing. If so, they were right, for the full evolutionary

implications of the social complexity of duck sex are only beginning to be unpacked.

On a cloudy winter day in 1973, when I was twelve years
old, I embarked on one of my earliest birding trips to the ocean.
I stood on the banks of the Merrimack River in Newburyport,
Massachusetts, just upstream from where it widens out into the
bay. With the proceeds from a paper route and mowing lawns, I
had just purchased my first spotting scope for watching distant
birds, and I was excited to be using it to observe ducks, gulls,
loons, and other waterbirds at this famous birding locality. It was
a cold February day, with chunks of ice on the riverbanks and in
some of the calmer eddies, but I was euphoric. I could see several
dense flocks of ducks churning away against a strong current on
the falling tide.

In my very first scan with the scope, I landed on a *lifer!*—a
flock of a couple dozen Common Goldeneye (*Bucephala clan-
gula*). The male ducks were crisp black on the back, snowy white
on the sides, belly, and breast, and crowned with a shiny, iridescent green head. On each glittering green cheek was a large round
white spot. As advertised, their eyes were brilliantly golden yellow. The females were drabber, with grayish sides and neck and a
brown head, but they shared the same yellow eyes.

For some reason—a reason that I would not understand
until years later—there were many more males in the flock than
females. Among the two dozen or so birds, there were only five or
six females. I was enjoying the scene, watching them as they dove
underwater to feed and then popped back up to the surface, when
suddenly a male thrust his head upward and then snapped it back
to touch his rump—a display known as the head throw. With his
head in this awkward position, he briefly opened his beak toward
the sky, after which he brought his head back to its normal position with a slight side-to-side waggle. Soon, other males joined in,
and the males in the flock were boiling over with bravado, jock-
eying for position around the females, and chasing each other. If

I had been closer to the action that day, I would have heard the raspy two-note call the male Goldeneye makes during the head-throw display. The male Goldeneyes performed various other displays, too, which have been given suitably nautical names like the bowsprit and the masthead. The bowsprit involves cruising around with the head and beak pointed up and forward, while the masthead is performed with the head raised, then lowered and cast forward along the surface of the water. Despite the freezing weather, this gathering of Goldeneyes was engaged in courtship displays. They would continue wooing the females with these displays throughout the winter months, before returning to their nesting grounds on wooded lakes in northern Canada.

That memorable outing was my introduction into the complex social world of ducks. Across the entire waterfowl family, males engage in similarly showy courtship behavior. The displays vary among species, but they generally consist of a series of highly distinctive postures and gestures, each lasting only a few seconds.

The head-throw display sequence of the male Common Goldeneye.

The males may repeat them over and over, but the basic elements are pretty simple, and because almost all duck displays take place on the water, they always involve a lot of churning, cruising, and splashing.

The display repertoires of some species of ducks are so outrageous that they can be quite comical. For example, the male Ruddy Duck (*Oxyura jamaicensis*) performs an especially impressive *bubbling* display. With his tail cocked straight up in the air, and his neck and breast swollen from air pumped into specialized pouches to either side of his esophagus, the male lowers his head rapidly and beats his blue bill against his rufous breast to make a low percussive *pop* sound. As he does so, his breast feathers create a frothy wake of bubbles on the surface of the water. He rapidly accelerates these chest-beating bill strokes in a crescendoing drumroll of ten or twelve *pops* that ends in a flatulent, groaning call that sounds like a breaking spring from a windup toy. The combination of feathers, postures, percussion, vocalization, and frothy bubbles makes for a very attention-getting performance.

A particularly extreme example of the duck display genre is that of the lovely, diminutive male Mandarin Duck (*Aix galericulata*) of Eurasia. Many ducks perform *sham preening* displays in which they ostentatiously preen their back feathers. But the Mandarin Duck combines the sham preening with a drinking display, which looks not so much like a courtship display as a flamboyant demonstration of drinking incompetence characterized by copious dribbling. The Mandarin Duck's sham preening is made more dramatic by the male's uniquely shaped, brightly colored reddish-brown inner wing feathers, which stand up vertically above the surface of his back. The "purpose" of these unusual feathers only becomes evident during the male's sham preening display when he reaches his head over his back (always on the side facing the female) and tucks his bright pink bill *behind* the upright planar feather, through which his eye is just visible to the female, as in a coy peekaboo game—or perhaps we should call it *beak-a-boo*.

I could go on and on. What all these rich and complex waterfowl courtship displays have in common is that they have evolved through female mate choice. Males go through all these antics

The peekaboo sham preening display of the male Mandarin Duck.

in their quest to be selected as mates by the extremely selective females. On the basis of her observations of male display, the female duck makes her choice about which one she wants to pair up with. In many species, like the Common Goldeneye, females choose their mates on the wintering grounds, and after they form a pair, they remain together for the rest of the winter months. There's no copulation during the winter months, because neither sex is ready. The annual cycle of sexual development in birds is a wild hormonal roller coaster, whose ups and downs are seasonally driven. Birds progress from completely asexual in the nonbreeding winter season to having gonads thousands of times larger only a few months later, in the spring, when it's time to mate. As mating season approaches, the pair migrates together to their breeding grounds. Once there, the male will continue to display, as well as to defend the female from other males. After much displaying, the pair will copulate on the water. The female signals that she is ready for copulation with a distinctive solicitation display, in which she extends her neck forward, holds her body horizontal, and raises her tail.

Why are female ducks so picky about whom they mate with? Because they *can* be. Remember how the female Common Goldeneyes I saw were surrounded by males who greatly outnumbered them? In most duck species, the sex ratio is highly skewed toward males, so females have plenty of mates to choose from. Given such a wealth of options, female ducks have evolved lots of elaborate mating preferences for colorful male plumage, extravagant displays, and complex, funky acoustic stimuli. And because many ducks begin courtship months before they reach the breeding grounds in spring, female ducks have ample opportunity to put the males through their paces in order to make a decision.

Sounds great for the females. Unfortunately, there is a dark side to duck sex, too.

Although some waterfowl, like the Canada Goose, Tundra Swan, and Harlequin Duck, form enduring, monogamous pairs in which both parents help to defend an exclusive nesting territory and raise the young together, most duck species, like the Mallards my fellow dinner guest was querying me about, do not. What distinguishes them from the pair-bonding waterfowl is that they are not territorial. They nest in habitats where their food supply is so highly concentrated and the populations are so dense that an exclusive feeding territory cannot be defended by any pair. And because they are non-territorial, their sexual and social relationships are quite different from those of the territorial species.

In these non-territorial ducks, the primary functions of the male of the pair, once they arrive at the breeding ground, are to have sex with his mate and to protect her from the sexual depredations of other males during the ten to fifteen days she is laying her clutch of eggs. He has a strong evolutionary incentive to do so, of course, because he is protecting his own paternity. But once the eggs have been laid, there isn't much for Papa Puddle Duck to do. The mother duck doesn't need him, because the building of the nest and incubating of the eggs are done entirely by her. And their ducklings won't need him either, because they'll be able to feed themselves soon after they hatch. If males aren't required to defend a territory against other members of their species, or to help with feeding the young, parental care in waterfowl consists mostly of trying to keep the ducklings from being eaten. This may actually be done better by one parent than by two, because more parental activity may only attract more predators, and the male's bright plumage colors act as a predator magnet. So just as McCloskey wrote in *Make Way for Ducklings,* in many non-territorial waterfowl the male of the pair abandons the female as soon as she begins incubating the eggs. At that point, with his paternity guaranteed, the male duck can no longer benefit evolutionarily from defending her, and she probably cannot benefit from his remaining with her. Which answers my dining companion's question: "What's with *that?*"

But now comes the shocking part about duck sex, the part that wasn't included in McCloskey's otherwise scientifically accurate children's story about puddle duck family life, the part that few would even think to ask about. McCloskey said nothing about the challenges the father duck might have faced in protecting his mate, or what might happen to her if his defenses were unsuccessful. Or where the male duck goes after he leaves. And this is where things can get very scary indeed in the world of the female duck.

Whenever there are a lot of ducks present in relatively small spaces, like the high-density ecologies of non-territorial puddle ducks, there are lots of opportunities for social interactions. For males, these social opportunities are also sexual opportunities. Because of the excess males in the population, many males end up unpaired. These unpaired males now have two reproductive options: they can wait another year and hope they have better luck; or they can try to coerce and force themselves on unwilling females. Thus, forced copulations are an alternative male reproductive strategy. Males whose mates have already begun to incubate may also pursue forced copulations when they leave their mates, which creates even darker implications to the Mallard drake's casual departure in *Make Way for Ducklings.*

"Forced copulations" is the term that ornithologists and evolutionary biologists now use to refer to rape among birds and other animals. The use of the word "rape" was routine in animal biology for over a century, but it was largely abandoned in the 1970s in response to multiple avenues of feminist critique. In particular, in *Against Our Will,* Susan Brownmiller built a powerful and effective argument that rape, and the threat of rape, in human societies functions as a mechanism for social and political oppression of women. Human rape is an act with such great symbolic and social impact that the term didn't seem appropriate in the context of nonhuman animals. As the ornithologist Patty Gowaty has written, "Because of the important differences between rape and forced copulations, those of us who study animal behavior agreed years ago to refer to 'forced copulation' in non-human animals, and to reserve the term 'rape' for humans."

I understand and agree completely with those concerns, but I think, unfortunately, that the shift to the term "forced copulation" in biology has contributed to a desensitization to the social and evolutionary impact of sexual violence in animal behavior. It has obfuscated the fact that forced copulation is a form of coercive sexual violence against the interests of many female animals as well, and it may have stunted our understanding of the evolutionary dynamics of sexual violence. (In chapter 10, I will further explore how this missed intellectual opportunity has held back our understanding of the impact of sexual violence in human evolution.)

Although I do not suggest that we return to the wholesale use of the word "rape" in animal biology, I think that the phrase "forced copulation" does an intellectual disservice to our understanding of sexual violence in nonhuman animals. Certainly, in the case of female ducks, it is scientifically critical to recognize that sexual coercion and violence are very much *against their wills* too.

Forced copulations are pervasively common in many species of ducks, which might suggest that there's something routine and ordinary about them, but they are also violent, ugly, dangerous, and even deadly. Female ducks are conspicuous in resisting them and will attempt to fly or swim away from their attackers; if they do not manage to escape, they mount vigorous struggles to try to repel their attackers. This can be extraordinarily difficult to do, because in many duck species forced copulation is often socially organized. Groups of males travel together and attack a single female in a form of gang rape. By attacking her in concert, males increase the chance that one of them will be able to overcome her resistance, and thwart her mate's attempts to defend her, than if they acted alone.

The cost to females of forced copulations is very high. Females are often injured, and not infrequently killed, in the process. So, why do female ducks fight back so vigorously? Female ducks absorb greater direct harm to their physical well-being by resisting forced copulations than if they acquiesced, so the intensity of their resistance seems difficult to explain from an evolutionary perspective. Nothing is more threatening to the ability to pass on one's genes than death, so why risk death by struggling?

This question delivers us to the crux of the complex interaction between the female acting on her sexual desire for beauty and the male using sexual violence to subvert her ability to choose her own mate. What is at stake in these attempts at forced fertilization is more than just the *direct cost* to a female's health and well-being; forced fertilizations will also create *indirect, genetic costs* to the female that may be even more important to the female. Why? Because females that succeed in mating with the males they prefer will likely have offspring that inherit the display traits that they, *and* other females also, prefer. These females will have the benefit of greater numbers of descendants through their sexually attractive offspring. This is the *indirect, genetic benefit* of mate choice that drives so much of aesthetic coevolution. Females that are forcibly fertilized, however, will have offspring that are sired by males that have random display traits, or traits that have been specifically rejected because they have failed to meet female aesthetic standards. Either way, the resulting male offspring will be less likely to inherit genes for the preferred male ornamental traits, and they will therefore be less sexually attractive to other females and less likely to obtain mates, which will result in fewer grandchildren for that female. This is *the indirect, genetic cost* of male sexual violence.

At the heart of the complex breeding biology of ducks is sexual conflict between males and females over *who* is going to determine the parentage of the offspring. Will it be females through mate choice based on the coevolved beauty of male plumage, song, and display? Or coercive males through violent forced copulation? In 1979, Geoffrey Parker defined sexual conflict as a conflict between the evolutionary interests of individuals of different sexes in the context of reproduction. Sexual conflict can occur over many aspects of reproduction, including who gets to mate, how often sex occurs, and the division of parental care investment and responsibilities. One of these sources of conflict is critical to the evolution of sexual beauty: the conflict over who will control fertilization, the purveyors of the sperm or the curators of the eggs.

Duck sex provides a premier example of sexual conflict over fertilization and allows us to investigate how Darwin's proposed "taste for the beautiful" creates the opportunity for the further evolution of sexual autonomy. A key insight is that both fundamental mechanisms of sexual selection in waterfowl—mate choice based on female aesthetic preferences for male displays, and male-male competition for control over fertilization—are occurring *and* in evolutionary opposition to each other.

This observation is actually quite subversive. As we've seen, ever since Darwin's publication of *The Descent of Man*, the mainstream, adaptationist, Wallacean view has considered all forms of sexual selection as forms of natural selection. Whether it's elephant seals or birds of paradise, this view holds that only the objectively "best" males will succeed at mating. But what happens when female mate choice and male-male competition operate simultaneously, *and* they are clearly running in different directions, as they do in waterfowl? The winners of these two distinct competitions cannot all be the "best." If the most sexually aggressive males are actually the best, why don't females prefer them? Clearly, the winners in mate choice and male-male competition cannot all be the same.

Rather, sexual violence is a selfish male evolutionary strategy that is at odds with the evolutionary interests of its female victims and possibly with the evolutionary interests of the entire species. By maiming and killing females, such violence lowers the population size of the species. And by further skewing the sex ratios, these violent deaths make sexual conflict even worse, because there will be more males losing out in the mate choice competition who will therefore be motivated to pursue this counterproductive strategy. Thus, sexual conflict in ducks demonstrates yet again Darwin's insight that sexual selection is not equivalent to natural selection.

One reason why duck sex is so exceptional is that unlike 97 percent of all bird species ducks still have a penis. The bird penis is homologous with the penis of mammals and other reptiles, but somewhere along the way the ancestor of most bird species lost his penis (more on that later in the chapter). Ducks and the other bird species that still have penises—including the nonflying birds the ostrich, emu, cassowary, kiwi, and rhea, and their close relatives, the flying tinamous—belong to the oldest extant branches in the avian Tree of Life. Among all the birds with penises, the ducks are the best endowed, in terms of the ratio of penis size to body size. In fact, one duck species is the best endowed of all vertebrate animals. In a 2001 paper in the prestigious journal *Nature*, the ornithologist Kevin McCracken and colleagues described the penis of the diminutive Argentine Lake Duck (*Oxyura vittata*). A duck that was itself only about twelve inches long and a little over a pound in weight had a forty-two-centimeter penis (about sixteen inches). The *Nature* paper, now cited in the *Guinness World Records*, was titled "Are Ducks Impressed by Drakes' Display?" McCracken hypothesized that female ducks may select their mates based on penis size. After all, what other possible explanation could there be for such an extravagant genital endowment?

However, we now know that penis size is not important in mate choice in most ducks because, believe it or not, the seasonal nature of the reproductive cycle means that the superlong duck penis is almost nonexistent during courting season, when the females choose their mates. The penis *regrows* every year as mating season approaches, but once mating season is over, it begins to

The record-setting 42 cm penis
of a male Argentine Lake Duck.
Photo by Kevin McCracken.

shrink and regress, until it's reduced to a small rudiment less than a tenth of its full-grown size.

Alternatively, McCracken also hypothesized that the male somehow uses his superlong penis to remove the sperm of other competing males from the female's reproductive tract. Proving once again that each scientific discovery merely opens up other unsolved mysteries, the paper concluded with the inquiry "How much of his penis does the drake actually insert, and does the anatomy of the females' oviducts [vaginas] make them unusually difficult to inseminate?"

In 2005, this question resonated with the interests of my new colleague Patricia Brennan. Brennan is Colombian but has lived in the United States for more than fifteen years. She is vivacious, enthusiastic, and scientifically unstoppable. She is not at all timid about working on, or talking about, avian sex. With two young children and a bit of gray hair, she still looks like the aerobics instructor she was during graduate school at Cornell. She is also a mean salsa dancer, which is to say still *una Colombiana*. Her Ph.D. was on the dinosaur-like, male nest care breeding system of the tinamous (Tinamidae). In the tropical rain forests of Costa Rica, Brennan came to know these extremely shy, chicken-like birds better than nearly anyone alive.

Once, when observing tinamous mating, Patty was shocked to see a fleshy spiral dangling down from the male's cloaca. The cloaca (a word that memorably derives from the Latin for "sewer") is the anatomical chamber inside the avian anus, which is a kind of one-stop business rear end that receives the outflow of the digestive, urinary, and reproductive tracts. In birds without penises, insemination takes place with a "cloacal kiss"—a poetical term for a chaste juxtaposition of orifices in which the male and female anuses come into contact, the male releases his sperm, and the female takes it up. The male does not enter the female, because he doesn't have anything that would allow him to. The tinamou penis had been described by Victorian anatomists who had performed dissections on natural history museum specimens, but these anatomical monographs were not inspiring enough to keep the topic alive scientifically, and the existence of the tinamou's penis had been almost completely ignored for more than a century. So when Brennan spotted the extrusion from the cloaca of the postcoital male tinamou, she was stunned. Her sighting was probably the first-ever observation of the tinamou penis in action.

When Patty first arrived in my lab in 2005, she was interested in continuing her studies of the tinamous, focusing on the anatomy and function of their penises. But tinamous are eminently edible, and they are heavily hunted throughout their range, which is why they are among the shiest of all the birds in the world, and therefore very hard to study in the wild. Whereas ducks also have penises and are comparatively easy to work with. So, Patty thought that ducks might provide an easier route to study the evolution of genital anatomy and function in birds.

This interest ultimately led her to a duck farm in the Central Valley of California in 2009. Although a duck farm is not an obvious place to pursue new frontiers of evolutionary science, the farm Brennan went to had some very special ducks. These drakes were trained to ejaculate semen into tiny glass bottles. This was done not to satisfy some perverse interest in duck sex but because the duck farmers wanted to create offspring that are a hybrid of male Muscovy Ducks (*Cairina moschata*) and female Pekin ducks (a captive breed of Mallard). In captivity, such hybrids show extraordinary vigor and put on weight rapidly—two qualities that are very attractive to duck farmers. But the Muscovy and Pekin ducks do not like each other, and if they are left to their own devices in a common pen, they will not mate at high enough rates to produce a commercially viable number of offspring. Modern agriculture's answer to this problem is artificial insemination, which requires some way of collecting the sperm. Hence the use of the little glass bottles.

All of which explains why one day the Latino workers who collected the sperm and performed the artificial inseminations at this farm were confronted with a lovely, well-educated, wise, and wisecracking Latina toting a high-speed video camera. As the videos showed, male Muscovy ducks will perform on demand—despite the little glass bottles, the scrutiny of the camera, and the glare of the lights.

The basic artificial insemination procedure goes like this: Male and female Muscovys are kept in separate pens to increase their sexual motivation. When it's time for the sperm collection to occur, the pair of ducks is placed in a narrow cage with their rear ends facing out of one open side. The male rapidly mounts the female and begins to tread on her back. The female becomes readily sexually receptive, as indicated by her reclining *precopulatory* posture: her neck extended forward, head lowered, rear end raised with the cloaca exposed, dilated, and secreting volumes of mucus. Soon, the male begins to lower himself toward the female's proffered rear. And then it happens.

Normally, the erection of the drake would take place into the female reproductive tract. During sperm collection, however, the farmworker prevents the male from actually entering the female and places what looks like a small glass milk bottle over the male's cloaca at just the right moment. The drake's penis then erects and ejaculates into the bottle. As in a discreet sperm bank, the sample is then passed through a little window into the hand of another worker who prepares it for the Pekin females who are waiting in the room next door. For Brennan's research observations, the farmers still prevented the male from entering the female but allowed him to erect and ejaculate into the air, or into the special glass contraptions that Brennan brought along on her next trip to the duck farm (more about those later).

Obviously, despite their ancient homology, the duck penis and the human penis are very different from each other. Like other reptiles, the duck penis is not external, but is stored, folded up, outside in, within the cloaca. It only emerges from the cloaca during copulation. Another difference is that unlike the erections of other reptiles, and of mammals, too, duck erections are powered not by the blood-fueled vascular system but by the lymphatic system. Inside his body on either side of the cloaca, the male duck has two muscular sacs, called lymphatic bulbs. When these contract, lymph squirts into the central hollow space within the penis, causing the penis to erect, rapidly unfurling out of the male's cloaca. It is difficult to envision, but the process generally resembles a cross between using your arm to evert a sweater sleeve that is inside out and unfurling the soft, motorized roof of a convertible sports car with a hydraulic drive—but much, much faster! The first part of the penis to be exposed is the base, and the rest unfolds in a wave toward the tip, with sperm traveling along an external groove on the penis from base to tip.

For ducks, the erection of the penis and its entry into the vagina are the *same* event. The duck penis does not become stiff and then enter the female, as in mammals and other reptiles. Rather, the penis is erected, or actively everted, *into* the female reproductive tract, and it remains flexible throughout the entire process. Furthermore, the duck penis is not straight, but spirals *counterclockwise* from its base to its tip. Over its twenty-centimeter length, the Muscovy Duck penis completes six to ten full twists.

The penises of ducks and other reptiles also lack an enclosed urethra, or tube, for the flow of semen. Instead, the duck penis has a sperm-carrying groove, called the sulcus, to transport semen. The sulcus runs along the entire length of the duck penis, rather like the seam in a shirtsleeve. But because the penis is coiled, the sulcus spirals counterclockwise as well. Those same Victorian anatomists who had described avian penises derided the sulcus as functionally ineffective—like a leaky, dribbling pipe. But they had clearly never watched the duck penis in action, and their armchair conjectures could not have been more wrong. As the high-speed videos of flying duck sperm would show, the avian sulcus may be a mere topological fold, but it works as well as any mammalian urethra.

Like a selection of sex toys from a vending machine in a strange alien bar (think perhaps of an X-rated *Far Side* cartoon by Gary Larson), duck penises come in ribbed, ridged, and even toothy varieties. These surface features point *backward* toward the base of the penis, and as the penis unfolds, they are rapidly deployed into the walls of the female reproductive tract to secure whatever inward progress the unfurling penis has made, like the pitons a mountain climber uses to maintain progress up a forbidding cliff face. Oh, and did I mention the duck penis's spiral twist? I did? Okay, well, there are so many odd things about a duck penis that it's hard to keep them all straight.

Although Brennan was well prepared by *years* of previous research on duck anatomy, even she was stunned by the duck penis in action. To be blunt, duck erections are "explosive," the very word we used in the paper we eventually published about our findings in *Proceedings of the Royal Society of London B:* "Eversion of the 20 cm muscovy duck penis is explosive, taking an average of 0.36 s, and achieving a maximum velocity of 1.6 ms-1."

That's nearly eight inches unfurled at three and a half miles an hour. In about a third of a second, the entire event is over, the male ejaculates, the penis begins to deflate, and the drake starts retracting it into his cloaca with a series of muscular contractions (color plate 16). Brennan's data show that it takes an average of two minutes for a male to complete the process of gathering his penis back inside his cloaca, or 190 times longer than it takes to erect it in the first place. Brennan was able to make these observations about speed because during her first trip to the California duck farm, she had filmed the high-speed duck erections in the open air to document the process of an unimpeded duck penile erection. This gave us the first measures of the velocity of erection and the first observations of the efficacy of the sulcus—the sperm-carrying groove that runs along the length of the penis.

After ejaculation and retraction, the farmers know that it will then be hours before the male will be able to perform sexually again—perhaps because that's how long it takes for a sufficient quantity of lymph to build up in the male's lymphatic bulbs to fuel another explosive erection. Whatever the reason, it takes a few hours for a drake to get his groove back.

When our duck-farm research was published, what was everyday knowledge to the farm workers turned out to be both scientifically notable and culturally irresistible. The videos themselves attracted tens of thousands of YouTube viewers in just the first few days—a veritable explosion of interest, shall we say.

Which brings us back to McCracken's question: How does the explosive, spiraling, ribbed, or even toothy duck erection function *within* the female duck? Why *do* some males evolve a forty-two-centimeter penis to fertilize a thirty-centimeter-long female duck? To find out, Brennan dissected the reproductive tracts of female barnyard ducks. What she found was, at first, wildly confusing. According to the textbooks, the avian vagina is a simple thin-walled tube that runs from the single ovary to the cloaca. But the textbook illustration didn't match up at all with what Brennan saw in the female duck's reproductive tract. The duck vaginas she examined had thickened, convoluted walls that were wrapped in a mass of fibrous connective tissue. To Brennan, they seemed at first like a complete and confusing mess. Then, surprisingly, in other specimens, she saw vaginas that *were* simple, thin tubes, just like those in the textbooks. Eventually, Brennan discovered that the simple tube specimens were from females outside the breeding season and the more complicated structures were found in females who were in breeding season. Turns out that the reproductive anatomy of the female duck follows the same seasonal rhythms as that of the male duck, with both of them redeveloping every year at breeding time.

Once Brennan was able to examine the vaginal anatomies of a number of breeding ducks, what she found instead of simple tubes were vaginas that had a series of dead-end side pockets, or cul-de-sacs, located near the cloaca at the bottom of the reproductive tract. Further up the reproductive tract, she saw a series of twists and turns in the vaginal tube. Interestingly enough, these twists were *clockwise spirals,* in the *opposite* direction of the counterclockwise-spiraling duck penis. Broadening the sample to include a comparative analysis of fourteen waterfowl species such as puddle ducks, diving ducks, mergansers, geese, swans, and "stiff-tailed" ducks, like the Ruddy, Brennan showed that the longer and twistier the penis, the more complex the vagina, with more dead-end pockets and upstream twists—and vice versa: the shorter the penis, the simpler the vagina.

But what was the cause of all this anatomical variation? The key insight was that there was a correlation between the more highly elaborated genital structures and the social and sexual lives of the species who possessed them. In monogamous, territorial waterfowl like swans, Canada Goose, and Harlequin Duck, males have a very small penis (about one centimeter) without any surface features, and females have simple vaginas without cul-de-sacs or spirals. But in non-territorial species, which frequently engage in forced copulations, like the Muscovy Duck, Pintail, Ruddy Duck, and, yes, even the Mallards in *Make Way for Duck-lings,* males have evolved longer, intricately armed penises, and females have evolved increasingly complex vaginal structures. A comparative analysis of penis and vaginal morphology showed that these two features—the longer and more elaborately struc-

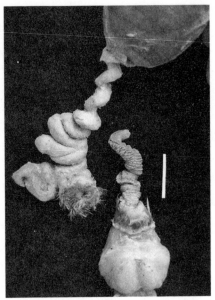

Coevolution of male and female genital morphology in waterfowl. (Left) The male Harlequin Duck has a very small, centimeter-long penis, and the female has a simple, straight vagina with no elaborations. (Right) The male Mallard has a long, corkscrew-shaped penis with hard ribs on its surface, and the female Mallard has coevolved a convoluted vagina with multiple dead-end cul-de-sacs, and several clockwise spirals. *Photos by Patricia Brennan.*

tured penises and the more complex and convoluted vaginas—had clearly *coevolved* with each other. But *why*?

We hypothesized that the coevolutionary elaboration of the duck penises and vaginas was the product of the *sexual conflict* between males and females over *who* is going to determine the paternity of the offspring. In cases like waterfowl, sexual conflict can create an ever-escalating war between the sexes, which is called sexually antagonistic coevolution. This process results in a kind of arms race between males and females, in which each sex evolves successive behavioral, morphological, or even biochemical mechanisms to overcome the evolved efforts by the other sex to assert control or freedom of choice over reproduction. That is, each evolutionary advance by one sex selects for a compensating counterstrategy by the other.

Male ducks had evolved penises that would enable them to force their way into an unwilling female's vagina, and the females in turn had evolved a new way—an anatomical mechanism—to counter the action of the explosive corkscrew erections of male ducks and prevent the males from fertilizing their eggs by force. Remember the duck penis is never stiff but unfurls flexibly in a counterclockwise spiral into the female's reproductive tract. It seemed to us that the cul-de-sac side pockets of the vagina, and its clockwise corkscrewed shape, could be blocking the drake's penis from progressing up the female reproductive tract during forced copulations. If the evolutionary advances in the female vaginal anatomy succeeded in foiling coercive fertilization, then males would evolve to counter female defenses with bigger, better-armed penises, and the females would in turn evolve ever more complex evasive anatomical structures, and so on and so forth.

The selection mechanisms at work in this dynamic coevolutionary process are complex. There is the sexual selection by mate choice that produces coevolution between male display traits and female preferences. In addition, male-male competition—another kind of sexual selection—is acting in the evolution of the coercive male behavior and in the evolution of the longer and more aggressively armed penis that allows males to succeed at fertilizing the females by force. Further, in response to the indirect, genetic benefit of autonomous mate choice (also a kind of sexual selection), female behavioral and anatomical resistance mechanisms evolve. Any genetic mutations that contribute to behaviors or vaginal morphologies that help females avoid forced fertilizations will evolve because those mutations will help females evade the indirect, genetic costs of sexual violence—that is, having unattractive sons that other females will *not* prefer.

On the face of it, this is a pretty depressing picture of duck social relations. It seems much more suitable for an apocalyptic dystopian sci-fi novel than a Caldecott Medal–winning children's bedtime story. The story, however, is not *all* depressing. There have been both escalations *and* reductions in this arms race in different lineages of ducks. Though some duck groups have evolved ever-longer and more elaborately armed penises and more complex vaginas, other lineages of ducks have essentially called off the arms race and evolved smaller penises and simpler vaginas. These reductions seem to be the result of external ecological factors that lower the density of breeding individuals, favor exclusive territoriality, and eliminate the social opportunity for male sexual coercion. In the absence of sexual conflict, both sexes seem to evolve away from these complex structures.

We wanted to test our hypothesis that female vaginal complexity functions in preventing forced fertilization. That required investigating whether there was something about the cul-de-sacs and spiral twists of the duck vagina that is specifically, mechanically designed to thwart the advance of the duck penis.

How could we test this hypothesis? It is impossible to get internal images of ducks during their sex act. Even if one could arrange for a male duck to forcibly copulate with a female in an MRI machine with the capacity to show a clear contrast between male and female tissues (and one definitely cannot!), it would be impossible to complete the imaging in the few tenths of a second during which penile erection is maximized and ejaculation takes place. Testing this hypothesis about sexually antagonistic evolution would take some creative thought.

Patty is nothing if not creative, however, and to test our hypothesis, she came up with the idea of creating four glass tubes that would help us analyze the interplay between the male and the female reproductive equipment. Two of the tubes would be designed *not* to challenge the progress of the duck penis in the vaginal tract. One would be straight; the other would be coiled counterclockwise to match the spiral of the duck penis itself. The other two tubes would be designed to act like a steeplechase obstacle course for the avian penis, mimicking the shape of the female reproductive tract in breeding season. One would be a tube with a hairpin turn similar to the female cul-de-sacs near the cloaca, and the second a tube with a *clockwise coil* like the upper reaches of the duck vagina. The diameters of all the tubes were to be the same; they would differ only in the shape of the interior space. We hypothesized that the duck penis would proceed without problems through the straight and counterclockwise spiral tubes. Conversely, we hypothesized that the tubes with the female-like hairpin turn and the clockwise spirals could frustrate erection and prevent complete entry.

Although glass tubes are nothing like the real thing, they have the advantage of providing a standard rigidity and uniformly smooth surface that would control for all mechanical factors other than the shape of the tube, which was the critical element of the hypothesis we wanted to test. The glass tubes would be unnatural but objective and fair. Plus, glass is clear, so we could observe and record on video the progress of the erecting duck penis down the tube.

To find someone to make the glass tubes, Patty and I went to talk to Daryl Smith at the Yale University Department of Chemistry Scientific Glassblowing Laboratory. The motto over the door read, "If not for glass, science would be blind." The display cases in the hallway leading up to the shop were filled with complex glass apparatuses with elaborate condensing coils, leading to flasks and bulbs leading to tubes with charcoal filters, and so on. Business was booming. Waiting outside the door was a line of students, each holding drawings of new designs they wanted to be made for their research, proof if any were needed that this classic art form is still a critical part of the science of chemistry. When our turn came to talk to Smith, we gave him a short introduction to the reproductive biology of ducks, to explain why we wanted him to make artificial duck vaginas in various shapes. We discussed the possible designs. Once we had decided on the final specifications, I asked Smith, "So, is this the weirdest request you ever had?" "Well," he responded, "I've been asked to make artificial vaginas before, but never for ducks!" We didn't inquire further about this previous request.

Brennan returned to the duck farm with new glass tubes in the male-friendly straight and counterclockwise spiral shapes and in the female-like hairpin and clockwise spiral shapes. When she placed the straight and the counterclockwise spiral glass tubes over the male Muscovy Duck cloacae, the penises succeeded at erecting completely 80 percent of the time, and they unfurled at the same velocity as a duck erection into open air. The few cases that did not erect completely only failed to unfurl at the very tip of the penis. In contrast, when faced with the hairpin and *clock-wise* spiraled tubes, the Muscovy Duck penises *failed* to erect 80 percent of the time. In each of these cases, the erection failure was complete. The penis became bottled up in the hairpin turn or in the first bend or two of the spirals and could not advance further. Sometimes, the penis proceeded to unfurl *backward* toward the opening of the glass vagina. These observations confirmed that the clockwise spirals of the duck vagina literally function as an anti-screw device.

To those who may feel concern about the feelings of the male ducks, they ejaculated just fine despite any and all mechanical challenges and seemed not to mind in the slightest. Turns out that because sperm travels down the sulcus, a duck penis can ejaculate regardless of how extensively it is erected. This observation might suggest that all the female's defensive structures are for naught. From the female perspective, however, the earlier the progress of the penis into the vagina can be impeded before ejaculation, the farther away from the ova the sperm will be when they are deposited, and the greater her chance of expelling the unwanted sperm with muscular contractions and preventing sexually coercive fertilizations.

The data from Brennan's glass tube experiments supported our hypothesis that the convoluted vaginal morphologies found in some duck species function to repel the explosively flexible duck penis during forced copulations. Further supporting these conclusions are real-life genetic data showing that these novel anatomical features are actually incredibly effective at preventing fertilization by force. By doing genetic paternity analyses, biologists can determine whether a female duck's offspring were fathered by her chosen male social partner or by other, extra-pair males. In several duck species, including Mallards, in which the forced copulations are a stunning 40 percent of the total copulations, *only 2–5 percent* of the young in the nest are sired by a male who is not the chosen partner of the female. Thus, the overwhelming number of forced copulations are unsuccessful. As a consequence of their elaborate vaginal morphologies, female ducks have indeed succeeded in maintaining freedom of choice for 95 percent of paternity despite persistent sexual violence.

But how is it, then, that the mate the female chooses can manage to overcome the twists and whorls of her defensive anatomy? How does voluntary sex differ from forced? We do not have any direct observations of the inner workings—again, MRI technology would need to take a huge leap forward *and* arrive in the barnyard to deliver such data. But, as mentioned above, Patty's duck-farm observations revealed that when female Muscovys were actively soliciting copulations, they assumed the conspicuously horizontal precopulatory display posture, dilated the cloacal muscles, and released copious amounts of lubricating mucus. It seems clear that females can make the reproductive tract a fully functioning and welcoming place when they want to.

To return once again to McCracken's question—what are the ridiculously long penises of these ducks *doing* inside the female's body? The answer turns out to be, "It depends." If the copulation is solicited, then clearly the female is in for the full ride. These penile structures can easily penetrate to the upper reaches of her reproductive tract if only momentarily. However, if the copulation is resisted by the female, then the penis's length and surface features are designed, evolutionarily speaking, to try to overcome the barriers imposed by female vaginal complexity. In the text above, I didn't use the metaphor of the forbidding cliff face lightly. It's clear that the ridges and hooks on the penis have evolved precisely for the purpose of helping it to claw its way through the various structures within the duck's vagina that are designed to keep it out. However, by being overwhelmingly successful at bottling up the penis during forced intromission, and preventing the vast majority of attempts at forced fertilizations, female ducks have managed to maintain the advantage in this sexual arms race. Even in the face of persistent sexual violence, female ducks have been able to assert and advance their sexual autonomy—their individual freedom to control paternity through their own mate choices.

This is a dark evolutionary tale with an amazing and profoundly redemptive outcome. What we learn from our investigations into duck sex is that despite the ubiquity of sexual violence in these breeding systems, female mate choice continues to predominate. Consequently, male plumages, songs, and displays continue to evolve. Beauty continues to thrive, even in the face of pervasive, violent attempts to subvert the freedom of mate choice that creates it. However, female sexual autonomy is *not* a form of female power over males. It is merely a mechanism for the assurance of freedom of mate choice. Female ducks do not exert sexual control over males, and they can always be turned down by the mates they prefer. Females do not, indeed cannot, evolve to assert power over others in response to sexual violence. Rather, females can only evolve to assert their own freedom of choice.

In this way, the concept of a sexually antagonistic coevolutionary arms race is really misleading because the "war of the sexes" is highly asymmetrical. Males evolve weapons of control, while females are merely coevolving defenses that create opportunity for choice. It's not a fair fight, because only males are really at war. However, as ducks show, female sexual autonomy can still win.

In March 2013, shortly after Barack Obama was inaugurated for his second term, negotiations between congressional Republicans and the White House over the U.S. federal budget broke down once again, and Republicans turned their attention to one of their favorite subjects: wasteful government spending. And that's how the research that Patty Brennan and I had done on sexual conflict and the evolution of duck genital anatomy became the focus of a mini-scandal about government excess, which propelled the topic of duck sex into the maelstrom of the political news cycle, where it was catchily dubbed Duckpenisgate by *Mother Jones*.

Our duck genital evolution research had been funded by a 2009 grant from the National Science Foundation (NSF), with money coming from the aptly named "stimulus" package—the American Recovery and Reinvestment Act (ARRA). For purposes of transparency, ARRA established an independent website,

Recovery.gov, which allowed citizens to "track the money" and see where their stimulus tax dollars were going. This is how, as I imagine it, some enterprising intern at Cybercast News Service (CNS), a conservative news website, came across our grant just a few months before it was due to expire. When a CNS news story describing our grant was posted on its blog, a conservative Twitter storm of outrage ensued. For example, the columnist Michelle Malkin tweeted, "Pass me the mind bleach. Blech." (Of course, why would you retweet a story you were supposedly so eager to forget?) The CNS story was quickly followed up on by Fox News, and the story went into heavy rotation for the week.

The Fox News anchorwoman Shannon Bream introduced a weeklong series of investigations into federal government waste with the following question:

> Did you know that $385,000 of your tax dollars were being spent to study duck . . . anatomy? You heard that correctly—$385,000 of your money to study the private parts of ducks. It's part of President Obama's stimulus plan, and it's just one example of the kind of spending decisions that have added up to massive debt and deficits.

The three-minute piece that followed was a tour de force of the tired genre of big-government lament. I never imagined it could be possible to combine quotations from Ronald Reagan ("Government is not the solution to our problems. Government is the problem!"), images of the Twin Towers burning, Barack Obama's teleprompter, and America's housing foreclosure and banking crises into an attack on our animal genital coevolution research program, but Fox News managed to accomplish just that. Never one to shy away from any antigovernment cause, Sean Hannity discussed the validity of federal funding of a Yale University study on duck genital evolution with Tucker Carlson and Dennis Kucinich later in the week in a segment titled "D.C. Wasteland."

Our duck penis research did have its strong defenders in the media, among them Chris Hayes on MSNBC, the science writer

Carl Zimmer, *Mother Jones*, the *Daily Beast, Time,* and Politi-Fact. After Patricia Brennan wrote an awesome defense of basic science research and funding for Slate.com, the storm appeared to be over.

Eight months later, however, when Senator Tom Coburn of Oklahoma published his *Wastebook* for 2013 and included our $385,000 grant as number 78 among the top 100 examples of federal government waste, the irresistible story of Duckpenisgate roared back to life. The *New York Post* headline read, "Government's Wasteful Spending Includes $385G Duck Penis Study."

Out of the $30 *billion* of waste reported in *Wastebook,* the *Post* headline focused on the 0.001 percent that went to our study. Somehow, the combination of money, sex, and power—your tax money, duck sex, and Yale's Ivy League prestige—made the story irresistible. And so it went, as the right-wing news outlets sought new ways to inspire the outrage that in an earlier era was reliably engendered by Ronald Reagan's Cadillac-driving "Welfare Queen" and the Defense Department's $700 toilet seats.

When repeddling this old story of government profligacy, news programs inevitably mentioned our research with a veneer of sexual titillation. So, when Sean Hannity sarcastically asked Tucker Carlson on Fox News, "Don't we really need to know about *duck genitalia,* Tucker Carlson?" his question belied the genuine human fascination with the topic. Like all the other attackers, he ignored the fact that we actually *do* have a tremendous amount to learn from the study of duck sex. There are important evolutionary findings, and perhaps even some of immediate practical value. If the pharmaceutical industry thought that Viagra was a big deal, just wait until duck developmental biologists unlock the secrets of the stem cells that allow the duck penis to regenerate itself every spring and to get bigger each year (which I think I might have forgotten to mention)!

Furthermore, our research has discovered that what the 2012 Missouri Republican Senate candidate Todd Akin said about rape in humans—that "the female body has ways to try to shut that whole thing down"—is actually *true* of ducks, but the reason it is true tells us something deeply important and new about the evolution of sexual autonomy in nature.

This chapter, like the research grant that had its fifteen minutes of infamy back in 2013, has focused on a group of birds in which female mate choice is threatened by male sexual coercion. What happens when mate choice is constrained, prevented, or denied by physical force? we asked. And as we have seen, the female ducks do not simply cave under the threat of violence or even death. Rather, their shared standards of beauty—even meaningless, arbitrary beauty—provide them with the evolutionary leverage to fight back against sexual coercion and reassert their freedom of choice over fertilization. Female ducks teach us a great lesson about the unexpected power of female sexual autonomy. In the words of the Eurythmics and Aretha Franklin song, they teach us that "Sisters are doin' it for themselves!" By doing so, females together become the agents of choice and the guarantors of their own freedom of choice. The evolutionary advantages of obtaining the mates they prefer—male offspring that will possess the traits they and other females have agreed are attractive—are so strong that they have reshaped female internal anatomy. Expanded sexual autonomy allows female waterfowl to continue to select for beauty in the form of male sexual display and everything that that involves—sounds, colors, behaviors, plumage, and so on. Even in the face of unrelenting sexual attack, female ducks have found a way to maintain the beauty in their world.

It is not an accident that these discoveries are consequences of the aesthetic view of mate choice. Only when we recognize that mate choice is a form of individual agency can we conceptualize sexual violence as a disruption of that agency. To paraphrase Susan Brownmiller, sexual violence is against the will of female ducks too.

The revelation of an aesthetic mechanism for the evolution of female sexual autonomy in waterfowl is a profoundly *feminist* scientific discovery. It is not feminist by accommodating the science to any contemporary political theory or ideology. Rather, it is a feminist discovery in that it demonstrates that sexual autonomy *matters* in nature. Sexual autonomy is not merely a political idea, a legal concept, or a philosophical theory; rather, it is a natural consequence of the evolutionary interactions of sexual reproduction, mating preferences, and sexual coercion and violence in social species. And the evolutionary engine of sexual autonomy is aesthetic mate choice. Only by acknowledging that these are real forces in nature can we make progress toward a complete understanding of the natural world. Of course, this should not be too surprising. As Stephen Colbert on *The Colbert Report* has observed, "Reality has a well-known liberal bias."

This discussion of duck genital evolution raises a broader question: Why *do* most birds lack a penis entirely? How did this happen? And what are the evolutionary *and* aesthetic consequences of the loss of the bird penis? Once again, the concepts of aesthetic evolution and sexual autonomy can provide interesting new insights.

Birds originally inherited the penis from their dinosaurian ancestors, but then it was lost some sixty-six to seventy million years ago in the most recent common ancestor of the group known as the Neoaves, which includes over 95 percent of the world's species of birds. We do not know anything about the ecology or morphology of the ancestral neoavian bird in which the loss of the penis occurred, so investigating this kind of event is difficult. But that doesn't mean we can't make some progress in thinking about it.

The penis could have been lost because it was no longer useful—like the eyes of cave fishes. But copulation is pretty important to reproductive success, so we have to ask what kind of selection could possibly select *against* the penis?

It's possible that the neoavian penis was lost because females explicitly *preferred* males without penises. Why? If one of the primary functions of the penis is to subvert female mate choice through forced copulations, as it is in waterfowl, then female mat-

ing preferences *against* intromission could have evolved to reduce
the threat to female sexual autonomy. The next two chapters will
focus in detail on how females can use mate choice itself to change
males both physically and behaviorally in ways that advance
female autonomy. But whatever the evolutionary mechanism, the
loss of the penis has had distinct consequences for sexual autonomy in birds.

Going penis-free means that active female participation is
virtually required for the intake of sperm into the female cloaca.
Although even in the absence of the penis males *can* mount a
female and forcibly deposit sperm on the surface of her cloaca,
they cannot deposit sperm *within* the female nor force her to
uptake their sperm by dilating her cloaca. In the more than 95
percent of bird species that are penis-free, females can eject/reject
unwanted sperm. For example, barnyard hens can eject sperm
after coerced copulations with unwanted males. Attempts at sexual harassment and intimidation do still exist in birds without a
penis, and the female birds may still suffer injuries by resistance,
but the loss of the penis has resulted in a nearly complete end to
forced fertilizations. Through the loss of the penis, female neoavian birds have essentially won the battle of sexual conflict over
fertilization.

What are the evolutionary consequences of this expanded
sexual autonomy? Interestingly, we can return to Darwin's observation in *The Descent of Man* with an entirely new perspective:
"On the whole, birds appear to be the most aesthetic of all animals, excepting of course man, and they have nearly the same
taste for the beautiful as we have."

Given that birds are among the few groups of animals that
have evolved a combination of complex sensory systems, cognitive
capacities, *and* expanded opportunities for mate choice thanks to
the loss of the penis, I do not think it an accident that birds have
also evolved into the "most aesthetic of all animals, excepting
of course man." The irreversible advance in avian female sexual
autonomy that occurred because of the disappearance of the penis
may be the most powerful explanation of the aesthetic evolutionary extravaganza among birds.

This evolutionary extravaganza, which is predicted by the Beauty Happens hypothesis, might in turn have contributed to birds' explosive speciation and aesthetic radiation, which could help to explain why penis-free birds are the most successful group of terrestrial vertebrates in terms of the number of species. Of course, there are many other factors contributing to avian evolutionary success, rapid speciation, and diversification, including the capacity for flight, their capacity for ecological diversification, migration, song, and song learning. But any future investigation into the question of the evolutionary success and diversity of birds should include the role of aesthetic evolution and the evolutionary loss of the neoavian penis.

Another striking observation about female sexual autonomy in penis-free birds is that it is strongly correlated with social monogamy, in which both the male and the female make substantial reproductive investments of time, energy, and resources into raising their offspring. The traditional explanation for the evolution of monogamy in these birds is that it was a "nonnegotiable" feature of neoavian biology. Unlike most other reptiles, neoavian birds have offspring that are helpless and entirely dependent on their parents when they hatch. These helpless baby birds—what ornithologists call altricial young—are so vulnerable to predation that they must grow up very fast to minimize the risk of being eaten in the nest before they learn to fly. Having two parents helping to raise them will protect them during this vulnerable period and will also speed their development and help them to fledge faster.

Intriguingly, however, we may have this evolutionary logic completely backward. Rather, the loss of the avian penis and the expansion of female autonomy might have had a decisive impact on the evolution of avian development, physiology, and social behavior, so that altricial young may be the result, not the cause, of the evolution of avian monogamy. All species of birds with penises have offspring that can feed themselves soon after hatching—ornithologists call them precocial young—who can be safely raised and guarded by only one parent. (Two-parent care may evolve in precocial bird species if territorial defense is required.) However, once the penis was lost, female birds might have evolved to use their expanded sexual autonomy to require *more* parental investment from males. Because penis-free male birds cannot force copulation, they are basically required to fulfill female mating preferences in order to reproduce. If females evolve to require greater investment in reproduction from their mates, then males will soon evolve to *compete* with one another to do a better job of providing resources for the offspring of those choosy females! The result will be evolution of a stronger, more extensive pair bond in which males are active participants and investors in parenting. This expansion of male reproductive investment could in turn have facilitated the evolution of helpless young, whose upbringing requires the kind of substantial investment that males evolved to make. Thus, expanded sexual autonomy that resulted from the loss of the penis has allowed neoavian birds to advance in their sexual conflict with males over parental investment, too.

The concept of sexual autonomy provides insights not only into the evolution of defenses against sexual violence and coercion but into the evolution of other, distinct paths to advance against sexual conflict. We will explore these ideas further, in birds in the next two chapters, and in humans too, in chapters 10 and 11.

So what have the females in the more than 95 percent of bird species that *lack* a penis done with all the sexual autonomy they have won? As our observations of bowerbirds and manakins in the next two chapters will reveal, they have pursued their aesthetic, and frequently arbitrary, mate choices, and by doing so have contributed to the nearly infinite varieties of colorful, tuneful, and exuberant avian beauty in the world.

CHAPTER 6

Beauty *from* the Beast

No description can really prepare you for the extraordinary architecture of the aesthetic structures created by male bowerbirds to
use as their courting arenas. Few creatures on earth lead a life
that is as thoroughly shaped by aesthetics as these birds, and their
bowers are their masterpieces, created with as much care, attention, and discernment as any artwork.

The aesthetic extremity of bowerbirds is the product of the
same evolutionary force we have been examining throughout—
female mate choice. We've seen how mating preferences exert
evolutionary pressure on ornaments and *coevolve* along with the
ornaments they prefer. And as we saw so vividly in the case of
ducks, when mate choice is infringed upon by sexual coercion,
the evolutionary advantages of maintaining freedom of mate
choice can drive the evolution of defensive strategies—including
behavioral and even *anatomical* mechanisms of resistance. In the
ducks, sexual conflict has resulted in a violent, costly, and self-
destructive antagonistic arms race between the sexes. Both sexes
invest heavily in arms and defenses, many females are killed or die
young, the sex ratio of males to females becomes more uneven, so
the sexual competition and coercion get worse, and the population size suffers as a result. Of course when ecological conditions

change, and make coercion less profitable, then sexual conflict is alleviated, and neither sex has to make these costly investments anymore.

But in the bowerbirds, we find a different and distinctive evolutionary response to sexual coercion. Instead of evolving separate evolutionary mechanisms for aesthetic mate choice and resistance to coercion, female bowerbirds have used the power of mate choice itself to transform male sexual behavior in ways that enhance and expand their sexual autonomy. As a result, females get the highly stimulating, exciting, and active males they prefer, but in a behavioral context that allows them nearly complete control over their mating decisions.

Bowerbirds provide us with a particularly vivid example of what I call *aesthetic remodeling*—the coevolution of female aesthetic preferences and male traits that enhance female autonomy. The result is a sexual partner that is both more pleasing to females and more amenable to female choice—in other words an attractive male that has to take no for an answer if the female prefers not to mate with him.

I vividly remember my personal introduction to the bowerbird clan on my first trip to Australia, when I traveled there with my wife, Ann, in 1990. Walking around the edge of the campground in Lamington National Park, which is located midway down the east coast of the continent near Brisbane, we encountered a male Satin Bowerbird (*Ptilonorhynchus violaceus*). The chunky male is the size of a small crow with a stout ivory-yellow beak, exquisite violet-purple irises, and a deep, lustrous blue plumage.

What makes the aesthetic expression of the Satin Bowerbird truly extraordinary, however, is not his plumage but his bower. Like the males of almost all the other species of the bowerbird family, the male Satin Bowerbird creates a courtship structure—a kind of bachelor pad, or crib—to attract mates. As Henry Alleyne Nicholson clarified in the first published use of the word "bowerbird" in his *Manual of Zoology* in 1870, the bowerbird's bower is *not* a nest but an entirely distinct structure built by a displaying

Types of bower architecture in Bowerbirds. (Below top) Display court of Tooth-Billed Bowerbird ornamented with green leaves and no bower. (Below bottom) Avenue bower of Great Bowerbird. (Opposite top) Maypole bower of MacGregor's Bowerbird. (Opposite center) Double-maypole bower of Golden Bowerbird. (Opposite bottom) Hut bower variation of the maypole bower of Vogelkop Bowerbird.

male for the sole purpose of attracting mates. The bower has no function beyond its use as a seduction theater—an ornamental stage for male sexual display.

Prior to the ornithological exploration of Australia and New Guinea by Western explorers and colonists in the mid-nineteenth century, the word "bower" referred only to a simple dwelling or hut (like a lean-to); to an interior chamber within a home, especially a lady's bedroom or *boudoir;* or to a shady recess with overarching branches and vines. As it happens, all of these traditional meanings seem happily appropriate when applied to the bowers created by male bowerbirds; however, bowerbirds extend these meanings in a whole new direction.

The male Satin Bowerbird's bower is located in a small clearing on the forest floor and consists of two parallel walls made of dry, upright twigs, branches, and straw, with a narrow passageway running down the middle of it (color plate 17). Hence the name given this kind of mating structure, the *avenue bower,* which is one of the two main forms of bowerbird architecture.

In addition to building the bower structure, the male Satin Bowerbird gathers objects with which to decorate it, all of them royal blue, and he piles them on a bed of straw in the courtyard area located at the front of the bower. Given his proximity to the trash from the campground in the national park, the first male that Ann and I saw had assembled a hoard of objects that included not just wild fruits, feathers, berries, and flowers but a mixture of man-made and relatively durable items like milk jug tops, pen caps, snack food wrappers, and other plastic packaging—all of them, from flowers to food wrappers, in the preferred shade of a medium royal blue. Though the Satin Bowerbird is highly discriminating in the color of the items he gathers for his bower, so long as they are of an appropriate blue, he is completely unpicky about the material properties or provenance of such items. A blue soda cap is just as pleasing as far as he's concerned as the most exquisite blue feather. The male attends his bower, keeps it in good order, gathers and curates his hoard of blue items. He also defends it from other males, who will take any chance they can get to pull his bower apart and rob him of his prized blue trinkets.

Of course, the entire function of this piece of architecture is to seduce a female to visit and to mate. Although I was never privileged (or patient) enough to observe a female visit, the display behaviors of the Satin Bowerbird have been well described. When a female arrives, she steps into the avenue between the bower walls and peers out at the male and his gathered materials. Like the horse stall at the starting gate of a race, the passage between the bower walls is narrow, affording her only enough room to face forward, where she can see the male, waiting for her. Once he has her attention, the male performs a series of highly energetic displays in which he suddenly fluffs out his body feathers and wings. He punctuates these displays with loud vocal squawks, bizarre, pulsating, buzzy electronic noises, and dead-on imitations of other local bird songs including that of the Laughing Kookaburra (well-known to us from Hollywood jungle sound tracks). Ultimately, the male will pick up an item from his collection of blue materials, or a twig or a green leaf, conspicuously display it to the female, and then replace it on the ground as he continues his vocal performance. If the female prefers him, she signals her interest with a low, crouching copulation solicitation posture. The male then enters the bower from behind and mounts her while she remains in the bower. If, however, the male attempts to copulate when the female is not receptive, then she can escape out the front of the bower and fly away to avoid his advances. In other words, the walls of the bower protect her from being jumped by the male.

Avenue bowers can differ quite substantially from each other. The simple avenue bower of the Satin Bowerbird consists only of a pair of parallel stick walls with a narrow pathway, or avenue, between them. But among other species, there are much more elaborate avenue "bower-plans" as well, including the double-avenue bower made by Lauterbach's Bowerbird (*Chlamydera lauterbachi*), which has two parallel paths on a raised platform, and the grand "boulevard" bower built by the Spotted Bowerbird (*Chlamydera maculata*), in which the central pathway is especially wide and the side walls are a transparent screen rather than a solid mass of sticks.

The decorations assembled by the male bowerbirds in the

areas in front of or behind the bowers also vary tremendously among species and sometimes even among populations within a species. In some species, the decorative objects are fruits, flowers, or leaves, while in others they include bones, shells, insects, or feathers. Different colors may also be preferred, depending on the species or population. Often the materials are laid on a bed of moss, straw, or pebbles.

Another avenue builder, the Great Bowerbird (*Chlamydera nuchalis*), has a wide distribution in dry open woodlands across the northern third of the Australian continent. In most Great Bowerbird populations, males collect and display light-colored pebbles, bones, and snail shells for their bowers. But the males of one population of Great Bowerbirds are particularly original in their choice of decorations, as I had occasion to observe in 2010, when I visited the Broome Bird Observatory in the northwestern corner of Australia. This preserve sits on the shores of Roebuck Bay, which is lined by steep, five- to twenty-meter-high cliffs of red clay and stratified rocks. About half a kilometer from the ocean cliff face, I observed a Great Bowerbird avenue bower with a surrounding courtyard decorated at both front and back with a vast pile of bleached, brilliantly white *fossil* clam shells (color plate 18). This bird's bower was a virtual paleontological museum, displaying fascinating examples of the earth's extinct biodiversity to attract prospective mates. Quite literally, this male's territorial calls meant "Do you want to come over and see my fossil collection?" The shells were so distinct in shape and color that it was easy to identify their source. At certain places along the red cliffs that tower over the bay, a brilliantly white layer of material about a foot thick is exposed. Closer inspection revealed that this was a layer of white fossil bivalves that had been deposited in abundance during an earlier epoch in the geological history of this corner of the ancient continent. As a museum curator myself, I felt a certain affinity with this bowerbird's paleontological passion.

The second major architectural style made by bowerbirds is the *maypole bower,* which consists of a pile of horizontal sticks placed around a central support, usually a sapling or a small tree. The stack of brown sticks is cone shaped, broadest at the base, and narrowing at the top to form a structure that is like a bottle-brush, or a bizarre, minimalist, postmodern Christmas tree. At the base of the maypole, the male clears a circular path, or run-way, which allows the male and the female to run a rapid circuit around the maypole during the courtship maneuvers. The court, which is located outside this circular runway, is decorated with materials the male has gathered, which can include flowers, fruits, beetle and butterfly parts, and even fungus. Some bowerbird species also adorn the twigs and branches of their Christmas-tree-like structure with decorative materials, such as regurgitated fruit pulp. (Okay, so maybe that's not so much like a Christmas tree.)

The first time I saw a maypole bower was during the same trip to Australia. A week after our sighting of the Satin Bowerbird, Ann and I traveled to the rain forest in the Atherton Tablelands in northern Queensland, where we hoped to see the Golden Bowerbird (*Prionodura newtoniana*) and its famous *double-maypole* bower. The Golden Bowerbird is the smallest of the bowerbird species. The male has dull olive-green body plumage and bright yellow patches on his crown, upper back, throat, and belly. I was familiar with its bower from a classic, multi-panel black-and-white drawing illustrating the diversity of bowerbird architecture that has appeared in every ornithology textbook since, appar-ently, the dawn of time. The double-maypole bower of the Golden Bowerbird was depicted in a panel adjacent to the simple avenue bower of the Satin Bowerbird and appeared to be about the same size. It never occurred to me to consider whether or not the struc-tures in the two panels were drawn to the same scale. So, as Ann and I headed down the rain forest trail scanning the forest floor for signs of the bower, I cautioned her in a whisper, "We have to be careful not to step on it!" In a few hundred meters, we rounded a bend in the trail and saw an *enormous* structure that was nearly waist high and more than a yard wide. It would have taken quite an effort to step over it, let alone step on it accidentally as I had feared.

After recovering from my shock at its size, I was equally stunned by the complexity of the structure. The double maypole consisted of two huge piles of horizontal sticks piled around a

pair of saplings but oriented in various directions. The two conical mounds merged together in the middle to create a saddle of
sticks. The Golden Bowerbird decorates the bower structure itself
but not the courtyard around it. This male had adorned one side
of his bower with many dozens of small flowers of an exact shade
of buttery forsythia yellow, and he had decorated the other side
with myriad tiny threads of a vivid fluorescent-green lichen. The
transplanted lichen threads were growing happily in their new
home, and the flowers were as fresh looking as those in a florist's
bouquet. Even at this cooler altitude, these flowers would clearly
not last for more than a few days, so the absence of any brown or
wilted petals was testimony to the male's constant and attentive
curation of his display.

Fifteen years later, I had the pleasure to visit Brett Benz, then
a University of Kansas undergraduate student, at his field site
near the village of Herowana in the central highlands of Papua
New Guinea where he was studying the MacGregor's Bowerbird (*Amblyornis macgregoriae*), which builds a single-maypole
bower. The maypole bowers of MacGregor's Bowerbird are sit-
uated high up on ridges that descend sharply under the dense
forest canopy. The male decorates his court and bower with a
remarkably diverse set of ornaments that includes fruits of various colors, a brownish fungus, and tiny, extraordinarily brilliant,
iridescent fragments of blue *Entimus* weevils. Brett had recorded
video of a male returning to his bower with a *living* blue weevil.
The male brutally pulled apart the writhing beetle on the court
floor and carefully placed pieces of it in his bower arrangement.
Stepping back after every such placement, he regarded each decorative possibility with a little cock of his head, like a fussy florist
checking on the arrangement he was creating. Perhaps the most
curious ornaments of all were the numerous stringy, threadlike
blackish clumps hanging near the tips of various horizontal sticks
in the bower structure itself, which turned out to be caterpillar
frass—or droppings. The list of found ornaments in the collages
assembled by this species was eclectic in the extreme.

Like other *Amblyornis* maypole builders, the male Mac-
Gregor's Bowerbird is mostly drab brown like the female, but

unlike other *Amblyornis* the MacGregor's male has a long, erectable crest of deep umber-orange feathers. During the courtship display, the male and the female stand on opposite sides of the circular runway, with their view of each other obscured by the maypole between them. Peering around the runway at the object of his desire, the male suddenly erects his brilliant orange crest feathers and flashes them at her, then quickly reverses course and peers at her around the opposite side of the maypole, and he continues to engage in a rapid succession of alternating glimpses in what is essentially an elaborate game of peekaboo. Sometimes the male makes a running dash toward her around his runway. If he approaches her too aggressively, however, she can scuttle to the side, keeping the maypole between her and her overeager prospective mate—or fly away.

There are several unique features of male bowerbird courtship behavior that require specific evolutionary explanation: the existence of the bower; the radical diversity of its architecture, which I've only begun to hint at in these brief examples; and the wildly eclectic nature of the items that the males gather to decorate the courtyards of their bowers. How did these extraordinary structures and behaviors come into being, and why? We must look to their evolutionary origins to find out.

The bowerbird family (Ptilonorhynchidae) includes twenty species in seven or eight genera that are endemic to Australia and New Guinea. Like manakins, the bowerbirds are frugivorous, and almost all species are polygynous. Unlike those of manakins, however, the male display sites are not spatially aggregated into leks. Instead, each solitarily displaying male builds and defends a bower.

We now understand the bower to be a component of the male bowerbird's *extended phenotype*. Coined by Richard Dawkins in a book of the same name, the phrase communicates that an organism is more than the proteins created by the expression of its DNA, more even than its anatomy, its physiology, and its behavior. An organism's complete phenotype includes all of the consequences of its genome's interacting with its environment, including its impact *on* the environment. Thus, the beaver dam, which can create major changes to the ecosystem through the creation of ponds that gradually silt in and become bogs, is part of the *extended phenotype* of the beaver. Entire communities of organisms can evolve to feed on or shelter in components of the extended phenotype of another species. All architectural forms created by any living organism—including not just bowers but bird nests, beehives, termite mounds, prairie dog burrows, and coral reefs—are manifestations of the extended phenotype of the species that build them.

As Dawkins implied in the subtitle of *The Extended Phenotype—The Long Reach of the Gene*—he views all components of the extended phenotype as further manifestations of adaptive evolutionary forces acting on selfish genes. As a confirmed neo-Wallacean, Dawkins believes that the extended phenotype is merely another, more expansive frontier in which to observe the pervasive effects of adaptive natural selection. However, when the extended phenotype becomes a form of ornamental sexual display, as with bowerbird bowers, it becomes subject to sexual selection. This is where the extended phenotype meets the Darwin-Wallace debate over the nature of mate choice, sexual selection, and natural selection.

Is the extended phenotype exclusively shaped by adaptive natural selection? Or can the Beauty Happens dynamic shape the extended phenotype, too? If so, what evolutionary patterns should we expect? Bowerbirds and their bowers offer a unique opportunity to investigate the "long reach" of the neo-Wallacean paradigm into the realm of beauty.

Fortunately for students of evolution, the bowerbird family has enough diversity, including sufficient examples of extant species exhibiting transitional forms of bower architecture, to "capture" some of the critical stages in the evolutionary origin of this unique behavior. The earliest branch in the phylogeny of the bowerbirds includes the three species of catbirds (*Ailuroedus*). Like the vast majority of all birds—but unlike any other birds in the bowerbird family—the catbirds are monogamous, have two-parent care, form enduring pair bonds, and don't have any display court or bower. Moreover, as documented by Clifford and Dawn Frith, a tireless team of bowerbird enthusiasts from Queensland, Australia, nest construction in catbirds is carried out *exclusively* by the females. Thus the existence of the catbirds at the base of the bowerbird family tree provides evidence that ancestral male bowerbirds had no experience or interest in any of the fundamentals of home building or improvement. The extraordinarily advanced architectural capability of the male bowerbirds is a later evolutionary development, unrelated in any way to nest construction behaviors, and driven entirely by aesthetic female mate choice.

But how do we *know* that bower design and ornamentation perform an exclusively aesthetic function? Well, we know that the bower serves no physical purpose other than as a location where courtship takes place. It's a stage set with props, created for a performance that is evaluated by females during courting season. Over the past thirty years, the fundamental role of bower structure and ornamentation in female mate choice has been well established through a long-term research program led by Gerry Borgia at the University of Maryland. Borgia has conducted decades' worth of observations and experiments on multiple species of bowerbirds, focusing especially on the Satin Bowerbird of eastern Australia. In a pioneering use of eight-millimeter film and later video technology, Borgia set up numerous cameras at multiple bowers, aiming electronic eyes down the central avenues of the bowers to trigger the cameras and to record the details of all activity, including any female visits, that took place there. This made it possible for Borgia and his students to observe and measure female mate choice behavior and the variable mating success of different males over the course of many years.

Borgia's juggernaut research program has produced much of what we know about mate choice in bowerbirds, establishing conclusively that the specific features of the bower and its decorations are critical to female mating decisions. As Borgia's students Albert Uy and Gail Patricelli documented in the course of tracking mate choices made by 63 females who visited a total of 34 male bowers, the females visited between 1 and 8 males apiece, averaging out at 2.63 males. Most females visited a number of males over a series of days and then returned to revisit a smaller number before finally selecting one of those for their mate. Their choices were strongly skewed toward those males with better-constructed and more highly decorated bowers. These revolutionary data are a strong indication that female bowerbirds make their aesthetic mate choices based on a pool of interactive, experiential data, rather than in response to a simple, hardwired cognitive stimulus threshold. So, there is direct evidence in support of the role of sexual selection in the evolution of bowers.

Turning now to the evolutionary history of ornamentation, we can look to another still-living member of the bowerbird family—the Tooth-Billed Bowerbird (*Scenopoeetes dentirostris*). Also descended from an early branch of the bowerbird family tree, the Tooth-Billed Bowerbird is a polygynous species in which females do all the parental care. But despite being members of this family, male Tooth-Billed Bowerbirds, like the catbirds, do not actually build a bower. However, unlike the catbirds, they do create a court, clearing a patch of ground that is about two yards wide and then decorating it with a dozen or more large green leaves, carefully spaced out over the bare ground. This primitive court with its simple ornamentation of leaves give us some insights into the origin of bowers and their decorations. We can see that the collection of court decorations is common to *all* polygynous bowerbirds and that it evolved *before* bowers came into existence. It is one feature of bowerbird life that has not been evolutionarily lost in *any* of the bowerbird species—a further indication of the importance of decorations to female mate choice.

What has changed over time, of course, is the nature of those decorations. The specific materials that males collect and the many ways in which they are laid out have continued to evolve among species, and sometimes even among populations within a species. It is astounding to consider the breadth of materials that the various bowerbirds employ as bower decorations—from fruits to fungi, flowers to feathers, berries to butterflies, seedpods to caterpillar poop, not to mention candy wrappers and clothes-pins. Some avenue bower builders even "paint" the interior walls of their bowers with masticated blue, green, or black plant materi-als. By any scale, this is an extraordinarily broad aesthetic palette.

The collection of these ornamental objects and materials is the result of *male* aesthetic preferences that have coevolved with female mating preferences. To please the females, the males have evolved a whole new class of behaviors and preferences of their own. In the process, they've made themselves into animal artists who vie for the attentions of their aesthetic patrons.

As with any artist, their use of materials is far from random. As we have seen from the paleontological treasures collected by the Great Bowerbirds of Roebuck Bay, and the campground detritus collected by Satin Bowerbirds, the bower decorations are partly determined by what's available in the immediate environment. But the role played by aesthetic choice is also very important, as demonstrated by pioneering work done by Jared Diamond in the early 1980s on the bower ornamentation among populations of the Vogelkop Bowerbird (*Amblyornis inornata*) in western Irian Jaya, the westernmost portion of the Indonesian half of the island of New Guinea. Diamond discovered that males of the Fakfak and Kumawa Mountains build a straightforward maypole bower decorated exclusively with drab-colored materials like bamboo, bark, rocks, and snail shells. In contrast, the males of the nearby Arfak, Tamrau, and Wandammen Mountains, which are only 50–150 kilometers away, build an elaborate hut bower with a maypole at its center and an outer court that is decorated with colorful fruits, flowers, insect parts, fungi, and seedpods (color plate 19). These differences occur even though males in all five of these mountain populations have access to the exact same materi-als in their environments. There was even differentiation within immediately neighboring hut-bower-building populations. Arfak and Tamrau Mountain birds included white ornaments in their display, for example, whereas Wandammen Mountain birds did not. The birds are extremely picky about what they use.

To further establish that bower decorations are the result of specific male preferences, Diamond did experiments in which he offered male Vogelkop Bowerbirds from the Wandammen Mountains—which build elaborate hut bowers with piles of diverse and colorful fruits, flowers, and other materials—a choice of different-colored poker chips. When males gathered the poker chips, they demonstrated significant preferences for specific colors, especially for blue, purple, orange, and red (in descending order of preference), and on their bower courts they grouped them with similarly colored flowers, fruits, or feathers. By marking the specific poker chips that made their ways into individual male bowers, Diamond was also able to establish that many of the poker chips were later stolen by other males to be incorporated into their bowers. The rate of theft reflected the same differential color preferences, with blue being stolen most often, red least often. In similar tests, males from the Kumawa Mountains—who build simpler maypole bowers with uniformly drab ornaments—rejected all colors of chips.

Decades later, Albert Uy repeated these ornament color choice experiments, and he expanded upon them with simultaneous measures of female mating preferences. Working with two of the populations Diamond had studied, Uy confirmed that maypole builders from the Fakfak Mountains avoided the bright colors and preferred brown, black, and beige tiles, while hut builders from Arfak preferred blue, red, and green tiles. Using automatic video cameras aimed at sixteen hut bowers in the Arfak population, Uy was also able to show that female mating preferences were highly skewed toward a small subset of males and that the mating success of these males was significantly correlated with both the size of the total area they had covered with blue decorations and the size of the hut—the bigger the better. Thus, in the Arfak population, female mating preferences were closely *coevolved* with the male extended phenotype—a preference for blue ornaments and for constructing hut bowers of considerable size.

Because the populations of the Vogelkop Bowerbirds are found in very nearby mountain ranges, isolation among these populations must be very recent. Thus, the differences in bower ornaments and architectural styles among them are likely to have evolved in a very short period of evolutionary time. Crucially, multiple aspects of the female mating preferences have coevolved with these variations in male extended aesthetic phenotype. This striking pattern of rapid differentiation among populations in display traits and preferences is exactly the pattern predicted by the Beauty Happens hypothesis.

But might there be another explanation? Could the decorations that male bowerbirds choose to gather be indicators of male genetic quality? Well, it's possible that the choice of objects could indicate male quality if the collection consisted of rare objects that required an investment of lots of time, energy, and skill to find. But Jared Diamond established that the mountain forests of these Vogelkop Bowerbirds all had the same materials available, so black fungus and red flowers were not rarer on one mountain than they were on another. Furthermore, Joah Madden and Andrew Balmford conducted an explicit test of the idea that ornaments provide honest information about search costs in a study of three populations of Spotted Bowerbirds (*Chlamydera maculata*) in Queensland, Australia. They found no support for the idea that the bower decorations that were favored were rarer than any others. Quite the contrary. Snail shells and white stones were preferred in those populations where they were *more* common, not less. Objects that were associated with sexual success were also more common, not rarer, than others. Moreover, the Spotted Bowerbird males preferred to display those fruits that degraded *less rapidly* than other fruits that were available, which further reduces the work (and therefore the cost) required to produce a consistently attractive display. Thus, there is no compelling evidence that bower decorations are costly, honest signals of male quality. Rather, they appear to vary like any other aesthetic styles among species.

More recently, the evolutionary biologist John Endler and colleagues have discovered a fascinating new wrinkle in the aesthetics of bower decorations in at least some populations of the Great Bowerbird. In eastern Queensland, Endler has documented that successful Great Bowerbird males create displays in which the size of the objects gradually *increases* the farther away from the bower they are. They hypothesize that the male is creating an optical illusion known as forced perspective. In this case, with the objects getting bigger in proportion to their distance from the opening, the result is a flattening of the visual space so that when viewed from inside the bower, the objects tend to appear more uniform in size. Endler and colleagues make various speculations about why this particular trick of the eye should appeal to female bowerbirds. Interestingly, however, the optical illusion is *not* in the appropriate direction to make the male himself appear bigger to the female, so the illusion cannot function as a strategically dishonest signal about male size.

Whatever the reason for it, there's nothing accidental about the effect that the males are creating. By experimentally rearranging the stones in the opposite order, creating a negative gradient in object size, Endler and colleagues were able to observe that the males noticed the rearrangement, that they were not happy about it, and that they moved the objects back into place in appropriate order to restore the optical illusion. Laura Kelley and John Endler have subsequently shown that males with stronger illusions have higher mating success.

That still doesn't answer the question of why the preference for this visual illusion evolved. Endler has proposed that a male's ability to create this illusion could provide honest information to females about the cognitive capacities of their prospective mates— that is, the better the illusion, the better the male's brain, and the better the male's genes. Regardless of what these exercises in perspective may or may not be communicating, the implications for this finding are amazing. Endler notes that techniques to create forced perspective in human arts did not arise in Western culture until the fifteenth-century Renaissance. Assuming that this behavior has been present in bowerbirds since before the fifteenth century, Endler asks, "Why did perspective evolve in bowerbirds before humans?"

Of course, the human invention of perspective occurred first in art. I think it's interesting that humans developed perspective in the service of art, long before we made any practical use of it. Why shouldn't we expect the same of bowerbirds? As we have seen, aesthetic evolution can be an excellent source of evolutionary innovation. Endler himself seems to acknowledge this by comparing "bowerbird art" to human art. In a *New York Times* interview, he stated that the optical illusion "is evidence that bowerbirds are actually creating art" and that female mating preferences and male aesthetic construction preferences "can be regarded in an aesthetic sense because judgments are made."

Back to the question: Why have bowers evolved at all? And why do they continue to diversify among species and populations of bowerbirds? In 1985, Gerry Borgia and Stephen and Melinda Pruett-Jones hypothesized that the building of bowers, and the male's ability to protect his own against theft and destruction by other males, were indicators of male status and quality. But those hypotheses could not account for the complex variations in architectural structure and ornamental preferences that have developed among different populations and species. Blue berries are no easier or harder to defend than white pebbles.

Beginning in 1995, however, Borgia proposed a compelling and novel hypothesis for the evolutionary origins of the bower. Borgia had observed that the intense, energetic, and often violent displays of male bowerbirds frequently startle or frighten visiting females. Whenever the female perches on the court to observe the male and his decorations at close range, she is exposing herself to the threat of sexual harassment and forced copulations. But it's a different story when she is inside the bower. Borgia hypothesized that the building of bowers evolved through female preferences for being *protected* from sexual coercion, physical harassment, and forced copulations. He cited lots of natural history evidence from the field in favor of this "threat reduction" hypothesis. For example, there are many observations to document that if a male attempts copulation with a female at an avenue bower before she has signaled that she is receptive, she flies out the front of the bower when the male tries to mount her from behind; if she's visiting a maypole bower, she can hop to the side of the circular runway, keeping the maypole structure securely between her and the male.

As further support of his hypothesis, Borgia described the extremely abrupt courtship of the Tooth-Billed Bowerbird, whose simple, open, leaf-decorated display court has no bower and thus nothing to protect the female. When a female Tooth-Billed Bowerbird arrives on the male's court, she is immediately and aggressively mounted by the male. The longest observed female visit to a Tooth-Billed male's court was 3.8 seconds.

Because the female Tooth-Billed Bowerbird has no opportunity to observe the male *or* his ornaments at close range before she lands on his court, she has to make her choice of mates based on observation of the male and his ornaments from a safe distance many yards away. At that distance there is no opportunity to discern any aesthetic complexity, so the male has no evolutionary imperative to develop a more elaborate display. By the time she arrives at the court, it is already too late to make a more deliberately informed decision. By contrast, Satin Bowerbird females often sit within the avenue of the bower observing a male's displays at extremely close range for several minutes at a time. Protected by the architecture of the bower, females have the opportunity to choose their mates after evaluation from only a few inches away, and the displays are intricate enough to merit such close inspection.

Borgia and his students have conducted several very creative tests of the "threat reduction" hypothesis of bower evolution. For example, Borgia and Daven Presgraves investigated the function of the unique "boulevard" bower of the Spotted Bowerbird, in which the avenue is broad and the walls of the bower are not a solid mass of sticks but see-through screens of lighter twigs and straw. Because of the width of the avenue and the transparency of the walls, females can sit sideways inside the bower and watch the males displaying *through* the thin screens of straw. Borgia and Presgraves observed that the greater physical protection to the female was correlated with louder, more energetic, and more aggressive male displays than those of other bowerbirds. The display repertoire includes a rapid, running rush toward the side of the bower, with the males sometimes even banging their bodies against the bower. When they experimentally destroyed one random wall of each male's bower, males continued to display, and females continued to observe them, through the remaining *intact* side wall of the bower rather than across the newly open side. This result supported the hypothesis that this unique architectural innovation functions to increase the female's sense of physical security while she watches the hyperaggressive male display behavior. Furthermore, it is clear that the extra-aggressive and hyperstimulating display repertoire of the Spotted Bowerbirds has *coevolved* with the enhanced security offered by their distinctive bower architecture.

Borgia's threat reduction hypothesis is genuinely revolutionary. It suggests an entirely new dimension to the complex behavioral interactions between the sexes, one that has rarely been raised anywhere in the sexual selection and mate choice literature. According to Borgia, the behaviors and structures he observed in the male Spotted Bowerbird evolved as the solution to a psychological conflict experienced by the females; the bower innovation solves the problem that females are frightened by the aggressive male displays they actually prefer.

However, I think that it is more likely that the threat reduction response evolves through a more profound sexual conflict rather than merely psychological conflict. To explore this idea, let's return to the male Tooth-Billed Bowerbird and his simple court decorations, which consist of large leaves scattered around a court. On the basis of what she sees from a distance of a few yards, the female Tooth-Billed Bowerbird decides whether or not to visit his court. If she does visit, she will be immediately mounted and copulated with. At some point, because Beauty Happens, the females might evolve preferences for more elaborate or specific court decorations. But, pleasing as these aesthetic innovations may be, females that prefer them will face a new challenge. More complex court ornaments require her to approach *closer* to the male's open court in order to actually evaluate them before making her decision about whether she wishes to mate with their creator. But if she gets too close, the rapid-attack mating style of the Tooth-Billed Bowerbird male will expose her to the risk of forced copulation, whether she has decided she wants it or not.

Forced copulation will result in her having offspring that do not inherit the male display traits she, and *other* females, prefer. Male offspring with these less preferred display traits will be sexually unpopular. As we know from the example of the waterfowl, this is the indirect, genetic cost of sexual conflict.

But unlike ducks, male and female bowerbirds did not end up in a costly arms race with each other. Instead of evolving defenses, females selected on male *aesthetic traits* that advance female sexual autonomy and reduce the threat and costs of sexual coercion. This distinct evolutionary response to sexual conflict is an example of the process I call aesthetic remodeling—the aesthetic coevolution of sexual displays and preferences that result in greater freedom of sexual choice.

In bowerbirds, aesthetic remodeling has taken the form of innovations in male court structures. Like all such changes, they would have started haphazardly and evolved gradually. Perhaps in the course of decorating his court, an early ancestral bower builder gathered a few sticks in addition to his standard repertoire of green leaves. Simple variations in stick placement could have ended in the creation of a rudimentary screen that could have served as a refuge from sexual harassment for the female. This stick-gathering bird would therefore have proved popular with the ladies, because his proto-bower afforded them greater opportunity for evaluation and choice. The sexual advantages of providing females with the aesthetic structures they prefer would lead to the evolution of bower construction by ever-increasing numbers of males. Over time, the distinctive avenue and the maypole bower architectures evolved, each providing expanded sexual security to the females in a different physical way. Females who visited males who built such bowers would have been able to safely spend a longer amount of time on their evaluations of males and their courts. The greater the opportunity for subjective sensory experience and judgment, the stronger the force of sexual selection based on the physical and display behaviors of the males themselves and on the architectural and ornamental features of the extended phenotypes they have created. Consequently, both male displays and bower constructions and decorations would have coevolved with female mate preferences to become more elaborate and complex and more diverse among species.

Like adaptive mate choice, the process of aesthetic remodeling proceeds through a correlation between male display and an aspect of the male phenotype. In aesthetic remodeling, however, the correlation is *not* to good genes or direct benefits but to the expansion of female sexual autonomy. Imagine a population in which 50 percent of the fertilizations are determined by female mate choice and the other 50 percent are determined by coercive male sexual violence. If some aspect of male display arose that resulted in a lower efficacy of sexual coercion—say the pile of sticks in the form of a proto-bower that I hypothesized above— females will evolve to prefer this new display. This preference will evolve in the population because any increase in the frequency of this display trait will result in a larger proportion of fertilizations determined by female mate choice—a larger proportion of females avoiding the indirect genetic costs of sexual coercion. In this way, aesthetic remodeling breaks down sexual conflict by using mating choice to transform male coercive behavior into a more socially amenable, aesthetic form.

Are bowers aesthetic structures? Absolutely. Are the bowers protective? Yes, indeed. And it is precisely because bowers are protective that they have also evolved to be so aesthetically complex and diverse. Essentially, the evolutionary function of the bower is to provide a setting for aesthetic evaluation that also protects the female from "date rape." Once females have secured their freedom of choice, they have free range to pursue aesthetic preferences for ever more diverse and complex forms of beauty.

Because bowers function as both *objects* of choice and *enhancements* of the freedom of choice, they create a new kind of ever-escalating aesthetic evolutionary feedback. Once females have secured their sexual autonomy, their aesthetic preferences will continue to *coevolve* with male displays and decorations, resulting in the creation of ever more complex, aesthetically integrated structures and performances. Like grand opera, bower performances engage and stimulate multiple senses simultaneously, offering song and dance in a theater with colorful sets and props, and even a comfortable front row seat from which the female can watch the show, with easy access to a "fire" exit if things get too hot. As we see from the Spotted Bowerbird, the evolution of aesthetic/physical mechanisms that protect the female from coercion can also allow for the coevolution of ever more aggressive and stimulating displays, because the female can enjoy them without being physically or sexually threatened. In bowerbirds, freedom of choice has greatly enhanced the process of aesthetic radiation.

The aesthetic remodeling of male display and behavior provides a whole new way to evolve sexual Beauty from the coercive male Beast. It is important to emphasize, however, that this evolutionary process does not occur because females prefer less aggressive males that they can physically or socially dominate. At the moment that they exercise choice, females actually *have* autonomy and are not evolving preferences for wimpy males. Rather, female bowerbirds have evolved preferences that facilitate the capacity of all females to exercise full freedom of choice based on the gratification of all their aesthetic desires.

As a graduate student with Gerry Borgia, Gail Patricelli developed a fascinating and unique research program to investigate the threat reduction hypothesis. Looking at videotaped visits by female Satin Bowerbirds to male bowers, Patricelli and Borgia observed that females are frequently startled or frightened by aggressive male displays, and it seemed to them that by crouching low in the bowers, the females appeared to be able to communicate their level of discomfort to the male. They further observed that those males who modulated their displays accordingly were *more* sexually successful.

To test these observations, Patricelli created a remotely controlled, robotic, stuffed female bowerbird model, which she dubbed a "fembot." The fembots produced such natural-looking standing, crouching, head-rotating, and wing-fluffing movements that they completely fooled male bowerbirds, as Patricelli's videos of males copulating with the fembots demonstrate. By placing the fembot in the bower and regulating its posture and movements,

Patricelli was able to confirm her hypotheses that (1) female Satin Bowerbirds are communicating their comfort levels to displaying males by crouching, (2) some males modulated their display intensity to put the females more at ease, and (3) those males who can regulate their display intensity to keep females more comfortable are ultimately the most successful at attracting mates.

Why should female Satin Bowerbirds be less threatened by aggressive displays performed by more attractive males with more appealing bowers? If what is at stake is the indirect, genetic cost of sexual coercion—that is, male offspring who will be less appealing to females and therefore less likely to perpetuate her genes—then, evolutionarily speaking, females *should* be more comfortable with the risks posed by an attractive male. Forced copulations from less attractive males will create the same risk of physical harm, that is, the same direct costs. However, the more attractive mates provide a lower risk of the indirect, genetic costs of sexual coercion. Thus, Patricelli's fembot experiments provide strong support for the idea that bowers function to protect females from the indirect costs of sexual coercion.

From Patricia Brennan's artificial duck vaginas to Gail Patricelli's fembots, the science of mate choice takes us down some creative paths! And like the ducks, the bowerbirds teach us a whole new way in which to understand the freedom of choice. Here, sexual autonomy is an evolutionary engine of beauty.

CHAPTER 7

Bromance Before Romance

It's extraordinary enough to realize that female mate choice has produced the explosion of beauty we've seen in male manakins and bowerbirds. It's even more amazing to think that female mating preferences could have had a profound impact on male social relations and that this happened even though, as I'll discuss in this chapter, much of the resulting male behavior is something the females themselves never witness. But in the case of the manakins, this is exactly what has occurred over the course of evolutionary time. The social relationships among males in a manakin lek have evolved into a virtual bromance—long-term, socially engaged relationships that sublimate and moderate competition—and it's all come about, I think, because of the female pursuit of sexual autonomy.

Recognizing females as the active agents in the origin of lekking goes against most of the traditional thinking about why the lek-breeding system evolved. But we will see that entertaining this possibility provides a productive new way to understand the complexity and diversity of the highly unusual behaviors of male manakins and of the variations in lek social organization as well.

Although there are fifty-four species of manakins, and therefore fifty-four variations on their breeding and social relationships, we can make a few general observations about manakin leks. To recapitulate the basics: Leks are groups of sexually displaying males. Within the lek, each male defends a specific territory of his own, but the territory includes nothing of value except the opportunity to mate. From species to species, there can be a lot of variation in the size and spatial distribution of these territories and in the number of territories within a lek (from a few to dozens). In some species the territories can be as small as three to fifteen feet or so wide, in others as large as thirty feet or more. In some the territories are closely packed and adjacent, in others more widely dispersed. In a few species, males defend "solitary" lek territories that are so far apart that they are outside visual and acoustic contact with one another. The males may occupy their territories for anywhere between four and nine months of the year, with some populations being on lek nearly the entire year except when the males are molting their feathers. Outside the manakins, leks have evolved in a wide variety of other birds, in various insects, fishes, frogs, and salamanders, and in a few ungulates and fruit bats.

Confusion about the nature and function of leks dates back to Darwin himself, who was divided in his assessment. He discussed avian lek behavior in several sections of *The Descent of Man*. In "The Law of Battle" section he interpreted it in the context of male-male competition, which is how most evolutionary biologists have discussed it ever since, down to this very day. But in the "Vocal Music" and "Love-Antics and Dances" sections he wrote about lekking birds in the context of female mate choice. For over a century, Darwin was unusual in considering even the possibility that leks could have anything to do with female choice.

In the absence of a working theory of female mate choice or sexual autonomy, it's not surprising that theorists of the evolutionary origin of leks have generally conceived of lek organization as a purely male-male competition phenomenon—a product of the struggle for male dominance or control. The traditional hypothesis is that the males within a lek fight it out ritualistically in order to establish a hierarchy, and the females acquiesce to mate with the dominant male. Females would thereby win the male who was by definition "the best," because he had fought his way to the top of the hierarchy. This fit in well with the Wallacean notion that all sexual selection is a form of adaptive natural selection.

The male competition concept of lekking might have reached its most extreme expression in the popular ornithology textbook that I used in college, *The Life of Birds* by the Beloit College professor Carl Welty. Welty compared avian leks to the medieval European droit du seigneur (or lord's right), in which the lord of the realm had the right to have sex with any virgin bride in his realm before her wedding. With his colorfully inapt analogy, Welty managed to equate a possibly mythical human cultural institution, which embodies the ultimate denial of female sexual autonomy, with an avian social system—the polygynous lek— that, as we will see, may be the premier example of female sexual autonomy in action.

In an influential 1977 paper, the behavioral ecologists Steve Emlen and Lew Oring espoused the traditional male-male competition theory of lekking, describing leks as "a forum for male-male competition," which made it possible for "females [to] choose primarily on the basis of male status." Recognizing the apparent evolutionary problem with this theory—after all, what's ultimately in it for males to join a lek, because most of them will lose the competition?—Emlen and Oring came up with a plausible-sounding explanation. They hypothesized that by gathering together to pool their advertisement signals, the males will be able to produce a louder invitation that will reach farther, and attract a larger total number of females per male, than they could with individual advertisements. However, the animal behaviorist Jack Bradbury soon demonstrated that males cannot actually gain by pooling their visual or acoustic display signals. Even though larger groups of displaying males do make a total advertisement that is louder than that produced by a smaller group, the volume increase is only incremental—in direct proportion to the number of males. This means that additional males that join a lek do not increase the *per male* effective broadcast area of the lek. Joining a lek will not result in any net gain in an individual male's broadcast efficiency or in the number of females he is likely to attract.

If males do not benefit by pooling their displays, then what other reason could there be to join a lek? Bradbury and others proposed several possible models based on advantages they believed lekking might be able to offer to the male. For example, the "hotspot" model predicts that males who aggregate in areas of high traffic for foraging females will be able to maximize their encounter rates with females. Then there's the "hotshot" model, which predicts that males who establish territories near other particularly attractive males—the "hotshots" who attract higher than average numbers of females—could benefit because some of the females may end up mating with one of them instead.

However, the evidence for both the hotshot and the hotspot hypotheses is mixed at best. With the use of exciting new scientific tools and techniques, including radio tracking and molecular fingerprinting, in combination with good old-fashioned, high-efficiency nest finding, a number of recent studies suggest that these theories are just flat-out wrong. For example, Renata Durães and colleagues found that some Blue-crowned Manakin (*Lepidothrix coronata*) leks were indeed located in areas of high female traffic, but contrary to the hotspot hypothesis these leks were smaller, not larger, than leks in low-traffic areas. In a subsequent study, Durães captured and analyzed the DNA "fingerprints" of a population of the male and female Blue-crowned Manakins. She found an incredible sixty-six active nests and obtained molecular fingerprints of the nestlings to identify their fathers and then determined how far the females had traveled from their nest sites to find a male to mate with. Durães found that most females did not choose a mate from the nearest lek, but on average selected a mate from the *third*-nearest lek, which contradicts the hotspot model. Durães concluded that female mate choice was not consistent with either the hotspot or the hotshot model.

In the 1980s, Jack Bradbury and the evolutionary biologist David Queller were among the first since Darwin to propose that the formation of leks had anything to do with female mate choice. In 1981, Bradbury proposed the revolutionary hypothesis that leks evolve because of female preference for male aggregation. Specifically, he suggested that females have evolved preferences for males concentrated in leks because having a number of males in proximity to each other enables them to make efficient comparisons of potential mates. Shopping for a mate is much easier and more convenient when there are a lot to choose from in a relatively small amount of space. It's sort of like shopping at a mall instead of having to travel greater distances from store to store.

David Queller went even further with the female choice idea and proposed a purely aesthetic, sexual selection model of lek evolution. Queller showed that leks can evolve if social aggregation is like any other male display trait—say, tail length. Once there exists a female preference for the trait, in this case aggregated males, male aggregations will evolve. Genetic variation for a mating preference for leks will become correlated with the genetic variation for lekking, and preference and trait will continue to coevolve with each other. According to this model, lek evolution is just another kind of arbitrary beauty, but one that pertains to male social behavior rather than to male physical features.

Bradbury and Queller both viewed leks as organizations that had evolved to provide a mechanism for female mate choice. Unfortunately, their emphasis on the female as the active agent was so far ahead of its time that their revolutionary models did not receive a lot of attention. And after a boom of interest in the evolution of lek behavior during the 1980s and 1990s, most of which—unlike Bradbury's and Queller's—focused on the many attempts to support the hotspot and hotshot models, research on the question slowed to a trickle.

The biggest weakness of the current models of lekking—both the male-competition *and* the current female-mate-choice models—is that they focus only on the lek as the place where mating occurs. They fail to take into account the fact that the lek is also a male *social phenomenon*. Leks are not merely convenient concentrations of territories where females can find their mates. Unlike clusters of competing gas stations and fast-food restaurants located just off the exit of a highway where motorists can

1. A male Blackburnian Warbler (*Setophaga fusca*) perched in a balsam fir on its breeding grounds in northern Maine.
Photo by Jim Zipp

2. A male Superb Bird of Paradise (*Lophorina superba*) displaying
to a female visiting his display log in the Central High lands of
Papua New Guinea. *Photo by Edwin Scholes III*

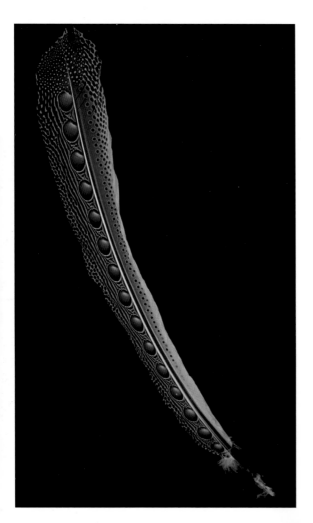

3. The fourth secondary wing feather of a male Great Argus (*Argusianus argus*). *Photo by Michael Doolittle*

4. Detail of the complex pigmentation pattern of the 3D golden spheres on the fourth secondary of a male Great Argus (*Argusianus argus*). *Photo by Michael Doolittle*

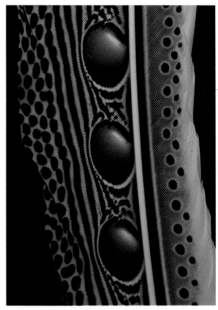

5. A male Guianan Cock-of-the-Rock (*Rupicola rupicola*)
in the lowland rainforest of French Guiana.
Photo by Tanguy Deville

6. A male Golden-headed Manakin (*Ceratopipra erythrocephala*) perched on its lek territory in the trees in northern Amazonia.
Photo by Juan José Arango

7. The male White-bearded Manakin (*Manacus manacus*) displays
on thin saplings around a cleared court on the forest floor.
Photo by Rodrigo Gavaria Obregón

8. The male White-throated Manakin (*Corapipo gutturalis*)
displays on mossy fallen logs on the forest floor.
Photo by Tanguy Deville

9. The male White-fronted Manakin (*Lepidothrix serena*) calls from a perch in the forest understory.

10. The male Golden-winged Manakin (*Masius chrysopterus*) has brilliant yellow wing patches that are usually hidden when the bird is perched, but prominently flashed during its log-approach flight display. *Photo by Juan José Arango*

11. The behavioral repertoire of the Pin-tailed Manakin
(*Ilicura militaris*) provides crucial evidence for analyzing the
evolution of display behavior in its close relatives, the
White-throated and Golden-winged Manakins.
Photo by Rafael Bessa

12. The male Club-winged Manakin (*Machaeropterus deliciosus*)
produces its tonal wing song by rapidly vibrating its inner
wing feathers from side to side over its back.
Photo by Tim Laman

13. The male Wilson's Bird of Paradise (*Cicinnurus respublica*) (below) displays the bald, featherless patches of bright blue skin on his crown to a visiting female (above). The female shares the same bald crown patches, albeit in a deeper blue hue.
Photo by Tim Laman

14. An orange male (left) and brown female (right) Guianan Cock-of-the-Rock (*Rupicola rupicola*) feeding on palm fruits. The crests of both sexes are composed of feathers that grow from the sides of the crown toward the midline.

Photo by Tanguy Deville

15. The plumage coloration of the Late Jurassic maniraptoran dinosaur *Anchiornis huxleyi* was reconstructed from analyses of electron microscope pictures of melanin pigment granules, or melanosomes, from its fossilized feathers.
Painting by Michael DiGiorgio; from Li et al. (2010)

16. After copulation, the corkscrew-shaped penis of a male Black-bellied Whistling Duck (*Dendrocygna autumnalis*) dangles down briefly before being retracted into the cloaca.
Photo by Bryan Pfeiffer

17. The male Satin Bowerbird (*Ptilonorhynchus violaceus*)
builds an avenue bower and decorates the court in front
of the bower with a plethora of royal blue objects
found in the environment. *Photo by Tim Laman*

18. Male Great Bowerbirds (*Chlamydera nuchalis*) usually decorate
their avenue bowers with bleached bones and sticks, but this
individual has decorated its bower with fossil clam shells.
Photo by Richard O. Prum

19. This male Vogelkop Bowerbird (*Amblyornis inornata*) in the Arfak Mountains of western New Guinea curates a collection of strange objects and materials on a planted garden of moss in front of its hut bower (clockwise from the upper left): globular red fruits; flakes of rotten wood infiltrated with green fungus; charcoal, black fungus, and rotten red fruits turned black; red flowers from *Freycinetia* vines; shiny black beetle elytra; blue berries; and gelatinous amber tree exudate. *Photo by Brett Benz*

20. In southeastern Brazil, a group of five adult male
Blue Manakins (*Chiroxiphia caudata*) performs a coordinated,
cooperative, cartwheel display to a visiting green female (left).
If she prefers the group's display, she will mate with the dominant
male of the group. *Photo by João Quental*

21. An Atlantic Puffin (*Fratercula arctica*) returning to its nesting burrow on Machais Seal Island, New Brunswick, Canada. During the breeding season, both sexes have identical, brilliant beak colors. *Photo by Jim Zipp*

easily find them, leks are highly *social* organizations in which a number of males gather, defend territories, fight, engage in often elaborate cooperative displays, and develop complex and enduring social relationships that can last a lifetime.

To understand just how elaborate these relationships are, we must take a look at the rather bizarre social lives of the males, which are in marked contrast to those of the females of the species. After hatching and fledging from the nest, female manakins live entirely independently. They have no social relationships with other adult females or with any adult males except during those few brief minutes of the year when they are visiting the displaying males, selecting a mate, and engaging in copulation. Their only other relationships of any kind are with their own dependent offspring, and those relationships end as soon as the chicks leave the nest.

The males are a different story entirely. Their relationships with females are minimal, as noted—confined to the short period of time they spend in the nest with their mothers, to the one- or two-minute visits they receive from various females visiting their territories during breeding season, and, if they are attractive enough and have sufficiently good displays to win the favor of one or more of these females, to brief mating sessions. But they do enter into complex, interactive, and long-lasting relationships with other males.

Once the young male manakins fledge and leave the nest, they roam around for a year or more (depending on the species), during which they must establish and defend a display territory within a lek of other males. Then they begin to develop the social relationships that are a characteristic of lek behavior. Because each manakin male typically defends the same territory in the same lek every breeding season for years on end, often for the duration of their lives, which can last a decade or even two, these relationships have the opportunity to develop over extended periods of time. So, social relationships between males on the lek consist of *daily* social interactions that typically continue for a decade or more.

———

So, why do males participate in leks? The best explanation is that males must aggregate because the females prefer it. In polygynous species like manakins, as we've seen, females do all the work of parenting entirely by themselves. They build the nest, lay the eggs, incubate the clutch, and feed and protect the nestlings until fledging. In exchange for all their efforts, females have gained control over fertilization. Males have no choice but to submit to female preferences because any renegade male that refuses to join a lek will lose any possible prospect for reproduction. The females are in charge, and male rebellion will lead to sexual irrelevance.

Is there any reason why independently living females who mate once a year would *not* prefer the kind of rich and complex aesthetic/sexual experiences afforded by a lek? Why not have sex the way you want it—amid a complicated, intense, stimulating circus of display activity? From the female's point of view, we can think of the lek as being like a brothel, but in reverse because it caters to females instead of males. Each male candidate for her sexual favor puts on an elaborate performance to woo her into choosing him. Even better, unlike the transactions that occur in a real brothel, the customer doesn't have to pay. Any male she wants is hers for the asking, free of charge.

The initial female preference for spatially aggregated males might have been a simple sensory/cognitive bias for greater, more intense sexual stimulation of the kind that occurs from observing multiple singing and displaying males in proximity. So it makes sense that lekking could evolve as a way of gratifying this kind of desire. But as noted earlier, leks are not just aggregations of male mating territories; they're also places in which males have developed elaborate social relationships with each other, which seems a very odd evolutionary development. After all, the males of almost all species are sexual competitors and frequently aggressive with each other. Evolving male cooperation is hard. In fact, *any* form of cooperative animal behavior presents a challenge to explain evolutionarily. Whether it's altruism in social insects, the development of human language, or the phenomenon of helpers at the nest, the evolution of cooperative behaviors always requires overcoming the substantial hurdle of the benefits of individual selfishness.

And make no mistake—this is a huge evolutionary challenge. The relative mating success of each male will be increased by aggressively interfering with the mating attempts of every other male. But such constant disruption would destroy the lek. Females would never be able to choose a mate if males are aggressively disrupting and fighting one another all the time. How, then, can leks ever evolve and survive if it is in the best interest of every selfish male to try to prevent every other male from mating?

The key to understanding this conundrum is to realize that male disruption of visits by females to other males within the lek is a form of sexual coercion against females, infringing on their sexual autonomy. In essence, one evolutionary mechanism, female mate choice, is being pitted against another, male-male competition. For female choice to prevail, manakins have to somehow get around male aggression.

How do they manage this? As in bowerbirds, female manakins have used their mating preferences to remodel male behavior in order to get what they want. In bowerbirds this reengineering takes the form of the bowers that protect the female from unwanted copulations while she's evaluating the male and deciding whether he's the one she will choose to father her offspring. Male bowerbirds are still highly aggressive to each other and even to visiting females, but the bowers that they themselves have built serve to mitigate the impact of much of that aggression on female freedom of choice.

In manakins, by contrast, the resistance to sexual coercion has been expressed not in architecture but in a fundamental reengineering of male social organization and behavior. The resulting transformation has greatly reduced male aggression and thereby maximized the female's chances of getting what she wants. It has also resulted in a lek-breeding system that is stable because it is not constantly disrupted by male aggression. Although fighting and disruption among males are not entirely eliminated, they are reduced to some tolerable balance between female freedom of choice and male competition.

Thus, I hypothesize that lekking is not an exhibition of male dominance hierarchy, and of female acquiescence in it and the adaptive benefits it offers, as was theorized for most of the twentieth century. Rather, leks are likely the result of female preferences for socially cooperative aesthetic gatherings of males.

What evidence is there that leks, particularly manakin leks, evolve as cooperative social phenomena? In fact, this evolutionary hypothesis is quite hard to test. Clearly, males in lekking species *are* much more spatially tolerant of each other than are other territorial birds. So, we know that manakin and other lekking males are socially distinct in some fundamental way. But it is hard to know whether female choice has been responsible for this transformation in male social behavior. Luckily, one highly unusual variation on lek display behavior that is very prevalent among manakins provides telling insights into the fundamentally cooperative nature of lekking.

In many manakin species, the social relationships among males may go well beyond mere peaceful proximity. Rather, male-male social relationships can extend to participation in highly elaborate coordinated displays among two or more males that can require years of fine-tuning to perfect. The specifics of such displays can vary dramatically depending on the species, but this kind of coordinated and cooperative behavior is a feature of many male manakins.

Although the aesthetic nature of these coordinated displays is highly diverse, when it comes down to a question of social function, there appear to be two classes: There are coordinated displays, which are performed by pairs of males, almost always when females are not present. And in the case of one particular genus of manakin (*Chiroxiphia*), there are what I call obligate coordinated displays, which are done by pairs or groups of males in the presence of females and which are an obligatory prerequisite of mate choice and mating. No male *Chiroxiphia* manakin can hope for a chance at mating with a female unless he participates in such a coordinated, multi-male display.

As dances, the coordinated displays are wonderfully variable. For example, as described in chapter 3, pairs of territorial male Golden-headed Manakins perform an elaborately choreographed series of maneuvers, after which they will sit side by side on the same branch, facing away from each other in a bill-pointing posture. In Blue-crowned Manakins and White-fronted Manakins, the males perform a coordinated version of the same display elements that they perform solo when females visit. These displays consist of "beeline" and "bumblebee" flights back and forth between saplings and chasing each other around a small court near the forest floor.

In the Golden-winged Manakin, pairs of males perform a coordinated version of the spectacular log-approach display described in chapter 3. The first male waits on the log for the second male to perform the log-approach display, and as soon as the second male arrives, he leaps up into the air and allows himself to be replaced on the log. Then the roles are reversed, and the second male waits for the first. In this case, the coordinated display is performed by pairs of males that may be made up of neighboring territory holders or of a territorial male and a younger, floating non-territorial male. None of the joint displays I've just described are performed during female visits to male territories; they merely function in male-male social relationships.

The ornithologists Mark Robbins, Thomas Ryder, and others have added to our knowledge of the social relationships among manakins with their descriptions of coordinated displays of males in the genus *Pipra*, which includes the Wire-tailed Manakin (*Pipra filicauda*), the Band-tailed Manakin (*Pipra fasciicauda*), and the Crimson-hooded Manakin (*Pipra aureola*). A territorial *Pipra* male displays with a number of other males, including both other territory holders and younger, non-territorial, "floater" males. Typically, a coordinated display consists of a territorial male waiting on his main display perch, while a second male performs an S-curved, swoop-in flight display, swooping below and then above the level of the display perch as he approaches it. When he lands on the perch, he produces a distinct call as he displaces the waiting male. The display is then repeated with the roles reversed,

Coordinated display of a pair of male Golden-winged Manakins. (Top) One male waits on the log in tail-pointing posture as another flies toward the log. As the flying male lands and rebounds off the log (dotted line), the waiting male leaps off the log (broken line). (Bottom) The two males cross in the air over the log and land facing one another in tail-pointing posture.

over and over again. Such coordinated flight displays can continue for several minutes without stopping. As with the displays described above, these joint displays are not usually performed for visiting females. They draw on the same vocabulary as intersexual communication—that is, the particular display elements are the same as those that a solitary male would perform for a visiting female—but they incorporate the elements into joint performances that are entirely male-male social behaviors.

All of the displays I've just been describing are of the first functional type—the simply *coordinated*. The second special class of this behavior, the so-called *obligate coordinated* display,

The coordinated swoop-in flight display of a pair of Band-tailed Manakins.

is unique to the blue *Chiroxiphia* manakins. *Chiroxiphia* males engage in the most extreme form of precopulatory male-male cooperation known in any animal. Pairs, or even larger groups, of males with long-standing relationships with each other perform coordinated displays that are a largely *obligatory* part of the courtship of females. Unlike other manakin females, *Chiroxiphia* females observe these coordinated performances and make their mating choices based on their evaluations of them. Once they decide which pair or group's performance they prefer, they have the opportunity to select the dominant, alpha male of the group.

To attract female visitors to the display site, *Chiroxiphia* males first sing loud, highly coordinated duets from perches high above the display perch—*Toleedo . . . Toleedo . . . Toleedo . . .* (or similar syllables). Then, when females visit, pairs or even groups of males perform an elaborate "cartwheel" or "backward leapfrog" display. In most species, the backward-leapfrog display

The obligate coordinated display of a group of male Blue Manakins for a visiting female (perched at right). As the male closest to the female leaps up and flutters back down the branch, the perched males sidle up the branch toward the female. The cycle is repeated dozens or even hundreds of times.

is performed by two males perched on a small, concealed horizontal branch located near the ground. In the Blue Manakin (*Chiroxiphia caudata*), the backward-leapfrog display is performed by a group of up to *four or five* males (color plate 20)! After the female has landed on the display perch occupied by the males, the male who is closest to her leaps upward and hovers in the air in front of her with his red crown fluffed. While hovering, the male gives a buzzy, snarling two-syllable call in flight and then flutters back down to the perch, taking a position farther away from the female. Meanwhile, the second male slides forward along the perch toward the female, then leaps up and performs the same display as the first. These leapfrog displays are repeated anywhere from *twenty to two hundred* times, depending on how much the female likes what she observes and how many bouts she's willing to watch. Eventually, the dominant, alpha male of the group gives a distinctive call, and the subordinate, beta male(s) leaves the perch. The alpha male does a few more distinct displays and then, if the female is still there, copulates with her on the display perch. The female may decide to leave at any point during the display sequence.

Considerable skill and coordination are involved in putting on these performances. Because the females are extremely discerning, their preferences select for males who have been in male-male social relationships that have lasted long enough to have allowed them plenty of time to practice diligently and iron out any kinks in their performances. Apparently, it can take years of practice to achieve vocal coordination between males that is good enough to attract mates. The ornithologists Jill Trainer and David McDonald have shown that the timing of the vocal coordination in the *Toleedo . . . Toleedo . . .* duet sung by male pairs of Long-tailed Manakins greatly influences their chances at sexual success.

This cooperative mode of display behavior has reorganized the entire breeding system of the *Chiroxiphia* manakins, resulting in a distinctly new form of lek. *Chiroxiphia* males do not defend individual territories, as other manakins do. Rather, each display territory is controlled by a team of males. The team consists of a dominant, alpha male who shares the territory with a subordinate beta—or in the case of *Chiroxiphia caudata,* with beta, gamma, and even epsilon males—all of them aspiring one day to succeed him as alpha male. The male partnerships within these shared territories are long-lasting and established over the course of years of interactions.

But the road to this kind of partnership is filled with challenges for the wannabe alphas. The young males must compete with each other as each of them strives to become an established beta male or alpha territory holder. And before they can even enter the competition, they must wait out the four-year period it takes for them to achieve mature adult plumage. At first, the young males look like green females, and each year they molt into a successively more male-like plumage. During this period, the subadult males consort with various groups and participate in rudimentary displays. Once they achieve adult plumage, males typically spend several more years displaying as floaters, trying to win the approval of an alpha male whom they can partner with. During this apprenticeship time they continue to work on improving the temporal coordination of their duetting songs and displays.

When at last a *Chiroxiphia* manakin achieves beta male status, what does he get for his time and effort? Well, he still doesn't get to mate with a female, because females choose only among the available alpha males. But at least he's now in a better position to inherit the alpha position if the alpha male dies or disappears— although that can take five to ten or more years to occur. If and when a male finally achieves alpha status, the struggle is not over, because there will still be constant competition with other alpha males in different display partnerships to succeed at attracting the most mates.

The intense, multi-tiered competition creates the strongest sexual selection measured in any vertebrate animal. For example, in his long-term study of Long-tailed Manakins (*Chiroxiphia linearis*) in Costa Rica, David McDonald was able to show that a very few males may obtain fifty to a hundred copulations per year for five or more years, while most males never have any opportunities to mate. The behavioral ecologist Emily DuVal found something similar when she conducted a remarkably exhaustive study of sexual selection in the Lance-tailed Manakin in Panama. By using the DNA fingerprints of chicks in the nest to establish paternity, DuVal documented that all young were sired by alpha males. Furthermore, out of a single age cohort of twenty-one males, only five of them became alphas, and four of those five were responsible for siring fifteen offspring, while the rest sired no chicks at all over a period of nine years. Clearly, *Chiroxiphia* females have such strong mating preferences that there will be far more losers than winners in the sexual competition. *Chiroxiphia* society is like a giant Ponzi scheme in which over 90 percent of males *must* lose.

So why do they do it? Obligatory male-male cooperation is a total loss for the vast majority of males. The only reason it can happen is that females are completely in charge. Males have no options because there is no other game in town. Like the judges of a male-pairs figure skating competition—or better yet, of a male-pairs pole dancing contest—females can be as picky as they want to be, or as picky as they have evolved to be. Perhaps unsurprisingly, the prize for extreme aesthetic expression goes to the Brazilian team! The Blue Manakin performs the cartwheel display in groups of three to five males and does so in the forests surrounding the Carnival capital, Rio de Janeiro, in southeast Brazil. There is no other show like it on the planet.

Male *Chiroxiphia* manakins are engaged in the most ruthless sexual competition known in nature. But this competition is not fought out with antlers or aggression. Rather, it is enacted entirely through a system of ritualized cooperative male dance. Extreme female choice has led to a transformation of males from aggressive competitors into disco slaves of fashion.

Traditionally, coordinated display behaviors have been interpreted (even by me in the 1980s) as a mechanism for males to ritualistically establish a dominance hierarchy. This view, however, is a hangover from the concept of leks as sites where males compete to establish a dominance hierarchy and females then acquiesce to the alphas. In fact, there is little evidence that male dominance per se contributes to sexual success among manakins. Another possible explanation for male coordinated display is kin selection. Perhaps males are displaying with close relatives to enhance the reproductive success of the genes shared with their half brothers or cousins? However, David McDonald and Wayne Potts have conclusively established that the males within displaying partnerships of Long-tailed Manakins are *not* more closely related to one another than by chance alone. These and other male-advantage explanations simply fail.

By contrast, the female choice/sexual autonomy model of lek evolution can explain both the evolution of lekking itself *and* the many varieties of socially coordinated manakin lek behavior. Coordinated display in manakins is an elaboration of the inherently cooperative nature of male lek behavior in general. It's another expression of the taming of the selfish aggression of individual males that makes leks possible in the first place. And it has likely evolved through the same mechanisms that created lekking—female mating preferences for cooperative male behaviors that contribute to female freedom of choice.

What's at first puzzling about this hypothesis is that the coordinated male-male display behavior of most manakin species is rarely, if ever, observed directly by visiting females. Thus, the evolutionary effect of female preferences on coordinated male social behavior must be indirect. If the females don't observe this behavior, why do they prefer males who engage in it? Basically, what's it to them?

It appears that by choosing males that get along socially with one another, females are selecting *indirectly* for males that perform coordinated displays. Males involved in such cooperative relationships are less likely to engage in violent mating competition, and females can thereby avoid harassment, which would waste their time and disrupt their mate choices. Thus, coordinated display evolves because these male-male interactions nourish the complex social relationships that females have forced upon them.

However, like other serendipitous consequences of aesthetic evolution, once cooperative display exists, it can become subject to sexual selection and lead to new forms of mating preferences. This mechanism could account for the evolution of the unique form of obligate cooperative display found in *Chiroxiphia* manakins. Perhaps coordinated male displays occurred so frequently in the ancestors of *Chiroxiphia* that females began to select specifically on this stimulating new form of male-male display, and their preferences then coevolved with these novel behaviors. Thus a once-incidental social behavior became an integral component of the display repertoire, yet another example of the evolutionary cascade of effects that can be created by aesthetic mate choice.

How can we test the hypothesis that coordinated display evolved as an essentially cooperative social behavior through female mate choice? Two interesting new data sets that employ an entirely novel perspective on manakin social relationships offer strong support for this idea. Recently, David McDonald has pioneered the use of network analysis to track the social relationships among male lekking birds. Network analysis is a way of describing the social interconnectedness of individuals via a graph of nodes (that is, individuals) connected by lines (that is, relationships). Law enforcement, security, and intelligence agencies are using network analysis tools to discover and track criminal and terrorist groups from cell phone records, e-mails, and metadata. This same technique can also help us investigate the role social relationships play in the sexual success of male manakins.

Using a ten-year data set from the obligately coordinated, cartwheeling Long-tailed Manakin, McDonald showed that the best predictor of a young male's future sexual success was the extent of his connectedness to the male social network. In other words, those young males with the richest social relationships— that is, the ones most consistently engaged in display relationships with lots of different groups of males—were *most likely* to ascend to alpha male status and to higher sexual success in later years. Similarly, Bret Ryder and colleagues have documented that the degree of social connectedness among young male Wire-tailed Manakins is a strong predictor of social advancement and subsequent sexual success.

These data demonstrate that rich male social relationships— bromance, not dominance and aggression—are the paths to male sexual success in the manakins. The loners and the antisocial males that cannot get along with others will be the sexual losers within a manakin lek.

Of course this raises the question of how female Wire-tailed Manakins know which males have the richest social lives, because they rarely if ever see the coordinated male performances and presumably don't have access to tallies of the number of Facebook friends each male has. However, while females can only select *directly* on male display behaviors that they see and evalu-

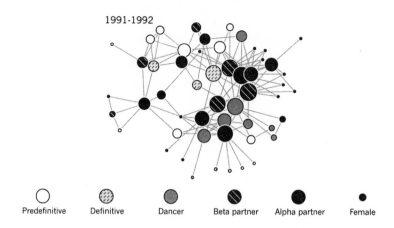

The social network of male Long-tailed Manakins for a single year identified by social status. *From McDonald (2007).*

ate, they also select *indirectly* for males with the most complex and sustainable social relationships. If practice makes perfect, perhaps by choosing the best displayers, females are also choosing those males who have engaged in the most varied, frequent, and enduring social collaborations. So if we ask, what goes into social and sexual success among manakins, the answer is probably a combination of genetics, development, and social experience.

In manakins, female mate choices have fundamentally reshaped the nature of an all-male world they rarely visit in order to advance both female sexual fancy and freedom of choice. The result has been the evolution of the lek itself and of the numerous and astonishing variations in coordinated male display found in so many species.

Nearly 150 years after *The Descent of Man,* we must wonder whether Darwin's statement—"Birds appear to be the most aesthetic of all animals, excepting of course man"—went far enough. If we measure the aesthetic accomplishment of an individual or a species in terms of the share of energy and investment dedicated to aesthetic expression, then manakins far exceed humans. All manakin males—half of the species—expend most of their time and energies in the rehearsal, perfection, and performance of a set of highly choreographed song and dance routines, in duet, group, and solo forms. By Darwin's criteria, the manakins and bower-birds beat humans by far!

CHAPTER 8

Human Beauty Happens Too

Charles Darwin's *Descent of Man* is basically a long book about the evolution of humans, with a few chapters about birds and other animals. Darwin included the birds (and other animals) in order to better support his hypothesis that sexual selection played a critical role in human evolution. This book takes a similar approach, but with the ratio of people to birds reversed. The mixed approach is as vital and productive today as it was then. By applying what we've learned about mate choice through our examination of the evolution of birds, we can gain a much fuller understanding of its role in shaping the appearance and the sexual behavior of our own species.

The forces we've witnessed in birds—Beauty Happens, sexual conflict, and aesthetic remodeling—play out in humans and their primate ancestors, too, and the chapters that follow will speculate as to how. I say "speculate," because human aesthetic evolution is a new science and most of the theories I offer here will need to be tested and analyzed with data from comparative studies and sociological investigations. But as we've seen with the birds, aesthetic evolution has great explanatory power, and what's more, it rescues us from the tedious and limiting adaptationist insistence on the ubiquitous power of natural selection.

And indeed, the study of human mate choice *is* currently dominated by such insistence, in the form of a field called evolutionary psychology. Contemporary evolutionary psychology has a profound, constitutive, often fanatical commitment to the universal efficacy of adaptation by natural selection. The application of the concept of adaptation to human biology is the *organizing principle* of the field. Evolutionary psychologists view human sexual ornaments and behavior as a cornucopia of honest advertisements and adaptive strategies. There is never any doubt what the conclusion of any evolutionary psychology study will be. The only question is how far the study will have to go to get there.

Where's the harm in this intellectual mission? What concerns me most is not merely that so much of evolutionary psychology is bad science. Bad science has a way of being corrected over time. What's worse is that evolutionary psychology is beginning to influence how we think about our own sexual desires, behavior, and attitudes. Evolutionary psychology teaches us that certain mate choices are sanctioned by science as adaptive (that is, universally good) while others are not, and these views are changing how we think about *ourselves*.

It matters to *me*, of course, whether female House Wrens prefer particular male songs because these songs are merely perceived as more aesthetically beautiful than others or because they signal superior male genetic quality or capacity for reproductive investment. But such ornithological debates are pretty narrow in their impact. However, when we misapply adaptationist logic to human bodies and our own sexual desires, as we will see, it becomes important to everyone to be sure that the scientific process has not been sacrificed to an intellectual movement.

Before we begin thinking about human sexual evolution, we need to place human beings and our sexual biology in historical and prehistorical context. The history of life, as we have seen, is a tree, and human beings belong to a distinct branch on this Tree of Life. Humans are apes—specifically, African apes. Apes are a lineage of old-world primates that includes gibbons and orangutans, gorillas and chimpanzees. The closest relatives, or sister group, of the apes are the diverse old-world monkeys, which include the vervets, macaques, baboons, mandrills, langurs, and leaf monkeys. Among the African apes, humans are most closely related to the chimpanzee (*Pan troglodytes*) and to the bonobo (*Pan paniscus*), also known as the pygmy chimpanzee. Together, humans, chimpanzees, and bonobos constitute the sister group to the gorilla (*Gorilla gorilla*).

Humans have had a complex evolutionary history that has resulted in dramatic changes to our species since our most recent common ancestry with the chimpanzees, about six to eight million years ago. Much more recently in evolutionary time, over the course of only the last fifty thousand years, the pace of change has accelerated, and we have undergone an enormous expansion across the globe, resulting in a great diversity of populations, languages, ethnicities, and cultures.

Because of this complexity, all hypotheses about human evolution have to be framed in the context of our evolutionary history in the Tree of Life. We can think about any evolved feature or evolutionary statement as belonging to one of four distinct evolutionary contexts:

1. evolution that occurred during our history of shared ancestry with various lineages of mammals, primates, and apes, or even further back;
2. evolution that occurred in the uniquely human lineage since our last common ancestor with the chimpanzees;
3. evolution that has occurred, and is continuing to occur, among living humans around the world today; and
4. processes of cultural change—or cultural evolution—that began relatively recently in humans and continue among and within human populations across the globe.

The observations that humans have evolved bones, four limbs, and hair but lack a tail are all statements about evolutionary events that took place at different points in time during evolutionary context 1. The assertion that humans have big brains and walk upright is a statement about evolutionary events that occurred in evolutionary context 2.

The observation that humans are still evolving is a statement about evolutionary context 3. The fourth evolutionary context is contemporaneous with the third but encompasses an entirely new

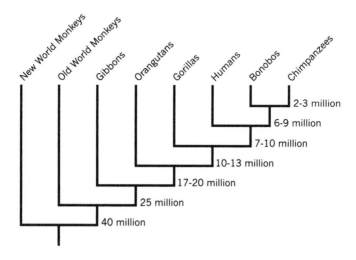

A phylogeny of monkeys and apes with estimates of the ages of the lineages.

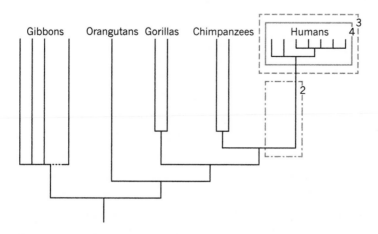

Phylogeny of the apes outlining four different contexts for statements about human evolution: 1. Evolutionary events shared with multiple other species (all lower branches leading to humans). 2. Evolutionary events in the unique lineage leading to humans since our common ancestry with chimpanzees. 3. Evolutionary events occurring within extant human populations. 4. Cultural evolution occurring within human populations.

phenomenon, human culture, which emerged at some point probably in the last million years. (Culture doesn't fossilize very well, so we have to remain vague on that date.) Operating alongside and sometimes interacting with biological evolution, culture has its own mechanisms of change, in the form of shared concepts, ideas, beliefs, and practices, which have sometimes profound effects on how humans think, behave, and *are*.

Because human sexual behavior is a kind of ape sexual behavior, it is important to understand what we have in common with the sexual and social behaviors of our ape relatives. But it is equally important to understand how we diverge. If we look at the behaviors of our fellow apes, especially the chimpanzees, this will help us to investigate what we've evolved to do differently since our common ancestry with them and then to ask the relevant questions about why. And, in our case, to ask whether any of those changes might have been a result of aesthetic evolution and the advancement of sexual autonomy.

Most primates, including humans, evolved to live in bands or troops held together by social relationships. Within the various primate breeding systems, there are many different kinds of sexual behavior, which are the result of differences in group composition, size, and social relations. Among those who are part of the African ape lineage—gorillas, chimpanzees, bonobos, and us—these differences are quite striking.

Gorillas live in bands of multiple females dominated by a single silverback male. The silverback controls the sex lives of all the females in his group, and there is little opportunity for female mate choice beyond the rare decision of what group to join. By contrast, chimpanzees live in larger groups consisting of multiple males and females. Male chimpanzees compete for social dominance within the group, and they use their dominance to assert reproductive control over estrous females. Female chimpanzees mate multiply with different males, but they sometimes form a temporary "consortship" in which a pair leaves the group for the duration of the female's fertile period.

Gorillas and chimpanzees are completely uninterested in sex

unless a female is in estrous. A female chimpanzee will have sex repeatedly during an intensive, two-week period of fertile estrous that occurs about every four years. The reason for the long gap between chimpanzee estrous and periods of sexual activity is that after copulating and conceiving, the female goes through seven months of gestation, followed by approximately three years of breast-feeding for each offspring. During breast-feeding, estrous and ovulation are suppressed. With the exception of the fact that ovulation suppression also occurs during breast-feeding in humans, the sex lives of women are obviously very different from those of the chimpanzees and gorillas.

Like chimpanzees, bonobos live in complex groups consisting of males and females. But male bonobos do not compete to dominate the group, and they exhibit very low levels of aggression both within and between groups. And unlike the other apes just described, both male and female bonobos freely engage in frequent sexual behavior with many individuals of their group—including those of the same sex—even when the females are not fertile, and they continue to do so throughout their adult lives, independent of reproductive season or fertility. During estrous, female bonobos mate with multiple males and exhibit variable mating preference.

Outside reproductive sex, bonobos engage in brief sex acts that are used to mediate social conflicts over food, reduce social tension, and foster reconciliation among individuals—regardless of their sex, relative status, or ages. Imagine, if you can, a tense negotiation at a bonobo business meeting in which two ace male deal makers suddenly pause to copulate or to rub their genitals together, after which they agree to a compromise. That is what bonobo sex is like.

It is important to recognize that such nonreproductive sexual behavior is still *sex*. Both reproductive and nonreproductive sex are primarily motivated by the sensory pleasure they provide. The consequences of these acts—whether social or procreative—are always downstream effects of the pursuit of the sensory pleasure of sex itself.

Like bonobos, humans have frequent sex outside the narrow period of female fertility. This is highly unusual, not just among apes, but throughout the animal kingdom. Nonetheless, in most respects we're not like bonobos either. Although we too have sex throughout our adult lives, independent of reproductive seasonality or fertility, we are highly discriminating about whom we have it with (at least by comparison to bonobos).

In trying to understand human sexual behavior, we must remember that many of our ideas about sexuality and gender are culturally influenced, or some would say "culturally constructed." Because all human beings are embedded within their own specific cultures, their attitudes and behaviors, sexual and otherwise, will necessarily reflect the ways in which their cultures have evolved (evolutionary context 4). Because an extraordinary diversity of linguistic, material, economic, ethnic, national, ethical, and religious cultures has developed among human populations around the world, there is a correspondingly great diversity of sexual beliefs and practices. Yet this fundamental truth does not obscure the fact that the biological processes of evolution (evolutionary contexts 1–3) remain deeply relevant to human sexuality, reproduction, and social behavior. The big challenge is to understand how our biological history and our cultural history interact to create the various expressions of human sexuality we see today.

Although the fullness of that complexity is beyond the scope of this book, I want to focus on a few issues within this biology/culture nexus that can be most productively understood through the study of aesthetic evolution. In particular, I will focus on the evolutionary changes in human sexuality that occurred between the time of our common ancestry with the chimpanzees and the invention of agriculture (and probably wealth) about fifteen thousand years ago (evolutionary context 2).

Even in this limited context, human sexuality is uniquely complicated. It has been shaped by interactions among multiple sexual selection mechanisms, often operating simultaneously. They include the following:

male-male competition
female-female competition
mutual mating preferences for ornamental traits that are
    common to both sexes
female mating preferences for male display
male mating preferences for female display
male sexual coercion
female sexual coercion
sexual conflict

Given the diversity and complexity of these sexual selection mechanisms, it is no wonder that our thinking about human sexual evolution can be so confused and convoluted. Where do we start? Because my goal here is to explore the workings of aesthetic selection in human evolution, we'll focus on a consideration of those ornamental features that are likely to have evolved through mate choice. Until now we've looked mainly at female mating preferences for male display traits, because in the birds we've discussed it is the females who are the drivers of sexual selection and the evolution of extreme beauty. But it's clear that in humans, as in certain bird species (like puffins and penguins), both sexes are involved in mate choice.

So, we will start with a look at those human sexual traits and preferences that evolved through *mutual* mate choice, which works in the same way except that both sexes have the same traits and preferences. Darwin proposed that nearly naked human skin—the evolutionary reduction in body hair—evolved as a sexually selected aesthetic trait. Alternatively, reduced human body hair could also have been an adaptation for better body-cooling efficiency for long-distance running. Regardless of whether the reduction of body hair is an aesthetic trait or not, it's clear that another unique trait—the *retention* of specialized patches of hair in the armpits, pubic region, scalp, and eyebrows—*is* ornamental. The fact that the retention of these patches of hair is the same in both sexes (what biologists call sexually monomorphic) strongly implies that it evolved through mutual mate choice, like the bright beaks and plumage of male and female puffins, penguins, parrots, and toucans. The hypothesis that underarm and pubic hair are evolved sexual signals is further supported by the observation that these patches of hair do not develop until puberty. These unusual patches of hair likely evolved for the purpose of pheromonal, sexual communication between mates, which is very common in mammals.

Underarm and pubic hair "cultivate" aesthetic sexual odors through a combination of skin secretions and microbes. Human skin provides a complex ecosystem for a great diversity of microbes, many of which have coevolved right along with humans. As the skin microbiologist Elizabeth Grice and colleagues have written, "Hairy, moist underarms lie only a short distance from smooth, dry forearms, but these two niches are likely as ecologically dissimilar as rainforests are to deserts." Indeed, some of these ecological differences are likely to be coevolved *aesthetic* features. (Future investigations of the microbiota of underarm and pubic hair might well focus on the contributions of skin flora to body odors—initiating an exciting new field of human coevolutionary microbial aesthetics.)

Rare among primates, male mating preferences for female sexual ornaments have clearly evolved on the uniquely human branch of the Tree of Life. The very fact that males *have* strong preferences would seem to be counter-indicated by one of the more tiresome evolutionary psychology truisms—the idea that because sperm are cheap and numerous, and eggs are expensive and rare, men are sexually profligate and women are sexually coy. The problem with this stereotype is how poorly it reflects human behavior. Despite the adaptive story of male profligacy and female coyness, the average lifetime numbers of sex partners for men and for women, at least in Western societies, are actually not that different.

Moreover, an open-ended desire for sex with random strangers can't have had much to do with human evolutionary history. Until only a few hundred human generations ago, when the development of agriculture resulted in greater population density, human groups were so small and dispersed that random sexual encounters would have been extremely rare outside periods of warfare. So male sexual behavior could not have evolved through specific selection to copulate with strangers. In fact, male sexual behavior has evolved in the opposite direction—toward choosiness.

We see evidence of this in our cultural depictions of sexual swashbuckling. The James Bond or Don Juan legends would be much less interesting if these famous Lotharios had sex with literally *any* woman they meet. But James Bond and Don Juan are sexual "heroes," that is, fulfillments of male sexual fantasy, because they are successful with many of the *most attractive* women, not just any women. In fact, Bond's sexual pickiness provides the humor behind his continual sexual disinterest in Miss Moneypenny, the attractive and endlessly available office secretary. Despite her loveliness, she is too available to fulfill the male fantasy of sexual selectivity.

In contrast to humans, all other male apes exhibit an openended sexual appetite that does not refuse *any* fertile sexual opportunity. Gorilla, chimp, and orangutan males will pursue *every* sexual liaison available to them. Men are conspicuously different. The sexual pickiness of human males is a derived feature that arose on the exclusively human branch of the ape family tree (evolutionary context 2). So, contrary to the evolutionary psychologists' eagerness to supply a reason for male sexual profligacy, we actually need an evolutionary explanation for the opposite quality.

There is indeed an evolutionary explanation for male sexual pickiness—quite a profound one, having to do with the unique qualities that make us human, as we will discuss in detail in chapter 10. For now it suffices to say that the pickiness is related to the fact that unlike other male apes, human males make substantial reproductive investments; that is, they dedicate resources, time, and energy to the protection, care, feeding, and socialization of their offspring. As soon as reproduction involves this kind of ongoing paternal care, males should be expected to evolve to be choosy about whom they want to reproduce *with*. And this is exactly what has happened: aesthetic male sexual preferences in human males evolved along with the increase in male parental investment (again, this occurred in evolutionary context 2). The result of this male sexual choosiness has been the coevolution of distinctly female sexual ornaments—like permanent breasts and distinctive body shape—which are completely absent in other apes.

Permanent breast tissue, a relatively narrow waist and broader hips, and fat deposits on the hips and buttocks are all evolutionarily derived in women since our common ancestry with the chimpanzees and therefore require evolutionary explanation. To be sure, basic versions of all of these features are under strong *natural* selection. Wide hips are necessary for birthing human babies, which have evolved to have heads that are larger than those of our ape relatives. Breasts are necessary for milk production and therefore essential to feeding infants. Efficient body fat storage is under strong natural selection when resources are limited or unpredictable, as they were for most of human evolutionary history. Yet each of these features has *also* evolved by male mate choice into ornaments that are exaggerated in specific ways that cannot be accounted for by natural selection alone, because they go well beyond their natural selection optimum.

Among the more than five thousand species of mammals on earth, permanent breast tissue is *unique* to humans. The mammaries of all other mammals increase in size only during ovulation and lactation, and they are not enlarged at other points in the life cycle. Human females, however, develop enlarged breasts with the onset of sexual maturity, and they retain enlarged breast tissue throughout their lives. Yet more than 100 million years of evolutionary history demonstrate that the original mammalian "as needed" breast design is perfectly suited to the successful nursing of offspring. This tells us that permanent breast development is not required for reproduction itself and has no naturally selected advantage. Rather, the existence of permanent breasts in women is likely an aesthetic trait that has evolved by male mate choice.

Similarly, the narrow waist, broad hips, and buttock fat in women might have been exaggerated beyond the proportions necessitated by natural selection alone. The distribution of body fat on the female body is distinctive. In particular, the fat on the buttocks accentuates the hourglass shape created by breasts, waist, and hips. Although there is little doubt that these features are sexually attractive to many, that does not mean that they have evolved, as evolutionary psychology would suggest, as adaptive indicators of mate quality. Even if a certain amount of body fat is an honest advertisement of either genetic quality or health, that does not explain the specifics of the way it is distributed on the female body. Yet there is an entire cottage industry of researchers dedicated to proving that large breast size and low waist-to-hip ratio are indeed signals of what evolutionary psychologists call mating value, a proposed objective measure of a particular person's adaptive genetic quality and condition.

One of the problems with the concept of mating value is that it rests on the assumption that there *must be* something of greater value in sexual attraction beyond mere sexual attraction, and it excludes even the possibility of the sexual appeal of arbitrary aesthetic traits. As we've discussed, evolutionary psychologists are like economic goldbugs in their conviction that there must be some extrinsic value behind every evolved ornament—a pot of evolutionary gold in the form of good genes or direct benefits. They assume that sexual attractiveness must involve *encoded meaning* and that the beautiful individual is in some way *objectively* superior. Despite the number of researchers working to prop up the idea of adaptive human mate choice, it turns out that the data to support its existence are surprisingly slim.

For example, though much effort has gone into trying to validate the idea that the supposedly universally preferred smaller waist-to-hip ratio is actually related to female genetic quality or health, the evidence does not bear that out. For example, one well-known study looked at a sample of Polish women and showed that larger breast size and a lower waist-to-hip ratio were correlated with higher peak levels of the hormones estradiol and progesterone during the menstrual cycle. Higher levels of these hormones have been associated with female fecundity, so the research was considered to support the adaptive hypothesis. But there was no indication that the hormonal variations documented in the study were large enough, or on the appropriate scale, to actually affect fertility. Nor, in fact, did the study find any significant effects of body shape on fecundity among the women, none of whom used contraception. So, the study actually *falsified* the hypothesis that body shape correlates with fertility. Yet it is still frequently cited as *support* for the very hypothesis that it falsifies. This is how a faith-based scientific discipline operates—looking for new reasons, however inadequate, to maintain belief in a theory that has failed.

Similarly, there is a large evolutionary psychology literature on facial "femininity"—that is, relatively small chin and large eyes, high cheekbones, and full lips—as an evolutionary indicator of female "reproductive value," or the remaining, individual, lifetime reproductive potential. This set of features is assumed to peak at puberty and decline with age. The problem with this idea is that youth is not heritable! Everybody starts young and gets older with time. So, male mating preferences for youthful mates with lots of future reproductive potential may be advantageous for males, but such preferences will not, by themselves, result in any evolution in females. The only plausible evolutionary response to mating preferences for indicators of youth is the evolution of traits that *lie* about age. So, to the extent that male mate choice focuses on reproductive value, we should predict the evolution of attractive, arbitrary traits that are dishonest about age. Thus, preferences for facial "femininity" are excellent evidence that mate choice is *not* adaptive but arbitrary.

Finally, although beautiful people do tend to have more friends, better jobs, and higher incomes, these facts are evidence of the *social benefits* of beauty, not evidence that beautiful people are actually objectively better than other people.

The antidote to this faith-based enthusiasm for the adaptive power of mate choice is to embrace the Beauty Happens null model. The Beauty Happens hypothesis proposes that human female sexual ornaments—like permanent breast tissue and the enhanced curves of the hips and buttocks—have arbitrarily coevolved with male sexual preferences for them and are not indicators of genetic quality or health. The Beauty Happens model does not preclude the possibility of honest advertisements; it merely requires that the existence of the evolutionary pot of gold behind beauty be supported by good science—that is, rejection of the null model—rather than propped up by ideological enthusiasm. So far, the Beauty Happens explanation is looking very good.

Oddly, there is a much smaller literature on female preferences for male physical attractiveness than vice versa. As the evolutionary psychologists Steven Gangestad and Glenn Scheyd have conceded, "Scant research has addressed female preferences for male body features." This lack of data is rather unexpected given how active the field of evolutionary psychology has been. If the costlier gametes of women are supposed to make them the choosier sex, then their greater selectivity *ought* to have resulted in the evolution of more extreme, finely tuned, and easy-to-measure mating preferences for many highly derived male ornamental traits. Scientifically speaking, studying female mating preferences should be low-hanging fruit, easy pickings.

Why, then, are the studies of women's mating preferences so rare? A few different explanations of this research gap are possible. Researchers may not find women's sexual preferences interesting to study, but I doubt it. Much more likely, I think, is that research on female mating preferences has simply failed to support the adaptive mate choice theory and therefore hasn't found its way into print. Because the mission of evolutionary psychology is to explain the ways in which human mate choice *is* adaptive, data sets that do not support the mission tend to languish unpublished in lab notebooks and hard drives. The scantiness of the published research likely points to reams of unpublished evidence that would support the Beauty Happens mechanism if they were ever to see the light of day.

Even the data that do exist in print are difficult to interpret as evidence of adaptationist views. For example, there is consistent evidence that females do not prefer the *most* "masculine" facial features, which have been characterized as prominent square jaws, wide prominent brows, thick eyebrows, and thin cheeks and lips. Numerous studies have shown that women instead prefer intermediate or even what some researchers call "feminine" facial features in men, and one study has shown that females prefer a light stubble over a more masculine full beard. According to a handful of disparate studies cited by Gangestad and Scheyd, these facial preferences seem consistent with the evidence on what women like to see in male bodies. They tend to like lean but somewhat muscular male bodies with broad shoulders and V-shaped torsos the most, and men with larger, more muscle-bound bodies the least.

These findings create a conundrum for adaptationists because masculine features are proposed to be indicators of strength and dominance, which every right-thinking, fitness-pursuing female *should* prefer. Of course, one possible explanation for why masculine features exist *despite* the fact that women *do not* prefer them is that they evolved through male-male competition for mates and social status, rather than through female mate choice. Evolutionary psychologists have also proposed that females may prefer men with less masculine features because such features signify men who will make bigger parenting investments in their children. However, they never explain why high-testosterone males with broad brows and prominent jawbones would make bad dads. It is just seen as obvious.

One reason evolutionary psychologists have so much trouble explaining away the apparent inconsistencies in female preferences is that they have drawn the concept of mating value too narrowly to capture the actual complexity of human mate choice. In a way, the very concept of mating value is a scientific expression of what cultural theorists have called the "male gaze"—a point of view that depicts women and women's bodies solely as the object of male erotic pleasure and control. Indeed, evolutionary psychology investigations of female mating value are almost universally conducted by having young men actually *gaze* at computer-generated images of women's faces and bodies. Is it really that surprising that the concept works so poorly as a tool to understand the sexual preferences of women? By reifying the male gaze as an adaptation, evolutionary psychology has enshrined sexist bias into human evolutionary biology, and notably failed to explain the mate preferences of the other half of the species.

What they have overlooked are the social interactions that are so critical to the mate choices we make. Indeed, social interactions are vital to how we experience sexual attraction, whom we have sex with, and how we fall in love. As new research from the field of experimental social psychology demonstrates, our social interactions with each other have the potential to override the information we take in only through our eyes. The psychologist Paul Eastwick's work has focused on how social interactions alter perceptions of sexual attractiveness. In a series of experiments and meta-analyses, Eastwick and colleagues have shown something that we all know from experience: our perceptions of sexual attractiveness change as we get to know each other.

Prior to any social interactions, people tend to agree on their initial (that is, superficial) judgments of the sexual attractiveness of others. But once they have opportunities to interact socially, they begin to diverge in their judgments and to notice features in other people's personalities that are specifically attractive to them. Ultimately, these *subjective* social perceptions have a much stronger effect on what they find attractive than does physical appearance. Paul Eastwick and Lucy Hunt write, "This idiosyncrasy will prove fortuitous, as it permits nearly everyone a chance to form relationships where both partners view each other as uniquely desirable." It is a happy thought that people are, by and large, built for finding social-sexual happiness with another despite the variations in physical attractiveness. "Mating value" is not a universal and objective measure; it is a subjective, relational experience.

Interestingly, Eastwick's studies also indicate that there is no difference between men and women in the degree to which social relationships influence their evaluation of attractiveness. The same guys that provide the data for evolutionary psychology studies on female mating value by gazing at computer screens are actually just as likely to be influenced by the qualities that emerge through social interactions as women are. Apparently, the male gaze is not a great recipe for male happiness either.

It is obvious that in the real world human mate choice occurs in a complex environment of individuals who vary not just in physical attributes but in personality and character. The bottom line is that the evolution of our ability to socially interact with each other in increasingly complex ways has affected the criteria that are involved in mate choice. With the origin of culture, material culture, language, and complex social relationships, a new dimension to the aesthetics of human attraction has come into being and greatly expanded—social personality. All the qualities that go into that—humor, kindness, empathy, thoughtfulness, honesty, loyalty, curiosity, self-expression, and so on—are now part of what attracts us to each other. In fact, it's likely that such traits evolved precisely *because* they proved to be attractive and helped reinforce the social stability of sexual relationships. Falling in love has become more and more elaborate, not to mention emotionally intense, enjoyable, and potentially heartbreaking, because it is the result of a coevolutionary process—millions of years of *aesthetic,* mutual mate choice. Even though they do have social personalities, I do not think that gorillas and chimpanzees can fall in love as we do, because these species have not gone through this coevolutionary process.

The evolutionary psychology concept of mating value suggests that we should be able to look at a picture of a potential mate and swipe left or right and make evolutionarily informed decisions accordingly. While this might be fun for a while, it usually fails as a long-term strategy because mating value cannot be defined on any objective scale based on superficial features. True "mating value" emerges only during the course of getting to know each other and falling in love, and it actually takes *time* to fall in love. For today's young urban dwellers, time is limited and sexual choices are nearly infinite. However, for most of the last few million years of human evolution, humans lived in very small populations with few sexual choices and all the time in the world. Human mate choice evolved to function in the latter context, not the former.

The real reason why there is an apparent paucity of morphological ornaments in human males is that female mate choice in human evolution has focused largely on social rather than physical traits. It makes sense that females, who until relatively recently on the evolutionary timescale were the ones charged exclusively with the care of their children, should care more about qualities that indicate the potential for relationship endurance. In the long run, women have evolved to want mates who will be good partners to them and good parents to their children. However, that does not mean that it doesn't take some shopping around to find that mate.

All that said, female mate choice has likely played a critical role in the evolution of one central feature of the male body—the human penis. We may not think of this vital piece of equipment as an "ornament," but like women's breasts the human penis has been shaped evolutionarily by simultaneous processes of natural *and* sexual selection, and it is worth asking which features evolved through which mechanism.

Darwin himself struggled to distinguish between the effects of natural and sexual selection on individual body parts. For example, he mused over whether the specialized grasping limbs used by certain male crustaceans to seize the female during copulation evolved as a result of natural or sexual selection. If the function of an organ was necessary for reproduction to occur at all, Darwin reasoned that it would evolve by natural selection. Yet any aspects of that same organ that were further derived through mating competition or mate choice would evolve by sexual selection.

The human penis is a fascinating example of the simultaneous action of these two evolutionary mechanisms. Given the fact of internal fertilization in mammals, we know that it is absolutely necessary for reproduction. So, the existence and maintenance of the human penis can be ascribed to natural selection alone. But various aspects of the morphology of the human penis, which are beyond those necessary merely to accomplish copulation and fertilization, are likely under sexual selection as well.

Among primates, the penis is one of the most variable of all organs. From species to species, there are radical differences in its length, width, thickness, shape, surface texture, and elaboration. All these variations are beyond what is required to accomplish reproduction. So why have the different species evolved penises that vary so dramatically from each other?

Here, of course, I will focus on the human penis. By any measure, the human penis requires a lot of explanation. It is substantially larger—both in absolute and in relative size—than that of any of the other apes, even though humans are intermediate in body size between gorillas and chimpanzees. The erect gorilla penis is only an inch and a half long. The chimpanzee penis is three inches long when erect, very thin, smooth, and finely pointed at the tip. The human penis is both longer—averaging about six inches when erect—and wider than the penis of other apes. The human penis is also characterized by a distinctly bulbous glans and coronal ridge at its tip. Similar structures have evolved in other primates, but they are not present in other African apes. We should also note that in contrast to their greater penis size and elaboration, humans have testes that are both relatively and absolutely smaller than those of our closest chimpanzee relatives.

In *The Third Chimpanzee,* Jared Diamond illustrates this genital variation in a memorable cartoon of what male gorillas, chimpanzees, and humans "look like to each other." The gorilla is a huge circle with tiny testes and an even tinier penis. The chimpanzee is much smaller in body size with huge testicles and a tiny penis. The human is between the gorilla and the chimp in overall size, but with small testicles and a huge penis. This mosaic of genital features has evolved under different sources of sexual selection in each species. So the variations tell a story about the dynamic evolutionary history of penis morphology—a story that lends itself to multiple interpretations, some more plausible than others.

Both testes size and penis size have frequently been hypothesized to evolve by male-male sperm competition. According to this hypothesis, when females have multiple mates, males will be under sexual selection to produce more sperm to outcompete the sperm of other males, which will result in the evolution of larger testicles. Chimps have a breeding system characterized by lots of multiple mating and high sperm competition, and thus they have huge testicles. Gorillas, on the other hand, have a breeding system characterized by male physical dominance over a group of reproductive females, with very little sperm competition or female mate choice, hence the tiny testicles.

The large human penis has also been interpreted as having evolved through sperm competition. The larger the penis, the closer it will be to the ova when sperm are released during sex, and the better the chances of fertilization. Or so goes the theory. Along the same lines, the prominent glans and coronal ridge of the human penis have been hypothesized as tools to displace the sperm of other males who might have previously ejaculated inside the female's vagina. The evolutionary psychologist Gordon Gallup and colleagues tested this hypothesis in experiments with artificial penises of various shapes, an artificial vagina (all purchased from Hollywood Exotic Novelties), and artificial ejaculate made of water and cornstarch. Not surprisingly, the realistic dildo with a prominent glans and coronal ridge displaced more of the cornstarch goo to the periphery of the fake vagina than did the smooth, sleek penis model. The human-penis-as-sperm-displacement-tool hypothesis was triumphantly supported.

Unfortunately, the sperm displacement hypothesis for the evolution of the size and shape of the human penis simply fails to line up with the evidence from the Tree of Life. The fact that there has been an evolutionary *decrease* in the size of human testicles since the time of our common ancestry with the chimps tells us that the sperm competition among human males has also decreased. So theories explaining human penis evolution through sperm competition and displacement mechanisms provide solutions to an evolutionary problem that has actually been waning over time. If larger penises with prominent bulbs at the tip function to remove sperm of previous males, why haven't chimps evolved them? Extracting ejaculate of another male with a penis would be a classic, nonaesthetic, mechanical function. Such a simple physical mechanism ought to have broad utility among all the species of primates that engage in sperm competition. Just like the beak of the finch, lots of primates should have convergently evolved the same tool for this job. Why, then, do chimps have relatively small, thin, smooth, tapered penises—basically the size of a human pinkie finger—despite their vigorous sperm competition? Sperm competition arguments for the evolution of human genitalia are plainly incongruent with the evidence from our primate relatives.

So where are the "honest penis" hypotheses when we need them? Oddly enough, evolutionary psychologists have not enthusiastically embraced the idea that penis size is an honest indicator of male quality. Although nearly every perceivable feature of the female body—waist-to-hip ratio, breast size and symmetry, facial symmetry and "femininity," and so on—has been scrutinized as a potential indicator of female genetic quality and mating value, the eminently measurable human penis has received little such attention. Perhaps male evolutionary psychologists are unwilling to submit their own anatomy to the same scrutiny they apply to the female body? Perhaps they lack the courage of their convictions?

Of course, it *is* rather hard to imagine that the size of the human penis could be an indicator of quality. After all, weighing in at an average of only about 4.3 ounces of flesh when flaccid, the average human penis, even if it were to double in size, would not be a costly investment, or Zahavian "handicap," because it would still represent just a tiny fraction of a man's body mass. If the penis were made up of rare, limited, biologically costly materials, perhaps such an increase in size would represent enough of an investment to signal superior quality. But the penis is not made up of anything special—just connective tissue, blood vessels, skin, and nerves. (Lots of nerves.) Nor are larger penises more costly to operate; for example, there's no evidence that erectile dysfunction is more frequent among men with larger penises.

Despite the overall lack of interest in the penis among evolutionary psychologists, there has been one aspect of the human penis that has attracted at least one honest indicator theorist, as we shall see, and that has to do with another biological innovation of the human penis. Human males are notably distinct from other primates in that they lack a *baculum*—also called the *os priapi*. The *baculum* is the mammalian penis bone, or the bone in "boner."

The *baculum* has been dubbed "the most variable of all bones." The prize for size goes to the bull walrus (*Odobenus rosmarus*), which has an *os priapi* that resembles a policeman's nightstick made of ivory. To take just one other example of the numerous variations in its size and shape, many squirrels have a *baculum* that is spatulate at the tip, with elaborately articulated tines like a tiny otherworldly pasta spoon.

Mammalogists have developed a mnemonic to help remember which mammals have evolved a *baculum;* PRICC is an acronym for primates, rodents, insectivores, carnivores, and Chiroptera (that is, bats). Although I assume few readers will be surprised to learn that humans don't have a penis bone, some may still be dismayed to learn that man is one of only two primate species—along with the spider monkey—to have been evolutionarily singled out to lack a *baculum.* The existence of a *baculum* in the other primates means that an erection is guaranteed by the presence of an ossified bone within the penis. However, there are many male mammals besides us that don't have a *baculum*—from opossums to horses, elephants to whales—all of whom achieve erections just fine without one. So we know the *baculum* must have functions beyond mere intromission, even though we don't know what they are. Actually, we do know that aside from producing erections, the *baculum* functions in retracting

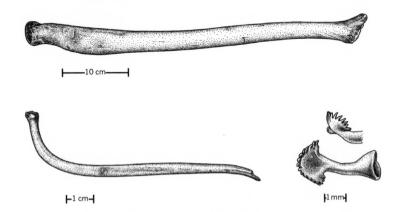

—10 cm—

1 cm

1 mm

A diversity of baculums from (top) a male walrus (*Odobenus rosmarus*), (lower left) a male raccoon (*Procyon lotor*), and (lower right) a spotted ground squirrel (*Xerospermophilus spilosoma*).

the penis between erections. What its other functions might be is still not clear.

But in the context of the current discussion I'm less interested in the question of why some mammals have a *baculum* than in why men have lost it. This is apparently not a new intellectual puzzle. Attempts to explain the mystery date back to the foundational text of Judeo-Christian culture—the story of the creation of Eve in the book of Genesis. In 2001, two well-respected academics—the developmental biologist Scott Gilbert at Swarthmore and the biblical scholar Ziony Zevit at UCLA—teamed up to investigate this question in a scientific paper titled "Congenital Human Baculum Deficiency: The Generative Bone of Genesis 2:21–23," which was published in the *American Journal of Medical Genetics*. Some twenty-five hundred years after the composition of the well-known Genesis creation story, Gilbert and Zevit propose that the story claimed that God had created Eve not from Adam's rib but from Adam's *baculum*. They maintain that the "rib story" would have been recognized as false by any ancient Israelite, based on the readily made observation that men and women have the same number of ribs. (Indeed, I remember counting my ribs and pondering this exact problem myself in Sunday school as a kindergartner.) Gilbert and Zevit further discredit the Adam's rib story as narratively lackluster—because ribs are *"lacking any intrinsic generative capacity."* Apparently, the Greatest Story Ever Told requires a more potent plotline than the King James translation has given us. Gilbert and Zevit provide some impressive linguistic evidence to support their radical hypothesis:

> The Hebrew noun translated as "rib," tzela (tzade, lamed, ayin), can indeed mean a costal rib. It can also mean the rib of a hill (2 Samuel 16:13), the side chambers (enclosing the temple like ribs, as in 1 Kings 6:5, 6), or the supporting columns of trees, like cedars or firs, or the planks in buildings and doors (1 Kings 6:15, 16). So the word could be used to indicate a structural support beam.

"Structural support beam" is a very succinct description of the *baculum*. Gilbert and Zevit then discover the smoking gun in this evo-scriptural mystery—unexpectedly clear anatomical evidence in the Hebrew Bible:

> Genesis 2:21 contains another etiological detail: "The Lord God closed up the flesh." This detail would explain the peculiar visible sign on the penis and scrotum of human males—the raphe. In the human penis and scrotum, the edges of the urogenital folds come together over the urogenital sinus (urethral groove) to form a seam, the raphe . . . The origin of this seam on the external genitalia was "explained" by the story of the closing of Adam's flesh.

In this interdisciplinary tour de force, Gilbert and Zevit took a fresh look at a very old story and arrived at a revolutionary new view of the Judeo-Christian creation myth. For some inexplicable reason, their paper has not yet received the firestorm of attention it deserves. It seems to me that everyone from the Vatican to feminist scholars should want to know about and debate this theory. Yet the paper has only been cited three times in fifteen years. Perhaps no one has time in our fragmented intellectual culture to ponder these questions? Shouldn't more people care whether the Hebrew God created Eve from Adam's penis bone? Inquiring minds should want to know.

If Genesis tells the story of the loss of Adam's *baculum* as an act of divine agency, how do evolutionary biologists explain it? Although there has been relatively little evolutionary theorizing about the human penis in general or its loss of the *baculum* in particular, one brave biologist does stand out in his eagerness to take on the task. Richard Dawkins hypothesized that the human penis evolved to be without a *baculum* in order that the penis could serve as—yes!—an honest indicator of health and genetic quality:

> A female who behaves like a good diagnostic doctor and chooses only the healthiest male for mate will tend to gain

healthy genes for her children . . . It is not implausible that, with natural selection refining their diagnostic skills, females could glean all sorts of clues about a man's health, and the robustness of his ability to cope with stress, from the tone and bearing of his penis. But a bone would get in the way! Anybody can grow a bone in the penis; you don't have to be particularly healthy or tough. So selection pressure from females forced males to lose the os penis, because then only genuinely healthy or strong males could present a really stiff erection and the females could make an unobstructed diagnosis . . . If you follow through the logic of my penis hypothesis, males are handicapped by the loss of the bone and the handicap is not just incidental. The hydraulic mechanism gains in effectiveness precisely because erection sometimes fails.

To be fair, Dawkins admits that this hypothesis "should not be taken too seriously" and that he only came up with it as a clever way to communicate Zahavi's handicap (that is, Smucker's) principle and its connection to good genes. However, when Dawkins admits that the idea is "less plausible than pleasing," he is actually making an unexpectedly revealing comment on the entire field of adaptive mate choice.

Dawkins's "tale of the lady doctor" reveals his priapic delight in the hypothesis that human erections are uniquely evolved symbols of male genetic superiority and condition. In his scenario, the ecstatic experience of male tumescence has been scientifically reified as an evolved indicator of individual male superiority. The adolescent male fantasy of erectile omnipotence has become an explanatory force in human evolution. In this way, Dawkins's "tale of the lady doctor" is a masterwork of phallocentric evolutionary biology.

However, as Dawkins admits, this scenario is not "plausible." Perhaps the main reason it's not is that for the average human male of mating age, having an erection—even "a really stiff" one—is no more a sign of some kind of superior health than growing a bone in the penis is for our primate relatives. More or less anyone, at least of a certain age, can do it—"you don't have to be particularly healthy or tough." Purely vascular, hydrostatic erections are not a challenge for mating-age males in practically any state of health. Most human erectile dysfunction is a result of senescence, and in the Pleistocene African savannas of our evolutionary past, most *Homo* were long dead by the age at which they would have had a problem maintaining an erection. No—despite the ubiquity of advertisements by pharmaceutical companies for erection-enhancing drugs, which would suggest some kind of epidemic of dysfunction, there is actually no shortage of human erections in the world today. How choosy would any woman be if, following Dawkins's scenario, she were to use male erectile competence as her criterion of mate choice? Only a relatively few aging geezers would be eliminated (ironically, along with their "good genes" for longevity). Thus, it seems unlikely that the loss of the *baculum* was an evolutionary answer to the female's need to assess male quality and health. Yet, notwithstanding Dawkins's own caveats, his penile handicap hypothesis for human *baculum* loss is actually taken seriously by evolutionary psychologists.

Implicit in Dawkins's hypothesis, however, is a much more plausible possibility—the thoroughly aesthetic proposal that the evolutionary loss of the human *baculum* occurred through female mate choice. An alternative to the honest advertisement hypothesis, and to the male-male competition theories, is that the loss of the *baculum*, the increase in penis size, and changes in penis shape all coevolved through aesthetic female preferences for penis morphologies that women found arbitrarily attractive. But why would female humans have evolved to prefer bigger, wider, and distinctively shaped penises? The answer, of course, is sexual pleasure, in all its many dimensions.

The human penis is a complex sexual ornament whose various features evolved to be experienced through two distinct sensory modalities: vision and touch. The aesthetic result is a visual ornament that doubles as a piece of interactive, personal, tactile sculpture. In other words, genital beauty happens too.

The convergence of these various features may have something to do with the fact that, thanks to the loss of the *baculum* and its penile retraction function, humans, unlike nearly every other primate species, have a penis that does not disappear when it is not erect. Instead, it *dangles,* and it dangles all the more visibly because it evolved to be bigger and longer than that of other primates. This suggests that the evolutionary loss of the *baculum* and the gradual increase in penis size in humans may be related and were the result of female mating preferences for a dangling genital display. Male genital dangle would have become an increasingly conspicuous display with the evolution of bipedality in the last five million years of human history.

An aesthetic function for the whole dangling human male genital package is further supported by the observation that the human scrotum is also more pendulous than that of other apes. Gorillas and orangutans have no prominent external scrotum. Chimpanzees have a truly pendulous scrotum and very large testes. Humans, however, have an even larger and lower-hanging scrotum than chimpanzees do. Paradoxically, this increase in the size of the human scrotum occurred simultaneously with the *decrease* in the size of the testes themselves, which are smaller in size in both relative and absolute terms than those of the chimpanzees. The exaggeratedly large human scrotum, which is far bigger than is necessary to house the testes, is indicative of a history of selection for an additional communication rather than a mere physiological function. That is, the scrotal sac might have gotten larger because females liked the way it dangled.

This is certainly not the only instance of sexual selection in the evolution of the scrotum. The co-option of the scrotum for sexual display purposes is well-known in various groups of mammals that see in color. These include the vervet monkey (*Cercopithecus pygerythrus*) and the mouse opossum (*Marmosa robinsoni*), both of which have a vivid, attention-getting, bubble-gum blue scrotum.

Of course, the human penis does more than just dangle, and the other derived features of the human penis have also likely evolved for sexually selected aesthetic functions. A dangling genital display will give females cues about penis size when erect. So why would females have evolved preferences for penises of a size far larger than those of any of our ape relatives? What benefit would these larger penises offer to females? Now that we've dispensed with the idea of penis size as an honest indicator of genetic quality, we can consider the aesthetics of the penis. The longer, thicker, broader human penis with the bulbous glans at the end is likely to have evolved through female preferences for male copulatory organs that produce greater pleasure. The first pleasure comes from observing the dangling penis at some distance, which was facilitated by the loss of the *baculum*. The size of the display would serve as an indicator of the potential tactile, sensory experience that sexual intercourse with that male would provide. This anticipatory pleasure is then succeeded by the pleasure of experiencing that penis directly during sexual interactions and copulation.

But does that mean that preferences for large penises are universal in women? Larger than chimps, yes, certainly. But not necessarily large by comparison with other human penises. Women's responses to the question "Does penis size matter?" are highly variable. Interestingly, male penis size is also highly variable. Could these two variations be related? Indeed, if penis size is an arbitrary aesthetic trait, then penis size, like many other aspects of human beauty, could be highly variable and responsive to a wide diversity of tastes, and so it is. Different strokes for different folks.

In contrast to the penis, which is highly visible, the size and shape of the glans are obscured by the foreskin while dangling, only to be revealed during erection and sexual intercourse. If, as I'm proposing, the shape of the glans also evolved through female choice because of the pleasurable sensations it provides, this suggests a mating preference for a feature that can only be evaluated *during* copulation, because it is otherwise hidden. Of course we usually think of copulation as something that takes place only *after* a mate choice has been made, but by the time sex is actually occurring, the mate-choice horse is out of the barn, so to speak.

It might seem odd that a mating preference could evolve for

a feature that is not experienced until copulation itself. But in humans, who mate—and mate repeatedly—without regard to season or fertility, mate choice does not have to end when copulation begins. It can even begin with it. Sex provides both individuals with a rich set of sensory stimuli that can be evaluated and used to influence *subsequent* mating choices, so all the fundamental features of aesthetic evolution can still apply.

Unlike the other apes, females of our species have evolved concealed ovulation, and therefore individual acts of sexual intercourse have a particularly low probability of leading to fertilization. Therefore, it would be better to think of humans as having *remating preferences*. Because these remating preferences may be partly based on the sensory experience of sexual intercourse itself, a fully aesthetic theory of human male genital evolution will encompass both those features that can be assessed before copulation—like penis and scrotum dangle—and those that can be experienced and evaluated during the act, including the size and shape of the erect penis itself. Interestingly, this evolutionary mechanism, which assumes female agency, directly contradicts the concept of the sexually "coy" female.

Female mate choice has had a profound effect on the appearance of human male genital "ornaments," which over the course of millions of years of evolutionary history have been reshaped so that they bear little resemblance to those of our ape relatives. But what we've discussed thus far takes us only to evolutionary context 2—evolution that took place in the human lineage since common ancestry with the chimpanzees—which falls short of taking into account more recent and continuing biological changes (evolutionary context 3) and the effect that culture may have on biology (evolutionary context 4). The role that human culture plays on mate choice—both male and female—is highly significant. What is considered sexy within one culture may be reviled in another. I propose that these arbitrary cultural preferences can reshape not just our social behaviors and relations but, over time, our actual bodies and their diversity.

When I lived in Brownsberg National Park in Suriname studying manakin display behavior in 1982, I paid a couple of dollars a
day for a bed in the bunkhouse that housed the workmen for the
park. They were all young Saramaccan men, members of a distinct ethnic group descended from African slaves who had escaped
from plantations on the coast in the early seventeenth century and
moved upriver into the forest to reconstitute new African Creole
cultures in the New World. Once or twice a week, tourist groups
would arrive to stay at the guesthouses in the park, and a few
young women would come from the local Saramaccan village to
clean the cabins and cook for the groups. From the porch and
windows of our cabin, the workmen subjected these women to an
endless stream of sexually suggestive verbal commentary as they
walked between cabins carrying sheets and towels, buckets and
mops, and the women would banter back laughingly. The young
woman who received the most attention was about five feet four
inches tall and well over two hundred pounds. Although she was
far from the ideal waist-hip ratio of any evolutionary psychology
textbook—a textbook that would have been written with Western ideals of beauty in mind—she was extremely attractive to the
workmen in camp, and she knew it.

If humans are the result of biological evolution, what accounts
for the great diversity in *ideas* about human beauty? So far, we
have focused on biological features of human sexuality that can
be credibly hypothesized to have evolved during the five- to seven-
million-year-long evolutionary history of humans since our common ancestry with the chimpanzees (evolutionary context 2).
Now it's time to look at the many unique evolutionary changes
that have happened more recently.

Humans evolved the capacity for spoken language, advanced
cognitive abilities, and complex social lives and social interactions. We expanded out of Africa multiple times—*Homo erectus,*
Neanderthals, and then modern *Homo sapiens*—and dispersed
around the globe. And with this dispersal across multiple continents, we continued to evolve and diversify genetically (evolutionary context 3). Thanks to the increasingly complex capabilities
and experiences that arose as a result, our cultures also continue

to change and diversify, probably at an ever-faster rate (evolutionary context 4).

The features that develop culturally are the result of interactions between the individual's social environment and the contingencies of human history. In other words, the distinct cultures that arise from human populations and subgroups that are geographically separated from each other are the way they are not just because of adaptations to their specific environments but also because of prior events in their history. The diversity of human languages is a great example of the arbitrariness of human cultural history. No one supposes that the differences between English, Japanese, and Navaho languages arose because of adaptations to the different environments in which these languages developed. With culture, who we become as individuals is greatly influenced by the history of the social groups, communities, and nations that we are born into and live within.

As with other culturally influenced features, our ideas about human beauty, our courtship and mating practices, and our sexual behavior vary depending on all of the above. Despite the evolutionary psychology belief in universal "mating value," there is no human sexuality without culture, and the only thing that's universal about culture is how variable it is. If we go back a few thousand years—a mere blip in human evolutionary history—we can see that immediately.

Classical Roman and Greek statuary depicts versions of female beauty so iconic that they were suitable for worship. Yet because of changing fashions, many of these faces and bodies would not be considered especially attractive in the contemporary Western world. Such changes in taste can appear not just over the course of thousands of years but in far shorter periods of time. In just a matter of decades, American culture has drastically changed its view of what men and women should look like. We have only to compare photographs of Marilyn Monroe or Rita Hayworth from the 1940s and 1950s with the comparatively emaciated, sometimes anorexic, female movie stars and fashion models of today to realize how rapidly cultural standards of beauty can change. Despite her legendary sexiness, the softly voluptuous

Marilyn Monroe would not make it into the first round of the reality television show *America's Next Top Model*. We have also changed our ideas about what we find attractive in male bodies. To stay at the top of the business, today's male movie stars must maintain well-defined, muscular physiques that are a far cry from the softer bodies of the 1940s and 1950s stars like Cary Grant, Clark Gable, and Gary Cooper.

Some cultures, unlike our own, revere and sexualize female obesity. In Mauritania and other parts of Africa, feminine obesity is regarded as so attractive that girls of normal body weight are sent to "fat camps" where they are force-fed enormous amounts of food to gain weight. Young Mauritanian men express specific sexual excitement over the stretch marks that appear on the skin from the young women's resulting rapid weight gain. In America, by contrast, we send young women to "fat camps" in order to *lose* enormous amounts of weight.

Even the most fleeting of sociosexual fashion trends can deeply influence human sexual desire and mating behavior. Some years ago, an anonymous man wrote a piece for the gossip blog *Gawker* about a sexual encounter he had had a few years earlier with a woman who, at the time of his report, was running for high political office as a Tea Party Republican. Several months after they had first met, he said, she and another acquaintance showed up at his apartment on Halloween night to invite him out to party. They all went to a bar and drank excessively, and then he and the future candidate returned to his apartment and ended up in his bed. What seemed like a predictable progression, however, came to an unexpected and sudden halt. As the kiss-and-tell blogger wrote, "When her underwear came off, I immediately noticed that the waxing trend had completely passed her by. Obviously that was a big turnoff, and I quickly lost interest." Even more unexpected to me than my sudden feelings of sympathy for the right-wing politician was my shock at the man's assumption that *his particular* sexual preferences were universally shared and approved of by his audience. Even though the anonymous tattler recognized that selective removal of pubic hair is a "trend," he still felt it "obvious" that any woman who wasn't a slave to this particular fashion would be a "turnoff" to any sexually well-adjusted man such as himself.

This anecdote is not just about the cultural variability of sexual taste, however. It also serves as another refutation of evolutionary psychology's contention that men are shaped by natural selection to be universally sexually profligate. In fact, men are quite sexually picky, and the form of their pickiness is greatly influenced by their cultural environments.

The reason that I've gone into some detail about differing cultural norms of beauty is that they have the potential to feed back upon biological, or genetic, evolution. When culture assumes a *causal* role in evolutionary processes, we call this a top-down effect.

One of the most striking examples of the top-down effect of human culture on genetics is the evolution of lactose tolerance in adults, which allows some to eat dairy products. Lactose is a special sugar found only in mammalian milk. All baby mammals digest lactose with the enzyme lactase, but mammals stop making lactase when they are weaned. However, during the last twelve to fifteen thousand years, various groups of humans domesticated sheep, cows, goats, and horses, and the ensuing widespread availability of milk—a rich, new source of calories and protein for adults—resulted in *natural selection* for genetic changes that produce the adult capacity to digest lactose in many populations of humans. Thus, the *cultural* practice of dairy herding exerted a top-down effect on human genetic evolution. In short, culture can shape biology.

In a similar way, I think that cultural ideas about beauty and sexuality could have top-down effects on the genetics of human appearance and behavior—via sexual selection. It would be very challenging to gather the comparative data that would be needed to test this idea. But in hopes of inspiring this sort of research,

Facing page: Variation in the frequency of adult lactose tolerance in human populations around the world. *Based on Curry (2013).*

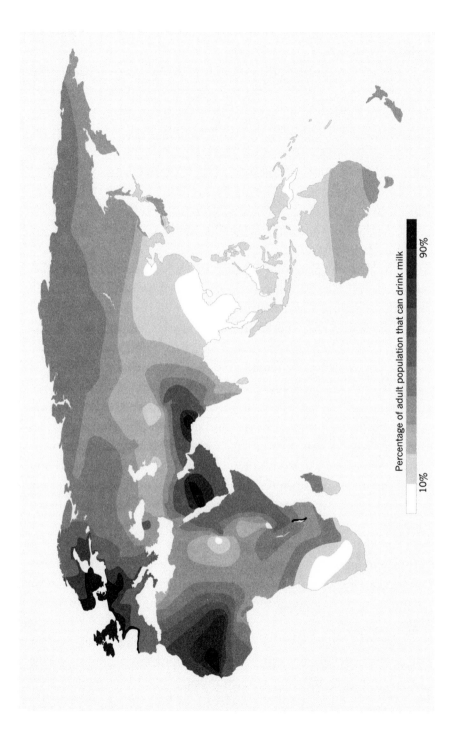

Percentage of adult population that can drink milk

10% 90%

I want to present a few highly speculative but plausible ways in which such a process could work.

Ethnic groups from different cultures can vary considerably in appearance, but few of these variations are likely to be under natural selection. For example, skin color variation is strongly associated with latitude, the probable result of strong natural selection for darker skin at equatorial latitudes to protect against skin cancer or (more likely) preservation of folate, and strong natural selection for lighter skin at higher latitudes to facilitate vitamin D synthesis. Hair and eye color often covary with skin color because these correlated traits involve many of the same genes for melanin pigmentation.

Most other variations in appearance among human populations and ethnic groups, however, are very unlikely to be under natural selection. These features include hair texture, hair length, nose shape and size, cheekbone shape, facial widths, lip size and shape, eyelid shape, ear size and shape, earlobe connectedness, breast size, patterns of female body fat deposition, the extent of male facial and body hair, and penis size. These features vary geographically among human populations and are strongly heritable, but there is virtually no possibility that such evolved variations among human populations are adaptations to variations in the environment. Although there are other possible explanations, I believe a strong argument can be made for the hypothesis that cultural ideals of beauty could produce top-down evolutionary changes in physical attributes.

As a speculative example, let's look at the Samoan and Hawaiian peoples, who have lived on their respective Pacific archipelagoes for approximately fifteen hundred years. By global standards, these populations are outstanding in their immense body size and mass. Traditionally, large bodies and high body weights have been seen as admirable and sexually attractive in these cultures. Their kings and queens have been famously tall, heavy, imposing figures. If cultural criteria for beauty mean that certain individuals within a society will be much more sexually successful than others, with higher numbers of offspring and perhaps more resources, it makes sense that the features that are culturally favored—in the case of the Polynesians large, voluptuous bodies—will become increasingly represented in the gene pool. In this way cultural *ideas* about attractiveness could propel a relatively rapid evolution in appearance.

Another example of the way this top-down effect could work can be seen in southern Africa, where the women of the ancient Khoisan ethnic group are well-known for large accumulations of body fat on their buttocks, creating a very distinctive, callipygian body shape. Not surprisingly, given that most cultures have strongly positive cultural associations with the features peculiar to them, the Khoisan men find this feature very attractive. Now, body fat storage in general should be favored by natural selection, but it is difficult to argue that such a specific anatomical shape would be favored in one specific environment and not in another. Rather, these particular body shape variations are likely the result of completely arbitrary sexual preferences. Among the Khoisan, it seems possible that cultural regard for a certain female body shape could have driven the evolution of heritable differences in body fat distribution. In other words, the cultural preference for this kind of female body shape likely helped to create it.

Using mathematical models that are very much in line with Fisher's runaway sexual selection model, the biologists Nathan Bailey and Allen Moore have documented that cultural mating preferences can create feedback loops that result in the evolutionary elaboration of certain traits that are deemed desirable but have no survival or fecundity value—only aesthetic value. These mating preferences are not merely the handmaiden of natural selection. Indeed, just as Fisher, Lande, and Kirkpatrick maintained genetically, a cultural runaway process is likely to erode any relation between beauty and honest indicators of quality, resulting in the evolution of traits that may even be counter-indicated for survival purposes.

This cultural-genetic evolutionary feedback could explain a lot of the aesthetic diversification in superficial appearance among human populations and ethnic groups. It is likely that human cultural diversity has begotten a great deal of our physical diversity. And this evolutionary mechanism would proceed entirely without adaptation by natural selection. Indeed, human culture makes it even more difficult for us to evolve honest sexual signals.

The possibility of arbitrary, aesthetic human mate choice stands in direct contrast to the adaptationist ideas about human mate choice that have so permeated Western culture. If this chapter has achieved its aim, I hope I have shown that we cannot automatically assume that variations in our appearance reveal anything about our inner genetic value. Before we can conclude that a given ornamental trait is adaptive, we must first reject the Beauty Happens null model. And when we fail to find evidence to reject it, we must accept that human Beauty Happens too.

CHAPTER 9

Pleasure Happens

In Greek mythology, the divine, reigning couple of creation—
Zeus and Hera—had a difficult marriage. Zeus was always running around looking for new ways to seduce beautiful young
women and father more children, and Hera was naturally in a
constant state of jealous rage over Zeus's frequent infidelities.
Because Hera's many titles included the goddess of marriage, her
husband's unwillingness to be true to her caused her not only personal pain but very public embarrassment. It was in the context
of this ongoing tension between them that Zeus and Hera had an
argument over which sex experiences greater sexual pleasure—
men or women. They both tried to defend their respective moral
positions on marital fidelity by claiming that the *opposite* sex
experienced greater sexual pleasure than their own. They decided
to settle the dispute by consulting the only authoritative source
they knew—a wise man named Tiresias.

Tiresias was what biologists today would call a sequential
hermaphrodite—an individual who changes sex during its lifetime (as happens in certain plants and animals). Tiresias was
born male and grew up in the land of Thebes. One day he was
walking in the countryside when he came upon two snakes copulating. He hit them with his staff and was instantly transformed

into a woman. Seven years later, Tiresias the woman was walking down this same path, when she observed the same pair of snakes copulating. Perhaps in the hope that the magical power would work in reverse, she hit the snakes with her walking stick *again* and was instantly transformed back into a man.

Hera and Zeus reasoned that Tiresias was the only human who had firsthand experience of the relative sexual pleasure of both man and woman. So they turned to him to resolve their debate. When Hera and Zeus popped their comparative pleasure question to Tiresias, he immediately responded that woman experiences *nine times* the sexual pleasure of man.

Why *nine* times the sexual pleasure? To the geometry-obsessed Greeks, the number 9 was a very special number indeed. Nine is 3^2. The number 9 tells us poetically that woman's sexual pleasure is not only greater in *magnitude* than man's but also greater in *dimension*. With a single symbolic number, Tiresias communicated that woman's sexual pleasure is a nonlinear, exponential increase over man's.

The myth of Tiresias reminds us that woman's sexual pleasure is possibly the most central and enduring mystery about sex. What is its purpose, and why does it exist? Yet even while attempting to deal with the evolution of female pleasure—including the female orgasm—the contemporary science of mate choice has been mute about the subjective experience of sexual pleasure. The theory of aesthetic evolution, however, has plenty to say about it, as do I, in this chapter. Viewing pleasure as the central, organizing force in mate choice, and mate choice as a major dynamic in evolutionary change, the aesthetic theory holds that women's pursuit of pleasure is at the very heart of the evolution of human beauty and sexuality.

The theory of aesthetic coevolution predicts that behind every elaborate sexual ornament, there is an equally elaborate, coevolved sexual preference. If the size and shape of the human penis evolved to fulfill an ornamental function, for example, then there must be a set of female preferences that coevolved with the evolutionary changes that occurred to the penis. As I proposed in the preceding chapter, those preferences had to do with the sensory experiences of enhanced sexual pleasure. And that leads us directly to the question of the female orgasm—its origins, its purpose—and finally to elaborate on the answer Tiresias supplied to Zeus and Hera, why it may be a more powerful and profound experience than the male orgasm.

Perhaps no topic in human sexual evolution has stimulated more scientific excitement and heated debate in recent decades than the origin of the female orgasm. The evolutionary explanation of the male orgasm has always seemed obvious; because the male orgasm is directly connected to the ejaculation of sperm, male sexual pleasure must have evolved, through natural selection, to motivate males to pursue reproductive opportunities. All in all, the male orgasm is a very tidy solution to the problem of how to keep the species going and in perfect keeping with the adaptationist point of view. In contrast, the origin and function of the female orgasm have been highly contested, with an abundance of theorists eager to supply possible explanations. What is surprising about these explanations of sexual pleasure, however, is how anhedonic they are.

In the early twentieth century, Sigmund Freud proposed a scientifically influential account of the female orgasm. He identified the clitoris as the location of infantile female sexual pleasure and the vagina as the appropriate location of mature female sexual pleasure. According to Freud, "normal" female sexual development required transitioning from the clitoral, masturbatory orgasm to vaginal orgasm achieved through heterosexual intercourse without clitoral stimulation. Women who failed to achieve the mythological transition were labeled as "frigid"—that is, sexually deficient, emotionally immature, not fully realized as "feminine."

Freud's hypothesis was influenced by the same autonomy-denying, anti-aesthetic intellectual tradition of Mivart and Wallace (see chapter 1), which viewed female sexual pleasure as merely an adaptive physiological stimulus to encourage and coordinate sexual behavior between the sexes and thereby ensure propagation of the species. Freud, Mivart, and Wallace all precluded the possibility that female sexual pleasure could be a goal in and of itself. As we have seen, Mivart was explicit in his antagonism toward female sexual autonomy. He was appalled by the very idea that "vicious feminine caprice" could have any evolutionary effects. Interestingly, Freud's failed theory of female orgasm might have been rooted in a similar anxiety about the consequences of recognizing the autonomy of women's sexual desires.

The modern scientific debate on the evolution of female orgasm began with Donald Symons's 1979 book, *The Evolution of Human Sexuality,* which proposed that the human female orgasm, like male nipples, evolved as a *by-product* of natural selection on sexual function in the opposite sex. The by-product theory holds that male nipples exist only because nipples are under strong natural selection in females; that is, they are necessary for nursing offspring. Similarly, the capacity for orgasm in females exists solely because orgasm is under strong natural selection in males; that is, it provides a mechanism for the delivery of sperm during copulation. Such by-products are able to arise because there is incomplete genetic and developmental differentiation between the sexes. Just as male nipples have the same evolutionary origin as female nipples, the female clitoris is homologous to the penis of males. So, Symons hypothesized, the female capacity for orgasmic sexual response is basically a happy accident—a by-product of natural selection for male sexual response.

Symons's by-product hypothesis was subsequently championed by the evolutionary biologist Stephen Jay Gould and the philosopher of science Elisabeth Lloyd. As Lloyd explained in an interview with the *Guardian,* "Male and female both have the same anatomical structure for two months in the embryo stage of growth, before the differences set in. The female gets the orgasm because the male will later need it, just as the male gets the nipples because the female will later need them."

The most persuasive evidence in favor of the by-product account is the simple fact that human copulation is by itself so ill-suited to eliciting female orgasm. There's also the fact that female orgasm is completely unrelated to female fertility. Women who have never had an orgasm during intercourse manage to produce babies just fine, as the by-product folks maintain, so orgasm cannot be seen as an adaptation to assist reproduction. The by-product account is further supported by the observation that female orgasm is broadly distributed in nonhuman primates— including stump-tailed macaques, chimpanzees, and bonobos. According to this model, there is nothing to explain evolutionarily in women. They come by their orgasmic capacity in exactly the same accidental way that other female primates do, and it has nothing to do with "adaptation."

Not surprisingly, the adaptationist sociobiologists of the 1980s and 1990s found the by-product account very unsatisfactory. In response, they proposed that female orgasm *is* an adaptation; that is, female orgasm *has* evolved by natural selection, its purpose being to support pair-bonding maintenance. Basically, this is the "good sex makes a happy marriage" hypothesis. However, pair-bonding hypotheses fell out of favor in the late 1980s when it was recognized that a female capacity for orgasm could be just as powerful a motivator for sexual liaisons outside the pair bond as within. This intellectual shift coincided with the discovery that many apparently "monogamous" birds are merely "socially monogamous"; that is, while they do form stable social pairs for parenting duties, they also mate extensively outside their social pairs. During the mid-1990s, this discovery led many of the members of an early generation of evolutionary psychologists to focus on the role of sperm competition in sexual evolution, which they eventually connected to theories about human female orgasm.

Positing that female orgasm has an important role to play in these "extra-pair" mating scenarios, they proposed that the uterine contractions that are part of the female orgasm are an adaptive mechanism that evolved in order to "upsuck [*sic*]" the sperm of genetically higher-quality males, making it more likely that the sperm of these superior males ends up fertilizing the ovum.

Who, then, are these higher-quality males whose sperm is so desirable? According to the standard evolutionary psychology scenario, this evolutionary mechanism acts because women are strategically and deceptively promiscuous; the woman's "social" mate is *not* the higher-quality male. Rather, the social mate is the one the female has chosen because he can provide the best direct benefits to her offspring in the form of resources, care, protection, and so on; he's the good ole reliable, but not very sexy, guy. It's the extra-pair mate she seeks out during her fertile period who is the higher-quality male—higher quality meaning that he's the sexy one and, being more attractive, the one she wants to father her children, because he can provide them with indirect benefits, that is, good genes. So the adaptationist theory is that a woman will have an orgasm *only* during sex with the more attractive, higher-genetic-quality man, because the upsucking mechanism of orgasm will give the advantage to his sperm and make it more likely that he will be the one to fertilize her eggs.

Elisabeth Lloyd presents a devastating case against the upsuck hypothesis in her book *The Case of the Female Orgasm*. She provides a comprehensive history of the contentious debate over female orgasm evolution, a review of the scientific and human sexology literature on human orgasm, and reams of data that make it clear that there is absolutely no support for the idea that female orgasm influences fertilization. Nor is there any evidence to suggest that males who induce women to orgasm are any more successful than other males at fertilizing the ova or that they are in any way genetically superior. If female orgasm has no effect on fertility or fecundity, and there's no correlation between male genetic quality and a man's ability to bring a woman to orgasm, then it is impossible to maintain that orgasm is a sperm-sorting adaptation to fine-tune the genetic quality of offspring. Lloyd goes on to document that critical papers in the upsuck literature rest on fundamentally flawed statistical methods and unjustifiable data manipulation and that many aspects of these studies have been influenced by the sexual biases of the researchers.

An important feature of the debate between the by-product and the upsuck accounts of female orgasm evolution is the way in which the *variability* of female orgasm is used as evidence by both schools of thought. In defending the by-product theory,

Lloyd proposes that the extreme variation among women in orgasmability during intercourse—with some women never having orgasms, others nearly always, and many others somewhere between these poles of experience—is profound evidence that orgasm is not under natural selection. If it were, natural selection would achieve more consistent results. If orgasm is not the result of evolved design, she proposes, then it should be viewed as an accident—albeit a very fortuitous one.

In contrast, the upsuck advocates maintain that variation is the very raison d'être of female orgasmic response—evidence in and of itself of female orgasm's adaptive function. As the evolutionary psychologist David Puts has written, the variation in women's orgasmability is a reflection of the variability in the "propitiousness of [their] mating circumstances." In other words, the higher the *woman's* mating value—that is, how sexually attractive *she* is—the higher the genetic quality of the male partners she can attract, and the more likely she is to orgasm during sex. More attractive females, who are of better genetic quality and condition, will attract more attractive males, who are also of better genetic quality, and these attractive males will more frequently induce those females to have orgasms, upsucking their higher-quality sperm to fertilize their higher-quality eggs. Thus, not only are beautiful women necessarily *better* (because they have better genes, health, status, and condition), but also they will be rewarded with greater sexual pleasure because of the higher genetic quality of the males they can attract as mates.

It would be hard to come up with an idea that would better reinforce the impression of a male bias in evolutionary psychology. The upsuck theory enshrines the fantasy of the superior male as the *proximate* and *ultimate* causal explanation of female orgasm itself.

A fundamental problem with the upsuck hypothesis is that it cannot explain why women vary in their *intrinsic* capacity to experience orgasm during intercourse—regardless of the attractiveness of the males they're having sex with. Recently, Kim Wallen and Elisabeth Lloyd published an article citing evidence for the possibility that frequency of orgasm during copulation may be related to women's genital anatomy. Based on their statistical analysis of historical data sets from the 1920s and 1940s—which, alas, are the only data available on this subject—Wallen and Lloyd propose that the closer the clitoris is to the vaginal opening, the greater a woman's capacity for orgasm during intercourse. This intrinsic, anatomical variability in women's capacity for orgasm during intercourse is congruent not just with the data they reviewed but with men's unscientific, anecdotal, personal experiences. After all, an individual man does not vary in genetic quality over time, but the frequency and ease with which the different women he has sex with over time experience orgasm during copulation *do* vary (no matter what he may say to the contrary). The upsuck hypothesis fails to explain this variation.

Another fundamental flaw in the upsuck theory is that it rests on the assumption of the importance of sperm competition which takes place only within a context of strategic female sexual promiscuity and deception. Upsuck theorists maintain that the female orgasm has evolved to meet the challenge of obtaining "good genes" when a woman is mating with multiple males of different genetic quality during her small window of fertility. If sperm competition does play a critical role in the evolution of female orgasm, as they posit, then the evolutionary elaboration of the female orgasm that has occurred in humans should be associated with increases in sperm competition. But this prediction is precisely the opposite of the story revealed by the comparative data. Testes size—the most reliable index of the evolutionary history of sperm competition—has significantly *decreased* in humans since our shared ancestry with chimpanzees, while the role of female orgasm in human sexuality has increased in importance. In contrast, chimpanzees have very large testes and strong sperm competition, and although chimpanzee females are capable of orgasm (as indicated by increased heart rate and rapid vaginal and uterine contractions), female orgasm apparently rarely occurs during sexual intercourse. Yet according to the upsuck theory, because female chimps mate with multiple males that vary in genetic quality, we should see female orgasm occurring as a sperm-sorting mechanism during chimpanzee copulations. But it does not.

Last, advocates of the upsuck hypothesis have surprisingly failed to think through the adaptive implications of their own model. If human female orgasm has evolved to mechanically increase the probability of fertilization, then human males should have evolved adaptive counterstrategies to *induce* these sperm-sucking orgasms during each and every copulation. What is human intelligence useful for if it cannot be applied by males to further their reproductive success? As a counterstrategy to female orgasmic sperm sorting, men should have evolved a universal, assiduous interest in women's sexual climaxes. As many a woman can attest, this has not happened. But the evidence goes beyond the anecdotal. Anthropological data from a range of cultures document that there are plenty of men who take little interest in women's sexual pleasure and orgasm. In many societies, men initiate sex with minimal foreplay and proceed to climax without ever concerning themselves with the woman's pleasure. In fact, in many cultures, men aren't even aware that it's possible for a woman to *have* an orgasm (or at least such knowledge was rare prior to the Internet). A 2000 survey found that 42 percent of college-educated Pakistani men did not know that women were capable of orgasm. Furthermore, many patriarchal cultures actively suppress women's capacity for orgasm through clitorectomy and other forms of female genital mutilation. The overwhelming indifference (not to mention frequent hostility) toward female sexual pleasure and orgasm by men in many of the world's cultures is a glaring explanatory failure of the upsuck theory.

There is still no resolution to the debate about the evolution of the female orgasm. The upsuck hypothesis has been thoroughly discredited. However, even though the fundamental data supporting the by-product account—that is, the genital homology between the sexes and the physiological similarities of male and female orgasmic response—are completely accurate, the question remains whether there is *more evolution* to be explained than the by-product account provides. Has human female orgasm evolved in its own right?

Interestingly, this issue has been raised by feminists who have argued that the by-product hypothesis marginalizes and trivializes the sexual agency of women, and I think they're onto something. Is the central place of sexual pleasure in many women's lives to be attributed to mere historical accident? Don't the prodigious qualities and potentials of female orgasm and sexual pleasure require a more *substantial* explanation than the by-product theory?

What has been missing from the debate is a genuinely Darwinian, aesthetic evolutionary perspective. There has been no direct intellectual engagement with the fundamental issue to be explained—women's subjective experiences of sexual pleasure. Both theories, in their different ways, marginalize and ignore female sexual pleasure as irrelevant to the historical, causal explanation of female orgasm.

It should not come as a surprise that science does such a poor job of explaining pleasure, because, as I discussed in the book's introduction, it's left the actual experience of pleasure out of the equation. The modern science of mate choice, in humans and other animals, has not been designed to address the question of sexual pleasure directly. Having grown out of the study of mate choice in other animal species, it simply can't. There is no way it can capture the pleasure that a female lyrebird experiences while listening to a male lyrebird's unremitting cascade of mimetic songs from his display mound or while watching the quivering veil of his gauzy tail feathers as he unfolds them over his body like half an umbrella. It cannot understand the aesthetic experience of a female Guianan Cock-of-the-Rock as she stands next to a screaming orange male sitting motionless on the bare dirt floor of his lek territory, the two of them surrounded by other screaming males lending their raucous cacophony to the courtship scene. The only thing we scientists can assess in these instances is the outcome—which mate did the female end up choosing? But by focusing solely on outcomes, biologists have obscured and ignored the richly pleasurable sensory and cognitive criteria that went into making the choice.

When it's our *own human* pleasures that we're investigating, however, we have an opportunity to understand sexual pleasure much more fully, because humans, unlike other animals, can tell us what they're experiencing. This ability to communicate can transform our analysis of the evolution of orgasm. It's time for evolutionary biology to embrace this opportunity. Fortunately, the theory of aesthetic evolution is uniquely well equipped to help us do so.

Aesthetic evolution explicitly addresses the subjective experience of the pleasure of mating preference. To understand the evolution of sexual pleasure, we need to create a corollary of the Beauty Happens hypothesis, which I will call the Pleasure Happens mechanism. In the Beauty Happens mechanism, the focus is on the coevolution of desire in one sex and the *physical objects* of desire in the other sex—in other words, the display traits. In the Pleasure Happens mechanism, we must focus on the coevolution of the *subjective experience of pleasure* with the features that elicit that pleasure. This means recognizing that the experience of mate choice is, in and of itself, pleasurable, something that is still rarely acknowledged in the scientific literature on mate choice. Darwin, however, proposed it.

Although Darwin was too proper, shy, or fearful of his audience's responses to explicitly discuss the sexual pleasure of humans in *The Descent of Man,* he did discuss sexual pleasure in animals, proposing that the sexual displays of animals evolve precisely because of the profound sensory pleasures they elicit. By the same reasoning, because female sexual pleasure and orgasm are fundamental components of the experience of mate choice in action—including all the physical interactions involved in sexual behavior—the exercise of sexual evaluation is inherently pleasurable. The pleasures that are part of it, including and especially the experience of orgasm, are the data upon which mate choice, or more to the point *remating choice* (see chapter 8), is made. Which leads us back to the question of how these pleasures evolved.

According to the Pleasure Happens hypothesis, female sexual pleasure and orgasm have evolved (that is, expanded in capacity and intensity since common ancestry with chimpanzees; evolutionary context 2) through indirect selection by women's mating preferences for those male traits and behaviors that they find sexually pleasurable. Because human mating preferences are largely remating preferences, based on repeated sexual encounters, female mate choice can encompass aesthetic evaluation of the physiological, sensory, and cognitive experiences of sex itself. As selection by female mating preferences gradually transformed male mating behavior, females' own capacity for subjective pleasure *coevolved* and expanded to become more complex, intense, and satisfying. To be as explicit as possible, the aesthetic proposal is that human female sexual pleasure and orgasm have evolved because females have preferred to mate, and *remate,* with males who stimulated their own sexual pleasure; females have thereby also selected indirectly for those genetic variations that contributed to the expansion of their own pleasure. By selecting on male traits and behavior that elicit orgasm more frequently, female mate choice has evolutionarily transformed the nature of female pleasure.

In the Pleasure Happens scenario, female orgasm is *not* an adaptation to accomplish any extrinsic, naturally selected function—sperm upsuck or anything else that adaptationists might come up with in their search for rhyme and reason. Nor is female orgasm merely a historical accident, second fiddle to male sexual pleasure. Rather, female sexual pleasure and orgasm are the evolutionary consequences of female desire and choice, and they are ends unto themselves.

The Pleasure Happens hypothesis of orgasm evolution is consistent with much of the evidence on female sexuality and sexual response—for example, its inherent variability. I agree with Elisabeth Lloyd's suggestion that the variability in female capacity for orgasm is an indicator that orgasm did not evolve by adaptive natural selection, because natural selection should result in much more reliable, highly functioning, and consistent experience. However, I disagree with the conclusion Lloyd then draws—that this means orgasm is simply a historical (but fortunate) accident. I think that human female orgasm is a highly evolved experience that is *about* something and has evolved *for* something. That "something" is pleasure, which evolves through the evolutionary action of their mate choices.

Although there is not yet enough comparative evidence about orgasm in various female monkeys and apes to support the conclusion that female orgasm has evolved or expanded in pleasure in humans since common ancestry with chimpanzees, I hope that proposing the Pleasure Happens hypothesis will lead to further investigations to test it. Until then, we can see that the Pleasure Happens hypothesis is congruent with lots of the current data. For example, the indirect sexual selection that drives the Pleasure Happens mechanism will be less efficient at evolutionary design than direct natural selection can be. In addition, female choice is not the only source of sexual selection in humans, so this mechanism may not predominate in determining the evolution of female sexuality. Thus, the Pleasure Happens mechanism is congruent with the fundamental variability of human female orgasm.

The hypothesis, furthermore, is supported by the existence of many evolved features of human sexuality that are different from our ape relatives and that can only be explained as expansions of sexual pleasure. For example, copulation duration in gorillas and chimpanzees is measured in seconds. On average, human copulation lasts for several minutes and of course can continue for far longer than that. (Much to the frustration of many women, however, the extensive variation in male copulation duration skews toward the short, chimpanzee end of the continuum.) These longer bouts of intercourse would enhance female stimulation and create greater likelihood of orgasm, but they would serve no adaptive function, because extending copulation duration cannot by itself increase fertilization success or make a male a winner at sperm competition. Any evolutionary explanation for longer copulation times in humans is inherently about enhancing the pleasurable sensory experience of sex.

Another piece of evidence that seems to suggest the primacy of female pleasure as the driving force in much of human sexual evolution is the diversity of copulatory positions. Male gorillas and chimpanzees generally mount the females from behind. Men and women are much more creative in their couplings, which is consistent with the aesthetic hypothesis that the evolution of our sexual repertoire is in service to the goal of expanding opportunities for clitoral stimulation and female pleasure. Likewise, the evolution of increases in copulation frequency, concealed ovulation, and the decoupling of sexuality from periods of female fertility all contributed to the expansion of the role of sexual behavior and sexual pleasure in the lives of human beings.

The aesthetic account is also completely consistent with the observation that female orgasm is unnecessary for procreation. Orgasm has no effect on female fecundity because it did not evolve for any adaptive purpose. The very fact that female orgasm is not *required* for anything is likely to explain both its variability *and* why it is so pleasurable. The female orgasm might have evolved to be so expansive and prodigious *because* it has no evolved function. It is sexual pleasure for its own sake, which has evolved purely as a consequence of women's pursuit of pleasure. In men, however, orgasm almost always occurs with ejaculation and is thus required for sexual reproduction. Consequently, the subjective experience of male orgasm is constrained by natural selection for a peristaltic pumping of semi-viscous seminal fluids up and down the vas deferens and out the urethra. Essentially, male orgasm is all about plumbing—moving stuff through tubes. And because of this ejaculation-orgasm connection, men need to replenish the seminal fluids produced by the prostate, the seminal vesicles, and the Cowper's gland before they can orgasm again. (Younger male readers may be alarmed to learn that this recovery period gets longer and longer with age.) Thus, the naturally selected physiological function of male orgasm places limits on the magnitude, frequency, and duration of male orgasmic pleasure.

By contrast, female orgasms are *not* constrained by design for any ancillary physiological function. Female orgasms do not need to deliver any goods or perform any task. The contractions of the vaginal, uterine, perineal, and abdominal muscles are all enlisted purely in the service of pleasure without the compromising constraints of fulfilling any other function. This helps to explain why many women are capable of rapidly repeated, multiple orgasms. Because women's orgasms do not need to accomplish anything beyond pleasure itself, women require no recovery period and have no limits on repeating the experience other than their own desire.

Thus, the aesthetic theory supports Tiresias's pronouncement. Because the female orgasm has evolved through a purely aesthetic evolutionary process of mate choice, women actually *do* have the capacity for greater sexual pleasure than men, and women's sexual pleasure *is* more expansive in quality as well as extent. When Beauty Happens, Pleasure Happens too.

The elaboration of the female orgasm in humans may be the greatest testament to the power of aesthetic evolution. And it may also be the premier example of the irrational exuberance of an aesthetic evolutionary bubble—evolution for no purpose other than the arbitrary pleasure of it. Fortunately, human orgasmic pleasure has not yet evolved to be so extreme that it has been countered by natural selection against having too much fun.

All this focus on women's sexual pleasure might have left the guys feeling left out, diminished, the magnitude of their own pleasure compared unfavorably with that of women, their orgasms denigrated to mere plumbing. But that doesn't mean that men don't have terrific sex. So why *are* men's orgasms so pleasurable? Recall that the male orgasm has always been explained as an adaptation to encourage males to pursue sexual opportunities. Natural selection for any behavior will often result in the evolution of physiological pleasure in that act. Animals need to eat, so eating when hungry has evolved to be rewarding, satisfying, and pleasurable. However, most men would agree, I think, that the pleasure of orgasm is far greater, more intense, and more rewarding than the pleasure of eating. So, I think it's fair to conclude that male orgasm is more pleasurable than it needs to be in order simply to ensure reproduction—that is, more pleasurable than natural selection alone can account for. This leads me to the conclusion that natural selection is not the only mechanism involved in the evolution of the human male orgasm and that aesthetic evolution has also played a significant role.

Although this is pretty speculative, I think it is clear that male orgasmic pleasure in humans has undergone an evolutionary expansion since the time of our shared ancestry with gorillas and chimpanzees. While other male apes pursue sexual opportunities with a fervor similar to men's, they certainly don't seem to enjoy sex as much as men do. The orgasms of male gorillas and chimpanzees do not appear to pack the same punch as those of human males. There is little foreplay, minimal touching, or even eye contact. After a brief moment of rapid thrusting, it's over and both male and female go back to sifting through the leaf litter. Consider also the fact that the length of time to orgasm in chimpanzees averages around seven seconds versus a few minutes in men. If the quality of orgasmic pleasure is correlated at all with the amount of time it takes to get there—a not unreasonable physiological conjecture—then men certainly experience more sexual pleasure than male chimpanzees.

If this is true, then we have to wonder why human male orgasmic pleasure evolved and how. The answer, again, is likely to be through aesthetic mate choice. Male chimps and gorillas are not sexually choosy, and they pounce on any sexual opportunity that arises. Without the involvement of mate choice, all evolutionary influence on sexual pleasure will be limited to the effects of natural selection alone. Humans, however, have evolved to be highly choosy. The history of mate choice in women *and* men, the evolutionary expansion of sexual behavior, copulatory frequency and duration, and so on have all created opportunities for the aesthetic coevolution and elaboration of male orgasmic pleasure as well. The evolutionary enhancement of men's sexual pleasure is a likely consequence of the fact that human males deviate from the evolutionary psychology stereotype of them as profligate purveyors of cheap sperm. It is only by eschewing some sexual opportunities in favor of others they prefer—in other words, it is only through the operation of mate choice—that human male sexual pleasure has been able to aesthetically coevolve beyond the baseline necessary for reproductive function.

The primary difference between the sexes may be that the evolution of male pleasure has been constrained by natural selec-

PLEASURE HAPPENS · 279

tion for plumbing functions, while female pleasure has not. In summary, human males and females are both a lot more sexually choosy than our close ape relatives, and the very fortunate evolutionary consequence of the choosiness we exercise in mate choice appears to be that we have evolved to experience a lot more sexual pleasure than they do.

Men and women are in this together, of course, and it seems probable to me that mutual mate choice, acting on many of the same pleasure-extending and pleasure-enhancing sexual interactions, has led to the elaboration of orgasm in *both* sexes. In his 2000 book, *The Mating Mind,* the evolutionary psychologist Geoffrey Miller also proposed a role for a Fisherian "runaway process" in the evolution of human orgasm. Perhaps out of discomfort with aesthetic thinking, however, Miller imagined the process as "a stimulatory arms race" between the penis and the clitoris. This unfortunately competitive and martial analogy obfuscates the expansive, pleasurable, sensory dimension of orgasm for both sexes. The changes in penis morphology and sexual behavior that have been driven by female desire have in no way diminished male sexual pleasure. Quite the opposite. Orgasm evolution is not the result of a war between the sexes; rather, it would be better compared to an aesthetic, coevolutionary lovefest.

Another way to describe the mechanism of mate choice is to say that aesthetic coevolution proceeds through the *sexual agency* of individuals. Thus, in a delightful and unexpectedly feminist fashion, the Pleasure Happens hypothesis identifies women as *the* active agents in the evolution of their own capacity for orgasmic pleasure. Women's orgasms are both the direct experiences *and* the evolved consequences of women getting what they want. In this way, every woman's orgasm is a celebration of the evolutionary history of woman's capacity to fulfill her expansive, and expanding, sexual desires.

Women's own sexual experiences might lead them to ask, "How could it be otherwise?"

CHAPTER 10

The *Lysistrata* Effect

We have all seen many *New Yorker* cartoons of a couple lying in a double bed. A bland piece of art hangs on the wall above the headboard, and matching lamps sit on the two bedside tables. From there, the details vary. Perhaps both people are wearing chaste pajamas, reading, and the sheets and blankets are perfectly smooth over their poignantly isolated bodies. Or the sheets are in disarray, their hair is tousled, and they are in postcoital reflection. Some couples are grinding through the later years of a difficult relationship. Others are young couples just negotiating their pair bond or engaged in a random hookup. In this moment, one of them makes a pithy, ironic, dreamy, poignant, exasperated, bitter, or wistful remark. The diversity of comments presents a microcosm of the cares, aspirations, obsessions, and desires of the modern (mostly white, heterosexual) couple.

The "Not tonight, . . ." cartoons form an entire subgenre:

She: Worse than a headache! I have three kids and a full-time job!

or

She: Not tonight, hon, I had a yogasm in class today.

The postcoital cartoons present an array of reflections upon intimacy, satisfaction, disappointment, infidelity, and the vagaries of desire. Some cartoons even parody the idea of honest Zahavian handicaps:

She: I faked my orgasm.

He: That's okay. This is a fake Rolex.

Others explore sexual disconnection. An attractive young couple lies separately in bed. He is looking at his iPad; she is wearing a fine negligee with her arms crossed.

She: Touch anywhere to begin.

Then there is the subgenre of infidelity cartoons. Woman lies in bed with another man when her husband in business suit walks into the bedroom.

She: Sorry, Burt . . . Outsourcing.

Like many good narratives, these cartoons embody conflict. These comedic scenes in couples' beds capture the primal human drama of sexual conflict. Of course, not all of the disagreements between partners are examples of sexual conflict in the evolutionary sense. We all have personal interests and desires that may be different from our partners'. However, it is easy to see that the explicitly reproductive dramas of sex, pairing, fidelity, child rearing, investment, divorce, and family life can be informatively understood as manifestations of the ancient and enduring evolutionary phenomenon of sexual conflict.

Sexual conflict occurs whenever the evolutionary interests of the sexes diverge in the context of reproduction. As in birds, human sexual conflict can occur over a wide range of issues including the number and identity of sexual partners, sexual fidelity, frequency of sex, types of sexual behaviors, control of fertilization, timing of reproduction, number of offspring, and how much each partner invests—in terms of energy, time, and resources—in the care of those offspring.

Of course, sexual reproduction is an intrinsically cooperative, self-sacrificing act at the genetic level. All sexual individuals must sacrifice half of their total genetic success by combining half of their genes with half of another individual's genes to produce each single offspring. This is the inescapable genetic cost of sexual reproduction. But the differences between the sexes, beginning with the difference between the size and number of the gametes and continuing through a whole cascade of anatomical, physiological, and behavioral characteristics that are necessary for sexual reproduction, create many opportunities for conflict.

Total reproductive success is a matter of how many offspring you have, how long they live, and how many offspring *they* will have, and so forth and so on. Of course, if sexual selection is occurring, then how attractive those offspring are can affect how many offspring they will have. For males or females, reproductive success may be maximized by having sex more or less frequently, by having more or fewer mates, by having more or fewer offspring, or by investing more or fewer resources in each one of them. It is easy to imagine how conflict between the sexes on all of these issues could arise.

Sexual conflict can result in sexual coercion—the use of force or intimidation to influence the outcome of sexual conflict. Sexual coercion is not limited to men, or even to one's own sexual behavior. At least some of the social conflicts that mothers- and fathers-in-law can create are sexual conflicts over the mate choices and other reproductive choices of their children. Humans are not alone in this regard. In colonial White-fronted Bee-Eaters (*Merops bullockoides*) in savannas of East Africa, sons often stay at home for a couple of breeding seasons to help their parents raise more siblings. In drought years, when their help is particularly needed, bee-eater parents will often harass and break up the attempts by their son to form a pair bond with a new would-be daughter-in-law so that the son will return to help at his parents' nest. This harassment includes disrupting their son's attempts to feed his mate and sitting and blocking the entrance to the new pair's nesting burrow. The result is a disruption of offspring sexual choice (that is, fewer grandchildren) for their own reproductive advantage (that is, more children).

As funny as the couple cartoons and mother-in-law jokes can be, in the real world sexual conflict is anything but humorous. The news is filled with dramatic and heart-wrenching stories about sexual violence, spousal abuse, genital mutilation, sex trafficking, child abandonment, rape, incest, and more. In this book, we've seen how mate choice has allowed females of different groups of birds to evolve various mechanisms that expand their sexual autonomy, reduce the efficacy of sexual coercion, and even reduce sexual violence itself. By exploring the history of sexual conflict in humans and our primate ancestors, we will discover that we have been shaped by a similar evolutionary struggle to resolve sexual conflict, overcome sexual coercion and violence, and expand human female sexual autonomy. Indeed, as we will see, the advance of sexual autonomy and the reduction of male sexual control might have been key innovations that made possible the evolution of many unique, complex features of human biology.

Now for a little duck sex redux.

Throughout this book, we have explored the coevolutionary "dance" between mate choice and aesthetic diversity. We have also seen how sexual coercion can challenge, constrain, disrupt, subvert, or undermine mate choice and how females have evolved means to advance their sexual autonomy in the face of persistent sexual violence and coercion.

In birds, there are basically two mechanisms at work in the evolution of female sexual autonomy. In many waterfowl, for example, females have evolved physical defense mechanisms to lower the effectiveness of forced copulation. Females with mutations for vaginal morphologies that prevent forced fertilization will have sons who inherit genes for their father's attractive traits. These females will therefore have greater reproductive success (that is, more grandchildren) because other females will be attracted to their sexy offspring. Or rather, they'll have such success *if* they're not grievously injured or killed.

Unfortunately, as we saw in chapter 5, the evolution of these elaborate vaginal morphologies has a big downside, because it has instigated a costly, ever-accelerating sexual arms race between the defensive capacities of females and the coercive tools and abilities of males. The reproductive success of the entire species suffers as a result.

Other birds, like the bowerbirds and manakins, have managed

to avoid a sexual arms race by using aesthetic mate choice itself to transform males in ways that facilitate female sexual autonomy. But it is important to note that the coevolutionary dances that have led to these restrictions on male coercive capacity have not resulted in female mating preferences for wimpy males that they can socially dominate or control. Instead, the females have continued to evolve preferences for high-energy males who perform dramatic, elaborate, complex, multisensory displays. From the female perspective, there is no evolutionary advantage to social control over males. The evolutionary advantage to females is the advancement of their freedom of sexual choice and the reproductive success, in the form of attractive offspring, that arises from that freedom. As we shall see, the same can be said, in a much more profound way, of the consequences of female sexual autonomy in humans.

Human sexuality has made a sharp break with the sexual habits of our primate ancestors. The "average" female old-world monkey lives a life of sexual subjugation with limited opportunity for real sexual autonomy. And unlike lekking bird species, where the females do all the work of incubating and raising the young but have evolved complete sexual autonomy, the female old-world monkeys get the worst of both worlds. Typically, the females make all the reproductive investment required for raising the young, while males invest exclusively in advancing themselves within the social hierarchy and, once they are in a dominant position, exploiting all the sexual opportunities they can.

Unfortunately for females, primate social hierarchies are inherently unstable. Younger, stronger males are always seeking social and physical opportunities to depose the dominant male within their social group. The results of this hierarchical instability for females are both shocking and instructive. When one male deposes the previously dominant male, he obviously gains new opportunities for advancing his own reproductive success through his newly won social and sexual control over the females. However, the new top guy cannot immediately capitalize on these reproductive opportunities, because at any given time most females in his group will be either pregnant or breast-feeding dependent young. Breast-feeding continues for months or even years, during which ovulation is suppressed and the females do not mate.

Consequently, males of many primate species have evolved to create new reproductive opportunities for themselves by *killing* all the dependent offspring of females when they gain control of the group. When a female's dependent child is killed, the fact that she is no longer breast-feeding will cause her to go into estrous, at which point she will resume mating. Infanticide is a selfish male solution to the problem of how to capitalize quickly on the advantages of having won the male-male competition. However, the results are devastating for female reproductive success and for the population as a whole. For example, in Chacma Baboons (*Papio hamadryas ursinus*) in Botswana, infanticide by males accounts for 38 percent of all infant mortality—as high as 75 percent in some years—and is more significant than any other cause of death.

While infanticide gives the new dominant male new opportunities for mating, the impact on the lifetime reproductive success of females is entirely and tremendously negative. Infanticide wastes all of the reproductive investment the female alone has made during the long period she spent gestating and breast-feeding that offspring. And because the maximum number of offspring she can have during her lifetime is fewer than ten, each child she loses to infanticide is a substantial blow to her ability to pass on her genes to the next generation.

Infanticide by males is a premier example of sexual conflict. It furthers the selfish reproductive interests of the dominant males at the expense of the reproductive interests of the females. However, this process is not just bad for the females of the species; it's inherently maladaptive, because the overall population of the species can be diminished as a result. Infanticide is not adaptive, because it does not improve the fit of the organism to its environment. Rather, infanticide evolves by male-male competition with each dominant male trying to gain advantage over other, previous males. But unlike the typical male-male battles like elk stags fighting it out with their antlers, infanticide is sexual conflict because it harms the evolutionary interests of females.

The bio-anthropologist and primatologist Sarah Blaffer Hrdy was among the first to theorize about female evolutionary response to infanticide in her 1981 book, *The Woman That Never Evolved*. At that time, female primates were often described as sexually and socially inert individuals that merely responded passively to male social dominance and hierarchy. Drawing from her years of work on langurs in India, Hrdy emphasized that female old-world monkeys are active, evolved agents in pursuit of their own social and sexual interests. Hrdy observed that in evolutionary response to infanticide, many female primates attempt to mate with multiple subdominant males during their estrous periods. Why? Hrdy hypothesized that a female mates multiply to convince other males that they *might be* the father of her offspring. Consequently, that male may be less likely to kill offspring that could be his. Thus, female monkeys evolved to mate promiscuously as a way of obtaining "insurance policies" against future infanticide should any of these males ascend to social dominance.

Like coevolved duck vaginal morphologies, this paternity assurance strategy proposed by Hrdy is a coevolved *defensive* response to sexual conflict. Primate females do not actually achieve sexual autonomy through multiple mating. Rather, they are making the best of a terrible situation. Females seek out multiple mates not because they prefer them but because females who mate multiply with socially ambitious males may prevent the future murder of their offspring. The primatology literature is filled with detailed descriptions of female strategies to deceive other males into imagining their paternity without threatening the dominant male's sexual control. But like the defensive vaginal morphology of the ducks, this defensive mating strategy also has a big downside, because it, too, initiates a violent sexual arms race. Dominant males will respond to a female's promiscuity with ever more aggressive efforts to control her reproductive life. These amped-up coercive strategies include mate guarding, violent physical punishment, and social intimidation. Sexually speaking, the average female old-world monkey is caught between a rock and a hard place. It cannot be that fun to be a female monkey.

Things have not improved much among most of our closest relatives within the African apes. Gorillas have similar male-dominated group structure, but usually with one large, dominant male in each multiple-female social group. Because the dominant male physically excludes all (or nearly all) other males from the group, there is little sexual conflict over mating. However, males still use violence to create an atmosphere of social intimidation that enforces their dominance. Thus, gorilla females who are newcomers to a group will receive *higher* rates of aggression from males, as one primate researcher has put it, as they "strive to develop new relationships with these new females." Some relationship!

If a new male gorilla takes over the group, or if a large group is broken up by a new male taking some of the females away into his own group, then male infanticide frequently occurs. How widespread exactly is very difficult to know, because unless you witness the actual murder, you can't be certain that an offspring that has suddenly gone missing or is observed to be dead was actually killed in a violent attack by the new male. Basically, when you study infanticide in apes, you are an infant homicide detective in a very leafy jungle where none of the witnesses will talk to you. It is a tough job. Still, knowledgeable estimates indicate that about one-third of all infant mortality among gorillas is due to sexually motivated male infanticide. This is a tremendous, maladaptive cost to the overall reproductive success of gorillas, which likely has a significant impact on their capacity for population growth.

Chimpanzees live in large groups consisting of both males and females that are subject to fission and fusion over periods of hours, days, or weeks. Within these social groups, there is a complex dominance hierarchy and extensive male sexual competition, which gives rise to sexual conflict over male paternity and female investment. When a female goes into estrous, her fertility is broadly advertised by a prominent perineal swelling. Multiple males initiate mating, and females acquiesce to all males. However, when her fertility reaches its peak on the tenth day of estrous, the dominant male increases his efforts to guard the female from other males and closely control her sexual behavior. As a result, even though females acquiesce in all mating attempts, the alpha males achieve about 50 percent of the fertilizations. It can also happen that a male-female pair temporarily leaves the larger social group to go off together during the female's estrous period, and this *consortship* may be an expression of female mate choice, but because males sometimes use violent attacks and intimidation to coerce females into joining them on these sexual sojourns, we don't know how many of them are truly freely chosen by the females. During the consortship other males cannot interfere with the couple, and paternity is thus assured.

Although forced copulation is essentially unknown in chimpanzees, it's not because females have sexual autonomy. Rather, it is because females effectively never limit sexual access to any soliciting male. As in gorillas, male violence toward females creates an atmosphere of sexual intimidation. In fact, when they are at the peak of their fertility, female chimpanzees associate most strongly with, and seek copulations more frequently from, those males that have been most aggressive toward them during the entire estrous period.

Infanticide by chimpanzee males is well documented, but most specific observations remain anecdotal. So, as with gorillas, actual estimates of infanticide in chimpanzees are tough to make, but male infanticide is clearly an omnipresent risk in the lives of chimpanzees and a serious challenge to the reproductive success of female chimps.

Like the chimpanzees, bonobos, or pygmy chimpanzees, also live in large groups consisting of multiple males and females, but their sexual behavior is very different from that of chimpanzees— indeed, from that of all other mammals. As we saw earlier, bonobos have evolved to use sexual behavior to mediate social conflicts, and they engage in sexual behaviors with individuals of both sexes across all age and social status categories. In general, males and females are social equals (or co-dominant) and share access to all ecological resources. Females have strong female- female social alliances or friendships. As a consequence, sexual coercion over fertilization is virtually nonexistent in bonobos, and there is no evidence of infanticide at all, nor of any other kind of extreme intra-group violence. However, as with chimpanzees and gorillas, it is the females of the species who do all the work of gestating and taking care of the young.

In summary, in our closest relatives—both species of chimpanzees—females are highly promiscuous (albeit for different reasons), only occasionally exhibit specific mating preferences, and contribute all the parental investment. However, only in chimpanzees do females face having their offspring killed by males.

Although sexual conflict and coercion are found in virtually every human society on the planet, they are very different in frequency, magnitude, and deadliness from what we see in the lives of our close relatives among the apes. The difference between us and most of our monkey and ape relatives is even more dramatic when we look specifically at infanticide by males. Viewed through the lens of human biology, the average male baboon, gorilla, or chimpanzee is an infanticidal maniac just waiting for his opportunity. Male infanticide accounts for 38 percent of infant deaths in baboons and approximately 33 percent in gorillas. What is common male monkey and ape behavior is almost unknown in any human society. Even though men are still responsible for the overwhelming majority of human violence, including the occasional death of children, human males simply *do not* murder young children for their own reproductive benefit. Actually, most of the anthropological literature on infanticide in humans is about infanticide by mothers.

The virtual elimination of male infanticide in humans constitutes a major evolutionary transition in primate biology. This transformation involved a reduction in male-male sexual competition and sexual coercion and a qualitative and quantitative advance in the sexual autonomy of females. How did this happen?

The question really is, under what conditions do males surrender their weapons? What evolutionary mechanism can counter the force of male-male competition that exacerbates sexual coercion? For us humans, the evolutionary stakes are high indeed. Most of the features that make us uniquely human—including intelligence, complex social awareness, cooperative social behavior, language, culture, and material culture—depend critically upon an extended period of child development and substantial ongoing parental investment. Growing a more complex brain to achieve all these innovative cognitive capacities takes more time and more parental investment. How could our human ancestors have evolved to invest *more* resources in each and every offspring if the most frequent cause of infant death is infanticide by male violence? The answer is that it could never have happened. An evolutionary solution to the infanticide problem was absolutely essential for the evolution of human biology.

The dominant view in evolutionary anthropology is that complex hominin social behavior evolved through an interplay between male-male competition and natural selection on foraging ecology—that is, more efficient and productive exploitation of food in the environment. For example, the evolutionary anthropologists and primatologists Brian Hare, Victoria Wobber, and Richard Wrangham have proposed that the distinctively mellow and cooperative behavioral temperament of bonobos has evolved through "self-domestication"—a process of ecological natural selection against aggression. They envision this process as being driven by the distinctive features of bonobo foraging ecology, such as the existence of higher-quality terrestrial herb food sources or a lack of competition from gorillas. Although the details are still unestablished, the idea is that more cooperative groups were more socially stable and were able to increase their overall ecological efficiency. In short, the "self-domestication" hypothesis proposes the evolution of social tolerance and cooperation as an ecological adaptation for the species rather than a transformation in the social and sexual behavior of males.

Hare and Michael Tomasello have taken the idea further, proposing that natural selection on foraging ecology could have favored lowered aggression and increased social tolerance in humans, too. Recognizing that bonobos and humans are historically independent in the way they evolved social cooperation, they argue that the social temperament of humans might have evolved by a similar mechanism of "self-domestication." However, Hare and Tomasello have a hard time documenting how human self-domestication could have actually worked. They conjecture that it could have involved cooperative aggression, in which multiple subordinates gang up to kill, ostracize, or punish overaggressive or despotic (male) individuals. But it is not clear why selection for cooperative aggression would not merely select for even greater aggression rather than disarmament. Furthermore, this mechanism for the origin of a cooperative social temperament *requires* the very cooperation they are trying to explain; that is, individuals need to be able to cooperate in order to gang up on someone whose aggression they are trying to contain. Last, they have not outlined the ecological circumstances that would have favored human self-domestication, upon which the hypothesis depends.

With few exceptions, human evolutionary biology has failed to incorporate the role of female mate choice, sexual conflict, and sexual autonomy into theories about human origins. Furthermore, it is critical to note that the evolution of human social intelligence and cooperation required the transformation of *male* aggression, *male* temperament, and the *male* behavior of infanticide specifically. Therefore, would it not make sense to explore those evolutionary mechanisms that explicitly focus on male violence and on those evolutionary agents who would benefit most from its transformation? In other words, females.

As with many of the fundamental questions concerning the evolution of human sexuality, we find again that the ancient Greeks did have some insight into this problem, which they expressed not in their scientific theories but in the genre of comedy. In Aristophanes's play Lysistrata (debuted 411 B.C.E.), the Athenian housewife Lysistrata enlists the women of the opposing city states of Athens and Sparta in a joint pledge to abstain from all sexual relations with their husbands and lovers until the men agree to negotiate a peace and put an end to the costly and harmful Peloponnesian War. The women's sex strike leads to a comic exacerbation of sexual conflict, followed ultimately by the men's complete capitulation to the women's terms. Peace is restored to Greece through the women's organized assertion of their own sexual autonomy.

Although the action in *Lysistrata* is not set in an evolutionary timescale, the play does make a few observations that are evolutionarily relevant. Women are far less tolerant of violence than men. Although more men than women die in this violence, women pay a high price in terms of their reproductive success because of their greater investment than men in raising the sons who die in war and other violence. Like infanticide, the loss of their children in war is a blow to their lifetime reproductive success. Furthermore, the comedy demonstrates that women's mating decisions can exert a powerful force to counteract the violence of manhood. The sex strike works because *all* the women of Athens and Sparta agree; it is consensus among women that gives them their strength. Lysistrata's mechanism to transform men is not merely sexual but explicitly *aesthetic*. In the drama, the women of Greece hold back from *choosing* to have sex until men transform themselves to be less aggressive. Lysistrata advises the women of Athens and Sparta that if their husbands force themselves upon them to not fight back and make sure the men enjoy it as little as possible. She proposes that men will soon become bored and will miss the full aesthetic engagement with consensual sex. Thus, the women seek to deny their men the coevolved *aesthetic* pleasure of sex if they are forced. Last, the women of Athens and Sparta are able to defuse male aggression *without* creating a costly, aggressive arms race.

So, in answer to the question "Under what conditions will males give up their weapons?" *Lysistrata* teaches us that the most efficient way to fight back against male violence is to hit men where they are most vulnerable—below the belt.

Which is exactly what I am hypothesizing as the evolutionary mechanism for lowered male aggression, cooperative social temperament, and social intelligence of humans. I think that these changes proceeded *not* by natural selection but by aesthetic *sexual selection* through female mate choice.

Here's how this could work: Imagine an ancestral hominid population in which fertilization is determined in part by violent male coercion and in part by female mating preferences on specific male display traits. As in bowerbirds or manakins, if a new female mating preference arose for a new version of a male display trait that coincidentally correlated with the expansion of female sexual autonomy—the protective bower for bowerbirds, the highly cooperative male social relationships in manakin leks—then these new mating preferences will continue to evolve because such traits and preferences would increase the frequency of uncoerced mate choices by all females in the population. In other words, female choice will further enhance female freedom of choice. Female preferences for these traits will erode the capacity of males to achieve fertilization through physical force and coercion, and an ever-greater proportion of fertilizations will occur through female choice. As we have seen with so many other physical and behavioral traits, the self-organizing mechanism of aesthetic coevolution will create a new feedback loop that reinforces the capacity of females to assert their mating choices in the face of sexual violence and coercion.

According to this hypothesis, human females transformed the nature of male social behavior by evolving to agree that male traits associated with aggression and sexual coercion were not sexy.

But if our ape ancestors lacked female mate choice, how did it ever get started in humans? Unfortunately, the origin of mate choice in humans would be very difficult to study because it likely occurred very early after our common ancestry with the chimpanzees. However, even though gorillas and chimpanzees didn't evolve much in the way of female mate choice, we can see in our ape ancestors the cognitive potential for it. People who are familiar with chimpanzees and gorillas, both in the wild and in captive situations, paint vivid portraits of their rich social personalities and their expression of strong personal likes and dislikes, which makes it clear that they are cognitively capable of recognizing and evaluating each other. In gorilla group fission or chimpanzee consortship, female apes can show some mate choice. Thus, female apes have the cognitive capacity for mating preference and choice, but they lack the social opportunity to act upon these desires. Regardless of the details of the ecological and social circumstances that would have made it possible for mate choice to emerge in our hominin ancestors, it is easy to imagine that early female hominins were capable of exercising mate choice as soon as they had the social opportunity to do so.

I have called this evolutionary mechanism *aesthetic remodeling* because it involves the use of aesthetic mate choice to transform, or remodel, males to be less coercive, disruptive, and violent. In humans, aesthetic remodeling involves a specific process of *aesthetic deweaponization*. Deweaponization is essentially the reduction of male armaments (which have evolved by the process of male-male competition) through female mate choice. Two primary examples of this process in human evolutionary history are physical traits—larger body size and elongated, razor-sharp canine teeth—that male primates use to assert violent control over each other and over females and their dependent young.

Although men still tend to be larger than women, our evolutionary history involves a tremendous reduction in sexual dimorphism in human body size—that is, a decrease in the difference in body mass between the sexes. Male orangutans and gorillas are gargantuan, averaging more than twice the size of the females of their species. Sexual dimorphism in body size is much less in chimpanzees and bonobos, with males being only about 25–35 percent larger than females. But the difference is even smaller in humans, with men averaging only 16 percent larger in body size than women. This amounts to a tremendous reduction in the physical advantage men would have in any conflict with women. Of course, men still *do* have a considerable advantage in physical conflicts with women due to body size alone. After all, boxing and wrestling weight classes, which are designed to ensure a fair fight, are based on body mass differences of only 2.5–5 percent. So, a 16 percent male body mass advantage is likely to be decisive in a physical fight.

Still, this notable reduction in human sexual size dimorphism is not merely accidental, because body size dimorphism typically gets *more* extreme, not less, as body size increases, and humans of both sexes have been evolving to be more massive since our common ancestry with the chimpanzees. (The observation that sexual differences in body size get even greater as body size increases is called Rensch's rule after the mammalogist Bernhard Rensch, who proposed it.)

Obviously, a female preference for greater equality in body size would have meant that males had less of a size advantage over females and that the females would have a better chance at resisting sexual coercion and other forms of violence. It is also possible that female mate choice for reduced size dimorphism resulted in correlated behavioral changes in males—specifically, a reduction in male aggression and an increase in male social tolerance. Interestingly, there is strong evidence in domestic dogs for this kind of genetic correlation between various *aesthetic* features such as curly tails, floppy ears, shorter snouts, and smaller teeth—precisely the things that humans find cute in dogs—and *behavioral* temperament, such as lower aggression, higher social tolerance, and heightened cognitive sensitivity to social cues. For example, a decades-long, Soviet-era experiment in the domestication of foxes that selected only for social tolerance ended up evolving foxes with the cute physical characteristics of domestic dogs. Closer to home, Hare, Wobber, and Wrangham point out that the evolutionary reduction in aggression in bonobos is associated with a host of other correlated changes in the species, including reduced sexual size dimorphism, infant-like pink lips in adults, slower social development, more passive responses to social stress, and greater sensitivity to human social cues (in captive experiments) than in the chimpanzee. So it is plausible that female mate choice on physical features of males—like body size—could also have had a strong evolutionary impact on the sexual and social behavior of males.

Another sexually dimorphic feature found in most old-world primates is an extreme difference in the morphology of the canine teeth (eyeteeth, or "fangs") of males and females. In macaques, baboons, orangutans, gorillas, and chimpanzees, male canine teeth are longer and have broader bases than female canines. These elongate canines are kept razor sharp by continuous honing against the third premolars of the lower jaw. As in old-world monkeys, the difference between male and female canines in orangutans and gorillas is extreme, indicating the importance of physical competition in male sexual success. Canine dimorphism is more moderate in both chimpanzee species, which is congruent with their smaller body size.

A simple glance at the smiling face of any man will document that a tremendous evolutionary reduction in male canine size has occurred since our common ancestry with other apes. The canines of men and women are virtually the same size, even though humans have increased in body size—another violation of Rensch's rule. The evolutionary decrease in canine dimorphism in hominins began soon after our common ancestry with the chimpanzees. The canines of *Sahelanthropus tchadensis* (7 million years ago) and *Ardipithecus ramidus* (4.4 million years ago) are less conical than those of chimpanzees and show no sign of the canine-premolar honing. By 3.2–3.5 million years ago, the time of the early hominid *Australopithecus afarensis*—

Variation in the sizes of canine teeth of a male lowland gorilla (left), chimpanzee (center), and human (right). *Photos by Shutterstock (left) and Ronan Donovan (center and right).*

the famous Lucy—canine dimorphism has diminished to what we see in modern *Homo sapiens*. Human paleontologists have traditionally tried to explain the reduction of male canines as an adaptation to chewing complex plant foods with side-to-side jaw motions in *Australopithecus afarensis*. However, it has recently become clear that canine reduction started much earlier in our evolutionary history and was already advanced in species like *Ardipithecus ramidus*, affectionately called Ardi, which entirely lacked the dietary specializations of *Australopithecus*. Thus, the absence of a solid adaptive, ecological, dietary explanation for hominin canine dimorphism reduction indicates that a new evolutionary hypothesis is needed—female mate choice.

The bottom line is that most male old-world monkeys and apes have deadly weapons in their mouths that females lack. Enlarged male canines are not ecological, foraging tools but social weapons that males use to assert sexual control. As Darwin hypothesized, these weapons have evolved not because of the advantages they provide in survival but through the sexual advantage they provide in aggressive control of female mates and other male rivals. Nonhuman male primates use these weapons in aggression toward other males, in violent coercion of females, and in infanticidal attacks on dependent offspring. A hamadryas baboon uses his extremely large canines to bite, or *threaten* to bite, estrous females he controls if any of them stray even slightly from his side or toward any of the roaming bachelor males of the band. Male mountain gorillas use canine teeth in male-male confrontations over group control and on dependent offspring of females in the group. In chimps, the repertoire of female-directed male aggression includes vicious biting.

Just like females' mating preferences for male body sizes closer to their own, female mating preference for deweaponized male canine teeth would have enhanced female freedom of choice. Reduction of male weapons would decrease the efficiency of male coercion and infanticide, giving females more opportunities to successfully choose their mates. Females that prefer males with smaller canines would receive the indirect, genetic benefits of having attractive offspring that other females would be more free to choose as their preferred mates. The result is an *aesthetic* expansion of female social and sexual autonomy.

Again, as with the reduction in size dimorphism, the aesthetic deweaponization would not result in emasculated, wimpy, or subordinate males. On the contrary, female mating preferences would continue to evolve for attractive male traits like male body proportions and vigorous sexual stimulation. There are no evolutionary advantages to individual female sexual control, only advantages to freedom of choice. Likewise, this entire process is not adaptive; that is, it would not lead to any better fit between the organism and its environment. Rather, it evolves because female sexual autonomy results in lower costs of male sexual coercion to females: that is, greater infant survival, lower direct harm to females, and enhanced population growth.

The aesthetic remodeling/deweaponization hypothesis of human evolution is speculative but plausible. The model provides an efficient explanation of many features of human evolution that still lack satisfying adaptive, ecological explanations, including the great reduction of sexual size dimorphism in humans, tremendous reduction in violent male sexual coercion including male infanticide, the expansion of female mate choice, and the evolution of male sexual ornaments. But can this model be tested? Is there any evidence to support or reject it?

The first challenge is to establish whether this hypothesis is feasible even in theory. Samuel Snow and I are working on a mathematical, genetic model of the aesthetic remodeling process showing that, given genetic variations in trait and preference, mutation for a display trait that incidentally advances female autonomy could indeed evolve. This model does not of course provide proof that such an evolutionary mechanism occurred in human evolution, only that it *could* have.

Probably the strongest contemporary evidence in support of the idea that female aesthetic remodeling did occur in human males comes from data showing that the average mating preferences of contemporary women do not skew toward those traits that are associated with male physical dominance. Rather, as we discussed in chapter 8, on average women prefer features in the middle of the "masculinity" spectrum—that is, slimmer, less muscular bodies, less prominent brows, and moderate amounts of facial and body hair. The fact that these more masculine traits persist among men indicates that other evolutionary forces, perhaps male-male competition, have favored more masculine features.

Other, more detailed tests of the aesthetic deweaponization hypothesis will become possible as evolutionary anthropologists bring the concepts of aesthetic evolution, sexual autonomy, and aesthetic remodeling to bear on their analyses of comparative behavioral ecology of primates, the fossil record of human evolution, evolutionary archaeology, and comparative human anthropology. What is clear now is that the dominant view of hominin evolution as an interplay between male-male competition and adaptive, ecological natural selection is insufficient to explain the key innovations that have occurred in the evolution of human cognitive, social, and cultural complexity. By including aesthetic female mate choice, sexual coercion, and female sexual autonomy in the evolution of humans, I think we arrive at a better account of how we have become human.

This exploration of human sexual conflict has focused so far on sexual conflict over fertilization—who will determine the paternity of the offspring. Yet sexual conflict can also arise over who will take care of those offspring after they are born and how much energy, time, and resources each parent will invest in the care of those offspring. After the evolutionary reduction of male infanticide, humans involved a *second* major advance in female interests, in the ongoing sexual conflict over parental investment. Among most old-world monkeys, orangutans, gorillas, and chimpanzees, males provide essentially *no* paternal care to their offspring. Even among the remarkably peaceful and egalitarian bonobos, males make no paternal investments beyond sharing food, which they already do freely with other individuals within the group. In all these species, females are losing the battle over having the males share in parental investment. Indeed, in these primate species, there doesn't even appear to be any overt sexual conflict over investment because females are making *all* parental investment themselves. Obviously, humans aren't like that. In virtually every human society and circumstance, men make substantial investments in their offspring in the form of food, economic resources, protection, paternal identification, and emotional engagement. Consistent biparental care would have been even more essential in our preagricultural, evolutionary past. Thus, the distinctively human pattern of collaborative child care is another major innovation in human reproductive biology that requires an evolutionary explanation.

It is possible that once our female ancestors gained enough sexual autonomy to substantially reduce or eliminate infanticide by males, they began to use mate choice to make additional gains in other areas of their ongoing sexual conflict with males. Specifically, female choice expanded from the immediately perceivable physical features of potential mates to encompass the broader social personality and social relationship experience, ultimately resulting in the evolution of male *paternal* investment. This transformation was accompanied by the aesthetic expansion of sexual intercourse itself to become more frequent, longer lasting, more variable and complex, more pleasurable and engaging, less related to reproduction, more obscured from paternity (through concealed ovulation), *and* entwined with new emotional content and meaning. Through female choice for more socially engaging, and interpersonally engaged, male partners, males gradually evolved to make new paternal investments—of food, protection, and a cooperative social relationship—in their mates and their mates' offspring. Ultimately, male reproductive investment evolved because of male competition—that is, competition to please choosy females and thereby gain enduring sexual access and social relationships that come with the pair bond.

Of course, male reproductive investment in offspring could provide decisive improvements in the health, well-being, and survival of a female's offspring, helping them to survive until they reach their own sexual maturity and reproductive years. It could also improve a female's survival, well-being, and fecundity, help decrease her interbirth interval (which is significantly shorter in humans than in other apes), and expand her lifetime reproductive success. This decrease is precisely why we humans managed to greatly increase our capacity for population growth over other apes. Thus, obtaining male parental care was an adaptive *direct benefit* to female mate choice.

At this stage in human evolution, mate choice evolved to advance through a series of mutual social and emotional interactions between people, during which we gained the opportunity to scrutinize and assess the social, emotional, and even psychological attributes that are important to us *individually* in our search for a suitable mate. For this evolutionary reason, developing an enduring sexual bond is not the result of a hardball, legalistic negotiation dictated by game theory. This is why prenuptial agreements are so unromantic and offensive. Rather, falling in love is a deeply *aesthetic* experience that involves mutual social, cognitive, *and* physical seduction.

This evolutionary model implies that the human pair bond did not evolve through the assertion of coercive male control over female reproductive freedom, as some cultural theorists propose. In other words, the human pair bond does not constitute a male harem size of one. Rather, it evolved through a distinct evolutionary advance of female interests in their sexual conflict with males over paternal investment. Ultimately, the human pair bond is an aesthetically coevolved social relationship through which females and males have advanced their mutual reproductive interests. Of course, human pair bonds have never been absolute or inviolate. This is not a theory of the evolution of monogamy, till death do us part. To have evolved, pair bonds need only persist long enough to have a decisive positive impact on offspring development and survival. At some point in the evolution of male reproductive investment, cultural evolution began, and a whole new set of social complexities and variations arose.

To put it plainly, the evolution of human paternal care is a really big deal. Male investment in parenting is rare in primates, and in mammals generally. Paternal care has been especially important in human evolution because human offspring require so much care and investment, take longer to mature, and have many more complex social, cultural, and cognitive developmental challenges than other primates face. After the infanticide problem was solved, I think the next most important evolutionary challenge in the origin of human cognitive and cultural complexity was the origin of paternal care. Interestingly, this second major evolutionary transformation also involved the expansion of female interests in sexual conflict.

I think a very powerful case can be made for the role of female mate choice in the evolution of the human species. Solving the evolutionary challenge of male sexual violence, coercion, and infanticide through an aesthetic remodeling of maleness would certainly have given females much greater sexual autonomy. But male deweaponization could also have been the key innovation responsible for the subsequent evolution of human social, cognitive, and cultural complexity. Less aggressive, more cooperative males living in ongoing relationships with females would have created an environment of greater social stability for their developing offspring, which in turn would have made possible the longer development times and greater investment in each offspring that were required for the evolution of all the qualities we prize as evidence of our humanness—intelligence, social cognition, language, cooperation, culture, material culture, and ultimately technology. This new view of human evolution requires much work to test, but the stakes couldn't be higher.

CHAPTER 11

The Queering of *Homo sapiens*

For decades, those iconic *New Yorker* couple-in-bed cartoons portrayed exclusively heterosexual couples. Like many American cultural institutions, however, *The New Yorker* has slowly begun to acknowledge the existence of gay and lesbian couples and to include them in occasional couple-in-bed cartoons. The first ones have been quite prim in comparison to the frequent depictions of postcoital heterosexual couples under rumpled sheets, and, indeed, one of the earliest gay-couple-in-bed cartoons insightfully explored the anxiety surrounding the awkward intercultural negotiation required to bring the simple fact of gay couples sharing a bed into the public conversation. In a brilliant 1999 cartoon by William Haefeli, two men, fully clothed in winter overcoats, are lying next to each other on a smallish, bare mattress among many in a big department store showroom. One man comments to his partner, "I still think we should get a queen-sized mattress—despite the obvious jokes it will invite among the sales staff."

Like traditional *New Yorker* cartoons, the previous chapters on the evolution of human sexuality in this book could be viewed as reinforcing a heteronormative concept of "human nature"—the idea that heterosexuality is the only "natural" human sexual

behavior, the only one that is somehow sanctioned by evolutionary science. However, diversity of sexual preference is a profoundly human characteristic that must be accounted for in any
natural history of human desire.

Sexual diversity poses distinct challenges to evolutionary
explanation. How can evolution account for sexual behavior that
is not directly related to reproduction—sperm meets egg? One of
the most exciting aspects of this emerging theory of aesthetic evolution is the possibility that it sheds light on this enduring mystery
of variation in human sexual desire. Understanding the origin of
variations in sexual desire requires that we focus specifically on
the evolution of the *subjective* desires of individuals—that is, the
individual aesthetic experiences of sexual attraction.

I will not be talking here about the evolution of sexual
identity—that is, the conceptual categories of heterosexuality, homosexuality, bisexuality, and so on. The idea that sexual
behavior is a marker or definition of a person's identity is actually a quite modern, cultural invention—perhaps only 150 years
old. Because we live in a society that is accustomed to conceiving
of sexual behavior in terms of sexual identity, we tend to think
that sexual identity categories are biologically real and, therefore, require scientific explanation. The problem is that scientific
research on the origin of "homosexuality" seeks to explain the
evolution of a social construct. As David Halperin, professor of
English at the University of Michigan, explained to me, "Proposing a theory about the evolution of homosexuality is like proposing a theory of the evolution of hipsters or yuppies!" Sure enough,
an ample scientific literature on the "evolution of homosexuality"
gets this issue mostly wrong and has undermined itself as a result.

Rather, here I will explore the biological and evolutionary history of human same-sex sexual behavior, or same-sex behavior
for short. Specifically, I want to investigate evolutionary changes
in the diversity of sexual desire and behavior in humans after our
common ancestry with the chimpanzees and before the modern
cultural construction of sexual identity (evolutionary context 2,
see fig. 2, p. 229). Throughout this discussion, however, it will
be important to remember that just like many nonreproductive

sexual acts—kissing, caressing, oral sex, and so on—same-sex behavior is still *sex*, even if it doesn't involve sperm meets egg.

Human sexual preferences form a continuum—from people who engage in exclusively same-sex behavior, to people who engage in such behaviors usually, sometimes, or rarely, to people who engage in exclusively opposite-sex behavior. As with many other complex human traits, any genetic influences on human sexual preference come from *many* different genetic variations at *many* different genes, which interact with each other and with the environment in complex ways during development. As a result, the breadth and specificity of the resulting sexual preferences, attractions, desires, and behavioral responses will vary greatly. Where individuals land on the continuum of variation will depend in part on the combined effects of these many small genetic influences, and many social, environmental, and cultural influences as well.

An even more basic problem with most of the current scientific literature on the evolution of human "homosexuality" is that it begins with the assumption that there *is* an evolutionary conundrum. However, prior to the introduction of modern concepts of sexual identity, it is not at all clear that same-sex preferences were associated with lowered reproductive success at all. Humans have evolved to engage in sex more frequently, for greater duration, with greater pleasure, and in a greater variety of ways than did our ape ancestors, and many of the resulting sexual behaviors do not contribute to reproduction directly, yet they are perfectly consistent with reproductive success. Do heterosexuals who engage in oral sex have lower reproductive success than those who don't? Obviously, that is a pretty silly question, and there is no reason to think so. But the issue is nearly the same when considering same-sex behavior. By trying to find an evolutionary explanation for something that is a cultural category, instead of investigating the evolutionary origins and maintenance of variation of the subjective experience of sexual attraction—sexual desire itself—much of this previous evolutionary research has simply missed the boat.

Until now, most theories of the evolution of same-sex behavior have tried to explain it by proposing adaptive solutions to the proposed loss of reproductive success. For example, it has been widely hypothesized that individuals with same-sex preferences could contribute to the survival and reproductive success of *other* related individuals from within their extended families. This *kin selection* hypothesis proposes that same-sex behavior persists because nonreproductive individuals with same-sex preferences contribute substantially to the care of their younger siblings, nieces, nephews, cousins, and so on. Because these "Helpful Uncles" or "Aunts" share genes with their kin, it's possible that copies of the genes that contribute to same-sex preferences will be passed on indirectly to the next generation through these other family members.

The problem with the "Helpful Uncle" hypothesis is that there is no obvious correlation between same-sex attraction and the inclination to help raise one's relatives' children. And the kin selection hypothesis entirely fails to account for the most salient fact requiring an evolutionary explanation—the variation in human sexual desire.

In short, there is no evidence that same-sex behavior per se contributes to making a reproductive investment in related offspring. A more direct route to such investment would be to evolve *asexual individuals* who engage in no sexual behavior at all—as in female worker castes in ants and bees. But the *absence* of sexual desire is the exact opposite of the phenomenon needing to be explained in same-sex behavior. Kin selection arguments fail to answer the core question of how variations in sexual desire itself could have evolved and persisted.

Here, I propose that human same-sex behavior, like many of the sexual traits and behaviors discussed in the preceding three chapters, might have evolved through female mate choice as a mechanism to advance female sexual autonomy and to reduce sexual conflict over fertilization and parental care. According to this aesthetic hypothesis, the existence of same-sex behavior in humans is another evolutionary response to the persistent primate problem of male sexual coercion. Although I think that *all* human same-sex behavior might have evolved to provide females with greater autonomy and freedom of sexual choice, I address the evolution of female same-sex behavior and male same-sex behavior separately because I think that their evolutionary mechanisms differ substantially in detail.

To start off, we need to understand that the social and sexual behavior of primates is greatly influenced by which sex leaves the social group into which it is born when it reaches the age of sexual maturity. The movement of young adults out of one social group into another is necessary to prevent genetic inbreeding. Many primate species accomplish this by the traditional mammalian pattern in which it's the males who disperse among social groups at sexual maturity, while the females stay at home in their natal groups. However, African apes and a few other old-world monkeys have evolved the opposite pattern, of female dispersal among social groups. Female dispersal is the ancestral condition for humans, and it continues in many human cultures in the world today. A fundamental consequence of female dispersal is that young female apes must disconnect from their natal social networks when they go out into the world to join their new social groups. Thus, all primate females within female-dispersal societies begin their sexual lives at a profound social disadvantage because of the lack of social support of developed social networks to help them combat male sexual coercion and social intimidation. After dispersal, females must forge new social networks to help them mitigate the various dangers of sexual coercion.

Even when females stay in their natal groups, they have to construct protective social networks. In baboons, for example, the primatologist Barbara Smuts and others have shown that male friends help protect the females' offspring from marauding infanticidal males. More recently, the bio-anthropologist Joan Silk and colleagues have shown that female-female friendships contribute to protection of each other's offspring against infanticide and other threats.

Because female primates use friendships to construct these

mutually supportive, protective social networks, I hypothesize that female same-sex behavior in humans evolved as a way to construct and strengthen new female-female social alliances and make up for the ones that were lost when the females left their original, natal social groups. Natural selection on female preferences for same-sex sexual behaviors would result in females who have stronger social bonds with each other, allowing them to exert more effective defenses against male sexual coercion, including infanticide, violence, and social intimidation. According to this hypothesis, female same-sex behavior is a defensive, aesthetic, *and* adaptive response to the direct and indirect costs of coercive male control over reproduction. It's defensive in that it functions to mitigate the costs of sexual coercion to female reproductive success directly. It's aesthetic because it involves evolution of female sexual preferences. And it's adaptive because it would evolve by natural selection on female preferences to minimize both the direct costs of sexual coercion, in the form of violence and infanticide, and the indirect costs, in the form of restricted female mate choice and coerced fertilizations.

Male same-sex sexual behavior in humans might *also* have evolved to advance female sexual autonomy, but by a different evolutionary mechanism, I think. I propose that human male same-sex behavior evolved through an extension of the process of the aesthetic remodeling of maleness that we discussed in chapters 6, 7, and 10. This aesthetic evolutionary proposal maintains that female mate choice has acted not only on male physical features but also on male social traits, in such a way as to remodel male behavior and, secondarily, to transform male-male social relationships. In other words, selection for the aesthetic, prosocial personality features that females preferred in their mates also contributed, incidentally, to the evolution of broader male sexual desires, including male same-sex preferences and behavior.

Once male same-sex behavior evolved within a population, it would advance female sexual autonomy in a number of ways. I suggest first that even if relatively few males within a social group had same-sex attractions, this could result in substantial changes in the social environment. As some males evolved same-sex sexual preferences, the increased breadth of male sexual outlets could lessen the intensity of male interest, and investment, in sexual and social control over females and diminish the ferocity of male-male sexual competition. Because male sexual competitors might also be sexual partners, this could further minimize their competitiveness with each other without necessarily producing any loss in their reproductive success. In fact, I'm proposing that the evolutionary changes in male sexual preferences occurred specifically because males with traits that are associated with same-sex preferences were *preferred* as mates by females. So, there is no reason to believe that their reproductive success would be compromised at all. Once the majority of human sexual behavior has evolved to be nonreproductive and unhinged from the confines of the female's brief fertile period, then same-sex attraction can be seen as just a further broadening of sexual behavior and its social functions.

Second, male same-sex behavior could have fostered the subsequent evolution of less aggressive, more cooperative social relationships among males *outside* the context of sexual behavior. These same-sex relationships could have contributed to the development of collaborative hunting, defense, and other mutually and societally beneficial behaviors—exactly the suite of social behaviors that the human "self-domestication" hypothesis was designed to explain (see chapter 10).

Third, as female aesthetic preferences continued to coevolve with male traits associated with broader male sexual preferences, the process of aesthetic remodeling could have resulted in a minority of males with predominantly, or even exclusively, same-sex sexual preferences. These males could have then contributed to supportive and protective nonsexual relationships with females. (Of course, the exclusivity of sexual preference prior to the invention of the concept of sexual identity is an open question.) If the genetic influences on sexual preference are, like other complex human traits, a result of many variations of small effect at many different genes, then some male offspring will inherit a larger than average number of the diverse genetic variations involved in the social behavior traits that females find attractive. These individuals would end up at one end of the sexual preference continuum with predominantly or exclusively same-sex preferences and would be available for nonreproductive, noncompetitive, and noncoercive social alliances with females in their social groups. In baboon society, male-female friendships function in this way to defend females from physical attack, prevent infanticide, and advance the social interests of the female and her offspring within the group social network. Thus, I suggest that social alliances between males with predominantly same-sex sexual preferences and females—what we would call gay-male-straight-female friendships—may be not an accidental, or purely cultural, feature of human variation in sexual preference but an *evolved function* of human sexual variation.

Any losses to male reproductive success resulting from the evolution of same-sex preferences do not create an evolutionary conundrum, because female mate choice necessarily results in variation in male reproductive success. There are always winners and losers in the mate choice game. Any possible losses in male reproductive success merely demonstrate that male same-sex preferences have evolved not as an adaptation for males but to advance female sexual autonomy.

In the previous chapter, I proposed that human evolution has been greatly influenced by female aesthetic remodeling of maleness in ways that advanced the progress of female freedom of choice. Here, I am suggesting that male same-sex behavior has evolved by an extension of this same process. Again, this hypothesis *does not* imply that male same-sex behavior evolved through female preferences for weaker, subservient, feminized, or emasculated males that females can socially or physically dominate, though these female mating preferences *will* reduce the ability of males to dominate *females* of future generations. Rather, this female choice mechanism will lower the total effectiveness of coercive male sexual control and thereby increase the proportion of future fertilizations that occur due to mate choice. Ever-lower rates of sexual coercion will be associated with an ever-higher likelihood of the success of female choice, which will result in a sexual autonomy snowball.

The aesthetic theory of the evolution of male same-sex behavior does not imply that men with a predominantly same-sex orientation have any physical or social personality traits that differ from those of other males. Exactly the contrary, in fact. The hypothesis maintains that there is nothing distinctive about such men, because the features that evolved along with same-sex preferences have become a typical component of human maleness in general. Therefore, individuals with exclusively same-sex sexual preferences are distinctive only in the exclusivity, not in the existence, of their same-sex desires.

These aesthetic theories of the evolution of human same-sex behavior are, of course, highly speculative. However, I think that this speculation is responsible and warranted because of the fundamental importance of the question, the failure of current adaptive explanations to address the evolution of same-sex desire directly, and the unfortunate impact the current adaptive theories have already had on the public and cultural discourse on human sexuality, especially by reinforcing the tendency to view ourselves merely as (flawed) sexual objects rather than as autonomous and deserving sexual subjects. Clearly, there is a need for a new evolutionary theory on this question. We can, however, put these aesthetic hypotheses to the test by examining both their plausibility and their congruence with current data on sexuality in both human and nonhuman animals. To begin, I will evaluate their plausibility first by examining their assumptions.

For example, these aesthetic evolutionary theories assume the existence of heritable genetic variations in sexual preference and in behavior traits related to sexual preference. Like many other social behavior traits in humans, there is good evidence that predominantly same-sex sexual preference—that is, self-identified homosexuality—is strongly heritable.

In the case of the evolution of same-sex sexual behavior in females, the plausibility of the evolutionary mechanism of natural selection for female social alliances is well established in general. So, this proposal merely requires the application of a well-known evolutionary mechanism in a new context.

However, the hypothesis that female mate choice can result in the evolution of male social behavior in ways that expand female sexual autonomy is a new idea. Sam Snow and I are developing a mathematical, genetic model that will establish the efficacy of the aesthetic remodeling mechanism as proposed in bowerbirds, manakins, and humans. Such models establish that an evolutionary mechanism could occur under certain realistic assumptions.

The aesthetic theory proposes that female mate choice can also transform male social behavior in ways that extend beyond males' social interactions with females, which is exactly the kind of process we've seen in lekking birds. Female mate choice in manakins has transformed the nature of male social competition so that *bromance* is key to success in *romance*. Same-sex behavior in human males may be another form of this female-driven aesthetic remodeling of male social relations, another evolutionary solution to the problem of male sexual coercion.

Strong evidence in support of the proposed anti-coercion, social functions of same-sex behavior in humans comes from the bonobos, who are among our closest relatives on the Tree of Life. Bonobos are well-known for having frequent, promiscuous, mostly nonreproductive sex, including extensive same-sex behavior. Sex in bonobos mediates social conflicts of various kinds (especially conflicts over food). As a result, bonobo society is remarkably egalitarian and peaceful. Bonobos are notable for the nearly complete absence of sexual coercion, even though male bonobos have a greater physical size advantage over females than do human males. Thus, bonobos demonstrate that same-sex behavior *can* function to undermine male sexual hierarchy and coercive sexual control in primates, that female same-sex behavior *can* strengthen female social alliance and reduce sexual and social competition among males, and that male same-sex behavior *can* lower competition and enhance group social cohesion. Despite these similarities in social function, however, same-sex behavior in bonobos has evolved independently of humans and by a very different mechanism.

The aesthetic hypothesis for the evolution of same-sex behavior is also congruent with what we know about the evolutionary elaboration of human sexuality *since* our common ancestry with bonobos and chimpanzees. Gorillas and chimpanzees pursue all available sexual opportunities with females, but *only* during their brief fertile periods; by contrast, human males are both sexually choosy *and* interested in sex outside the context of the narrow window of female estrous. Similarly, other female apes exercise little in the way of mate choice, but human females have evolved to be highly selective.

Human evolution has also involved many other changes to sexual behavior. There has been not only an increase in the frequency of sexual behavior beyond the limited period of female fertility but a broadening and deepening of its sensory and emotional content. As sexual behavior among humans evolved to serve social as well as reproductive functions, it could have expanded to function in same-sex relationships. The evolution of concealed ovulation and the expansive aesthetic evolution of sexual pleasure would also have furthered the process of decoupling sexual behavior from reproduction in humans.

Previous theories about the evolution of human same-sex behavior have either focused only on male same-sex behavior or lumped female and male same-sex behavior together as a single phenomenon. By contrast, these hypotheses propose that there are different evolutionary mechanisms for same-sex behavior in the two sexes. Because these mechanisms are distinct from each other, we can predict, accordingly, that there should be differences in the frequency and social function of these same-sex behaviors.

For example, because male same-sex behavior evolves via sexual selection for the advantages it provides to females, not to males, the possibility of evolving nonreproductive individuals is not an evolutionary conundrum but an expected outcome of sexual selection. In contrast, natural selection for alliance-building same-sex preferences in females should not result in any significant losses to female reproductive success. Accordingly, the frequency of individuals with exclusively same-sex preferences should evolve to be much higher among males than in females. Indeed, this prediction is borne out by the evidence that exclusive homosexual identity is about twice as frequent in men as it is in women. The aesthetic remodeling mechanism hypothesizes that the physical and social personality features that are associated with male same-sex preferences have evolved precisely because these traits are preferred by females. Consequently, even though the evolution of same-sex preferences could result in losses to individual reproductive success of some males, these losses will arise because of the exclusivity of their same-sex preferences, not because these males would fail to succeed in attracting female mates. As noted earlier, there is nothing distinctive about such males, because the features that evolved along with same-sex preferences will have become a typical component of human maleness in general. This prediction is consistent with the data (discussed in chapter 8), which indicate that women generally prefer men who have physical features that are somewhere in the middle of the "masculinity" spectrum. This prediction is also consistent with the observation that most men with predominantly same-sex preferences would be perfectly successful at obtaining female mates if they preferred them.

The sexual autonomy hypothesis also predicts that the capacity for broad, nonexclusive, same-sex sexual attraction should be quite common in humans, if not ubiquitous. This prediction is difficult to test because of the long history of moral and social condemnation of same-sex behavior in many cultures. We cannot yet say how most people would behave in the absence of such strong cultural discouragement. However, there is ample reason to believe that same-sex attraction is quite common. For example, in the 1940s and 1950s, Alfred Kinsey found that 37 percent of men and 13 percent of women, based on samples of over five thousand each, had some experience of same-sex behavior culminating in orgasm. We know that Kinsey's samples were not representative of the American population as a whole. Nevertheless, Kinsey presented clear evidence that same-sex attraction and sexual experience are much more frequent in occurrence than is represented by the relatively tiny percentage of people who identify as having exclusively same-sex preferences. The biological capacity for same-sex attraction appears to be broadly distributed in human beings of both sexes.

Furthermore, same-sex behavior is a common occurrence in certain cultures and institutions in which it is not condemned or suppressed. For example, fascinating work by the feminist cultural anthropologist Gloria Wekker in the urban, working-class, Creole culture of Paramaribo, Suriname, has documented that an estimated three-quarters of women have engaged in enduring same-sex sexual partnerships that were simultaneous with long-term sexual relationships with the men who fathered their children. The women in these relationships were deeply engaged in providing cooperative child care, emotional support, and sexual pleasure to their female partners. We also know that the frequency of same-sex behavior may go way up in same-sex populations such as one finds in prisons and boarding schools, in which the cultural sanctions against same-sex behavior may be loosened.

The aesthetic theory also hypothesizes that females can advance their sexual autonomy through their friendships and social alliances with males who have predominantly same-sex sexual preferences. It is hard to investigate this observation given the complex social construction of gender, sexual identity, and social relationships in contemporary human cultures. However, we do know that such friendships are commonly acknowledged in our culture as a special sort of social relationship in a way that "straight-male-lesbian" friendships are not. At the heart of the long-running NBC hit television show *Will & Grace* was the enduring friendship between the housemates Will, a gay lawyer, and Grace, a straight interior designer. This phenomenon is not unique to Western culture, however. The 1992 Japanese movie *Okoge* tells the story of a straight young female office worker's friendship with a gay friend and his lover. The title for the film comes from a Japanese slang use of the word for "sticky rice," which refers to straight women with close gay friends. The fact that this slang even exists suggests that this is as well recognized a phenomenon in Japan as it is in Western cultures.

I do not, however, know of any example of an iconic relationship between a lesbian woman and a straight man. *Will & Grace* has not been followed by a show with a contrasting pair—say *Rosie & Rocky* starring the garrulous housemates Rosie O'Donnell and Sylvester Stallone. Nor are there any evolutionary hypotheses about the advantages that might accrue to either the males or the females in such relationships. However, to pursue these observations further, we would need to conduct serious sociological and psychological research into the nature of gay-straight relationships and their roles in the lives of real people.

Last, the aesthetic hypotheses about the evolution of same-sex behavior as a mechanism for reducing sexual violence also predict that male same-sex behavior should be associated with lower levels of sexual coercion, sexual violence, and domestic partner violence than male heterosexual behavior. The data available on this question are promising. The 2010 National Intimate Partner and Sexual Violence Survey reported that lifetime incidence of all categories of sexual partner violence (including rape, physical partner violence, and stalking) were significantly lower for men in same-sex relationships than for women in heterosexual relationships.

The evolutionary models I'm proposing assume the existence of a genetic variation for same-sex attraction, preferences, and behavior. To many people, however, the mention of genetics and sexual preference conjures up the prospect of identifying the "gay gene" and the possibility of genetic screening by health insurance companies or expectant parents. However, given what we know about the genetics of other complex human traits, such fears are unfounded.

Genomic studies are establishing that most complex human traits—from heart attack risk, musical ability, social personality, and shyness to autism—are influenced by the interactions of *many* genetic variations with individually small effects at *many* different locations, or genes, in the genome. As a result, even though these complex traits may be highly heritable, each instance of these traits is the result of a unique combination of genes, gene interactions, and the developmental environment. For example, a recent study of thousands of human genomes has shown that 82 percent of the simplest DNA sequence variations (called single nucleotide polymorphisms, or SNiPs) occur at frequencies of less than 1/15,000, or less than 0.006 percent. An individual human would only need three or four of such variations to be genetically *unique* among all the world's seven billion people. However, your genome actually has many *thousands* of such variations. Thus, it is hard to overestimate how truly unique each human being really is.

In this way, modern genomics has discovered the overwhelming fact of *human individuality*. Because there are myriad unique and distinct genetic combinations that influence each of these complex traits, including sexual preferences, we can be highly confident that there is no such thing as the "gay gene." Any genetic influences on individual sexual preferences are likely to be virtually unique. Genetics will not produce a reductive science of human sexual attraction. Its causes are simply too diverse.

In summary, the hypothesis that human same-sex behavior has evolved through natural and sexual selection for the expansion of female sexual autonomy is congruent with a great deal of the evidence on variation in human sexual preference and behavior. However, this hypothesis may seem inconsistent with the observation that in many cultures—such as ancient Greece and various indigenous peoples in New Guinea—male same-sex behavior occurs along with highly restricted social and sexual autonomy of women. However, these cultural examples may be exceptions that prove the rule. Such cultures construct male same-sex behavior within highly age- and status-structured relationships, usually involving an active, penetrating, socially dominant, older male and a passive, receptive, socially subordinate, younger male. The rigid hierarchical structuring of same-sex behavior appears to be a cultural mechanism to co-opt same-sex behavior into a coercive male hierarchy and thereby control the inherently autonomy-enhancing impact of same-sex behavior.

Although these proposals on the evolution of same-sex behavior are still speculative, I think they demonstrate that there is a productive new research area to be explored at the interface of aesthetic evolution, sexual autonomy, and human sexual diversity. In surprising ways, the evolutionary hypotheses I've outlined are strongly consistent with, and supportive of, some of the basic elements of contemporary gender theory. For example, aesthetic theories of the evolution of human same-sex behavior support elements of both sides of one of the most important contemporary debates within the lesbian, gay, bisexual (LGB) communities. On the one hand, some LGB rights advocates have argued that LGB people are essentially just like heterosexuals except for their sexual desires and partners. This school of thought—represented eloquently by Andrew Sullivan in his 1995 book, *Virtually Normal*—has made a major contribution to the achievement of same-sex marriage in the United States and in many other developed countries. Aesthetic evolutionary hypotheses provide support for the *Virtually Normal* view, because they predict that same-sex attraction is an evolved feature that is broadly shared by a large proportion of humans. Homosexuals are indeed fundamentally "just like everyone else." They differ only in the *exclusivity* and *specificity* of their same-sex preferences, not in having them.

However, many LGB people take issue with this assimilationist perspective, for they view variation in sexual orientation, desire, and behavior as inherently—and healthily—*disruptive* to heterosexual society. This opinion, well represented by Michael Warner's *The Trouble with Normal* and David Halperin's *How to Be Gay*, maintains that there is something about same-sex desire that is intrinsically subversive to normative heterosexual culture, hierarchy, and power. Interestingly, the aesthetic explanation for the evolution of human same-sex behavior comes down strongly in *support* of the inherent subversiveness of same-sex behavior.

According to my proposals, the evolved *function* of same-sex behavior is, quite specifically, to subvert male sexual control and social hierarchy. Thus, the evolutionary queering of the human species likely proceeded through female sexual desire to escape coercive male control.

Furthermore, if same-sex desire evolved as a means of subverting coercive male sexual control, this could explain why many patriarchal cultures have adopted such fervent moral and social sanctions against same-sex behavior. From this perspective, prohibition of same-sex behavior constitutes another means of reinforcing male capacity to exert sexual and social control over women and reproduction.

I hope, therefore, that aesthetic evolution and sexual conflict theory will provide a productive new intellectual interface between evolutionary biology, contemporary culture, and gender studies. After decades of reductionist, adaptationist arguments from sociobiology and evolutionary psychology that either ignored same-sex behavior as an aberration or misinterpreted it as a form of nonsexual behavior, who could have imagined that evolutionary biology and queer theory could be on the same page about anything? Actually, I think there will be many more productive commonalities to explore in the future.

CHAPTER 12

This *Aesthetic* View of Life

John Keats ends his famous poem "Ode on a Grecian Urn" with the following lines—a message from the Urn itself:

"Beauty is truth, truth beauty,—that is all
Ye know on earth, and all ye need to know."

Although Keats, writing several decades before Darwin, certainly knew nothing of evolution, his concluding lines are an oddly apt slogan for the long tradition in evolutionary biology of equating beauty with honesty. Indeed, this may be the most succinct and memorable articulation ever written of the honest advertisement paradigm.

While this may be an immortal conclusion to a poem, it is a poor guide to understanding beauty in the world. Keats's aphorism is really a *flatitude*—a faux insight that acquires its supposed profundity by flattening the intellectual complexity of the world. It does damage while claiming to be a glorious solution.

By contrast, Shakespeare—predating Darwin by centuries rather than decades—portrays a character with a much richer perspective on truth and beauty. In act 3, scene 1 of *Hamlet, Prince of Denmark,* Hamlet encounters his beloved Ophelia, who has recently shunned him without explanation. Ophelia, whose actions are being directed by her father, returns Hamlet's love letters and claims she no longer values his poetry because "rich gifts wax poor when givers prove unkind." Hamlet is understandably hurt by her actions and suspicious of her motivation because he knows he is innocent of her accusations. Because Ophelia is both as gorgeous as ever and obviously lying, Hamlet instigates his own inquiry into the relationship between truth and beauty.

> HAMLET: Ha, ha! Are you honest?
> OPHELIA: My lord?
> HAMLET: Are you fair?
> OPHELIA: What means your lordship?
> HAMLET: That if you be honest and fair, your honesty should admit no discourse to your beauty.
> OPHELIA: Could beauty, my lord, have better commerce than with honesty?
> HAMLET: Ay, truly; for the power of beauty will sooner transform honesty from what it is to a bawd than the force of honesty can translate beauty into his likeness: this was sometime a paradox, but now the time gives it proof. I did love you once.

Here, wily Hamlet gives a far more skeptical account of the "commerce" between beauty and truth than Keats's Urn. Beauty, he says, can transform truth into a bawd—a whorehouse madam, a procuress of a false and superficial love. Indeed, Hamlet makes the decidedly Fisherian proposition that it is the *power of beauty* that actually subverts honesty. Hamlet's paradox is the challenge we all face in reconciling the seductive force of beauty with the great desire to see beauty as having a higher purpose, as being an absolute good, as reflecting universal, objective quality.

On the one hand, we have Keats's poem, whose lines are a perfect representation of our deep desire to see beauty as an "honest" signifier of quality, of some kind of superiority. On the other hand, we have Hamlet, whose life experience has taught him that beauty is not truth; it is merely beauty, signifying nothing but itself, and often in fact at odds with truth. On the one hand, an insistence on "meaning"; on the other, an acceptance that the arbitrary power of beauty can undermine truth. These conflicting views are at the very heart of the contemporary scientific debate I have been engaged in in this book.

This same intellectual divide was explored by Isaiah Berlin in the essay *The Hedgehog and the Fox,* in which he analyzed an ancient Greek aphorism as a metaphor for a contrast in intellectual styles: "The fox knows many things, but the hedgehog knows one big thing."

According to Berlin, an intellectual Hedgehog, in search of a "harmonious universe," sees the world through the lens of a single "central vision." The Hedgehog's intellectual mission is to propagate this great vision at every opportunity. An intellectual Fox, by contrast, has no interest in the seductive power of a *single* idea. The Fox is drawn instead to the subtle complexities of a "vast variety of experiences," which he does not attempt to fit into a single all-embracing framework. Hedgehogs are on a mission. Foxes play for the joy of it. And like children, whenever they want, Foxes drop their toys and start a new game.

The intellectual styles of Berlin's Hedgehog and Fox provide further insights into the co-discoverers of natural selection: Darwin the Fox, and Wallace the Hedgehog. Having each intuited the mechanism of adaptive evolution by natural selection, the two diverged radically in how they elaborated on this key insight. To deal with the diversity of phenomena he observed in nature, Darwin proposed additional biological theories of phylogeny, sexual selection, ecology, pollination biology, even ecosystem services (for example, in his study of the ecological impact of earthworms), and more. Each theory was subtly different, requiring new arguments, types of thinking, and data. Wallace, on the other hand, despite his empirical breadth, strove to establish a "pure Darwinism" in which all of biological evolution was distilled down to the single omnipotent explanation of adaptation by natural selection.

The conflict between the Hedgehogs and the Foxes in evolutionary biology continues unabated to this day. In recent decades, adherents of the thoroughly foxy, Darwinian subdisciplines of phylogeny and developmental evolution (a.k.a. evo-devo) have worked to restore their places in an evolutionary biology that has been dominated, indeed hijacked, by adaptationist Hedgehogs. In this book, I have argued that the Darwinian theory of aesthetic evolution should also be restored to evolutionary biology. Each of these Darwinian subdisciplines focuses on diversity itself—the "vast variety" of specific instances—rather than on law-like generalizations of adaptive process.

Darwin concluded *The Origin of Species* with an inspired and poetic evocation of the "grandeur in this view of life." Later, in *The Descent of Man,* he articulated an equally moving grandeur in an *aesthetic* view of life. It has been my goal to revive Darwin's theory of aesthetic evolution and to present the full, distinctive richness, complexity, and diversity of this *aesthetic view of life.* Here, I want to conclude by discussing how an aesthetic view of life can have a positive impact on science, on human culture, and on the development of a newly respectful and productive relationship between them.

In many ways, Darwin's idea that the aesthetic evaluations involved in mate choice among animals constitute an independent evolutionary force in nature is as radical today as it was when he proposed it nearly 150 years ago. Darwin discovered that evolution is not merely about the survival of the fittest but also about charm and sensory delight in individual subjective experience. The implications of this idea for scientists and observers of nature are profound, requiring us to acknowledge that the dawn bird song chorus, the cooperative group displays of the blue *Chiroxiphia* manakins, the spectacular plumage of the male Great Argus Pheasant, and many other wondrous sights and sounds of the natural world are not merely delightful to *us;* they are products of a long history of subjective evaluations made by the animals themselves.

As Darwin hypothesized, with the evolution of sensory evaluation and choice comes the emergence of a new evolutionary agency—the capacity of individual judgments to drive the evolutionary process itself. Aesthetic evolution means that animals are aesthetic agents who play a role in their own evolution. Of course, this fact would be unsettling to a Wallacean Hedgehog who believes that the power of the idea of natural selection lies in its all sufficiency—its ability to explain *everything*. However, I am afraid that, to quote another passage from *Hamlet*, "there are more things in heaven and earth . . . [t]han are dreamt of in your philosophy."

Richard Dawkins once described evolution by natural selection as the "blind watchmaker"—the impersonal, inexorable force that produces functional design from variation, heritability, and differential survival. This analogy is entirely accurate. But because natural selection is *not* the only source of organic design in nature, as Darwin himself was the first to recognize, Dawkins's analogy remains an incomplete description of evolutionary process and the natural world. The blind watchmaker cannot actually look at nature and see all the stuff that he hasn't made and cannot explain. Indeed, nature has evolved its own eyes, ears, noses, and so on and the cognitive mechanisms to evaluate these sensory signals. Myriad organisms have then evolved to use their senses to make sexual, social, and ecological choices. Although animals are not conscious of their role, they have become their *own* designers. They are no longer blind. Aesthetic mate choice creates a new mode of evolution that is neither equivalent to nor a mere offshoot of natural selection. The concept of aesthetic mate choice is at the heart of Darwinian aesthetics, and it remains a revolutionary idea to this day.

The aesthetic view of life reveals new ways in which evolutionary biology has been hampered by failing to recognize the aesthetic agency of individual animals. For example, we can see that much of the scientific study of sexuality has been characterized by a deep anxiety about the subjective experiences of sexual pleasure and desire—especially when it's a matter of *female* pleasure. A symptom of this anxiety is the great lengths that evolutionary biologists have gone to avoid engaging with sexual pleasure and desire altogether. After the rejection of Darwin's aesthetic view of mate choice, sexual desire and pleasure had to be explained away as mere secondary consequences of natural selection.

Unfortunately, the anxious remove of sexual science from sexual pleasure was built into the structure of scientific objectivity—into the discipline of science itself. To imagine animals as aesthetic agents with their own subjective preferences was considered anthropomorphic. Scientific "objectivity" came to require us to discount or ignore the subjective experiences of animals. Adaptive, anhedonic theories of mate choice were developed to explain animal mating behavior and reproduction, and these theories were proposed to be sufficient to explain the evolution of human sexuality. Sexual pleasure was not only laundered from scientific explanation; it was banished as an appropriate *subject* of science. The result was generations of antiaesthetic sexual biology—such as Zahavi's handicap principle, or the upsuck theory of female orgasm—which entirely ignored and denied the existence of subjective experience of sexual pleasure.

A scientific anxiety about sexual pleasure persists in much of the contemporary science of mate choice. The result is a sanitized sexual science that lacks the theory and vocabulary necessary to investigate and explain sexual pleasure in the natural world and ourselves.

A bizarre consequence of this traditional framework is an inexplicable inversion in the rationality of nature. Because animals are denied aesthetic agency, we conclude that animal choices reflect the universal and rational hand of natural selection. But, of course, we understand that humans can be highly irrational when it comes to sex and love. So, because animals lack the cognitive ability to escape from the brute laws of adaptive logic, dumb animals are more rational than we are. Ironically, in this view, human cognitive complexity only provides us with the novel opportunity to be irrational!

Another important implication of an aesthetic view of evolutionary biology concerns the painful history of political and ethical abuse during the twentieth century—eugenics. Eugenics was the scientific theory that maintained that human races, classes, and ethnicities have evolved adaptive differences in genetic, physical, intellectual, and moral *quality*. Eugenics was also an organized social and political movement to employ this flawed scientific theory to "improve" human populations through the social and legal control of mate choice and reproduction. Because eugenics specifically concerned the evolutionary consequences of mate choice, it remains deeply relevant to human sexual selection and aesthetic evolution.

For multiple reasons, evolutionary biologists are uncomfortable discussing eugenics. First, between the 1890s and the 1940s, *every* professional geneticist and evolutionary biologist in the United States and Europe was either an ardent proponent of eugenics, a dedicated participant in eugenic social programs, or a happy fellow traveler. Full stop. Few of us are eager to confront this embarrassing, shameful, and sobering truth. Second, eugenics provided a pseudoscientific justification for abuses of human rights at every level—from everyday racism, sexism, and prejudice against the disabled, forced sterilization, imprisonment, and lynching in the United States, to the Nazi-engineered genocide of Jews and Gypsies, and mass murder of the mentally handicapped and homosexuals in Europe. Eugenics is the most egregious example of the destructive misuse of science in all of human history. Science gone bad. Really bad.

Last, another uncomfortable truth is that much of the intellectual framework of contemporary evolutionary biology was developed during this enthusiastically eugenic period in our discipline. Most evolutionary biologists would like to believe that eugenics ceased to be an issue in evolutionary biology after World War II, when evolutionary biologists rejected eugenic theories of racial superiority. But the uncomfortable fact is that some core, fundamental commitments of eugenics were "baked into" the intellectual structure of evolutionary biology, and they contributed to the flawed logic of eugenics. Without providing a detailed analysis here, I want to illustrate how aesthetic evolution provides an essential antidote to this poisonous intellectual history.

Eugenics and population genetics were both developed during the period when mate choice was either entirely rejected *or* presumed to be essentially identical to natural selection. This was the same period when Darwinian *fitness* was redefined and expanded to subsume all of sexual selection. As we have seen (chapter 1), "fitness" to Darwin referred to the capacity of an individual to do tasks that contributed to one's survival and fecundity, just like *physical fitness*. In the early twentieth century, fitness was redefined as an abstract mathematical concept—the relative success of one's genes in subsequent generations. This new definition of fitness confounded variation in survival, fecundity, *and* mating/fertilization success into a single concept, obscuring the differences between the Darwinian concepts of natural and sexual selection. Despite this redefinition, the original connection of fitness to adaptation was retained. In this way, the rejection of Darwin's concept of arbitrary aesthetic mate choice was forged into the language of modern evolutionary biology, making it almost impossible to talk about reproduction and mate choice in anything other than adaptive terms.

The new broad definition of fitness meant that all selection is, and *should be,* about adaptive improvement. Arbitrary mate choice was essentially defined out of existence, which is why arbitrary mate choice has had such a hard time in the discipline ever since. This intellectual stance contributed directly to the logical inevitability of eugenic theory. If one accepted the facts of natural selection, human evolution, heritable variation within and among human populations, and variation in human "fitness" and "quality," then the logic of eugenics was practically inescapable. And in fact, no one in the discipline escaped it. What was missing from both the eugenic framework and all of evolutionary biology was the possibility of arbitrary, aesthetic mate choice.

Although I do not think that contemporary sexual selection theory or research is actually eugenic, I do think that evolutionary biology did not overcome its eugenic history—*our* eugenic history—merely by rejecting theories of human racial superiority during the twentieth century. Obvious and uncomfortable intellectual similarities remain between eugenics and current adaptive mate choice theory. Eugenic theory and social programs were concerned with both the presumed genetic quality of offspring (that is, good genes) and the cultural, economic, religious, linguistic, and moral conditions of the family as the locus of human reproduction (that is, direct benefits). The twin eugenic concerns for genetic and environmental quality are still echoed in the language of adaptive mate choice today. The contemporary term "good genes" actually shares the same etymological roots as "eugenics"—from the Greek *eugenes* for wellborn or noble (*eu,* good or well; and *genos,* birth). Eugenics was also explicitly antiaesthetic and anxious about the maladaptive consequences of the seductive power of sexual passion. In general, the eugenic commitment to the idea that all mate choice is, and *should be,* about adaptive improvement persists today in the language and logic of adaptive mate choice.

The adaptive commitment of most contemporary researchers makes it difficult to investigate the evolution of human variation in ornamental traits, because to do so would require judgments to be made among human populations as to genetic and material quality. One reason why evolutionary psychology is so focused on the evolution of human *universals*—that is, behavioral adaptations that are shared by all humans—is that applying the same adaptationist logic to investigate evolved variations *among* human populations would explicitly resurrect eugenic research.

To permanently disconnect evolutionary biology from our eugenic roots, we need to embrace Darwin's *aesthetic* view of life and fully incorporate the possibility of *nonadaptive,* arbitrary aesthetic evolution by sexual selection. This requires more than a tacit recognition of the mathematical existence of Fisher's runaway theory. It requires undoing the Wallacean transformation of Darwinism into a strict, adaptationist science and abandoning the default expectation that all mate choice is, or should be, inherently adaptive. To sever our historical connections to eugenics, evolutionary biologists should restore the Darwinian view by defining natural and sexual selection as distinct evolutionary mechanisms and conceiving of adaptive mate choice as the

DESIRABLE TRAITS IN WOMEN

SOCIALLY:

Beauty first

Delicate features

No "deep" intellect

Vivaciousness

Slim figure

Tiny waist

Small hips

Dainty wrists
and hands

Slender, soft
tapering limbs

Slim ankles

Tiny feet

EUGENICALLY:

Beauty unimportant

Strong features

High intelligence

Seriousness

Sturdy figure

Ample waist

Broad hips

Sturdy wrists,
strong hands

Solid, sturdy
limbs and
ankles; good-
sized feet

The explicitly anti-aesthetic goals of eugenic social programs can be seen in this illustration from a popular eugenic test, *You and Heredity,* by Amram Scheinfeld (1939). The illustration contrasts "socially and eugenically desirable traits in women." Sexual passion and desire were equated with the maladaptive consequences of unregulated mate choice.

result of specific, special interactions *between* these mechanisms. Accordingly, evolutionary biology should adopt the nonadaptive, Beauty Happens null model of the evolution of mating preferences and display traits by sexual selection.

The reincorporation of aesthetic evolution into evolutionary biology can permanently inoculate the discipline against the intellectual fallacies of its eugenic past. Adopting the Beauty Happens null model breaks the logical inevitability of eugenic thought by formalizing the expectation of a nonadaptive, or even maladaptive, outcome (see chapter 2). The resulting, genuinely Darwinian evolutionary science will allow anyone to pursue the investigation of adaptive mate choice in any animal, including humans, but the burden of proof on adaptive mate choice will be appropriately high. Evolutionary biology will be the better for this change. And so will the world.

A truly unexpected, personal consequence of adopting Darwin's aesthetic view of life has been the discovery of new insights into the evolutionary impact of sexual coercion and sexual autonomy. When Patricia Brennan first proposed to work with me on the evolution of duck genitalia, I thought to myself, "Well, I've never worked on *that* end of the bird before." I figured we would learn a lot of interesting anatomy, but I never imagined how the project would grow or that the results would transform my view of evolution so profoundly and raise so many surprising new directions and implications.

Of course, it has long been clear that sexual coercion and sexual violence are directly harmful to the well-being of female animals. But the aesthetic perspective allows us to understand that sexual coercion also infringes upon their individual *freedom of choice*. Once we recognize that coercion undermines individual sexual autonomy, we are led, inexorably, to the discovery that freedom of choice *matters to animals*. Sexual autonomy is not a mythical and poorly conceived legal concept invented by feminists and liberals. Rather, sexual autonomy is an evolved feature of the societies of many sexual species. As we have learned from ducks and other birds, when sexual autonomy is abridged or disrupted by coercion or violence, mate choice itself can provide the evolutionary leverage to assert and expand the freedom of choice.

In the later chapters of the book, I have proposed that the evolutionary struggle for female sexual autonomy played a critical role in the evolution of human sexuality and reproduction and was a critical factor in the evolution of humanity itself. But if this is true, why aren't the women of the world enjoying the proposed fruits of this evolutionary process—universal fulfillment of sexual and social autonomy? The ongoing existence of rape, domestic violence, female genital mutilation, arranged marriage, honor killings, everyday sexism, economic dependence, and political subservience of women in many human cultures might seem to be direct evidence to falsify this view of human evolutionary history. Are we forced to acknowledge that such behaviors are an inescapable part of "human nature"—a part of our evolutionary legacy that humans will never overcome? I think not, and sexual conflict theory can help us to understand why.

Sexual conflict theory tells us that female aesthetic remodeling is not the *only* evolutionary force at work (see chapter 5). Males are simultaneously evolving through the force of male-male competition (another form of sexual selection), which can work simultaneously to maintain and advance sexual coercion. This process happens because there are limits to the effectiveness of female mate choice. It can expand female sexual autonomy, but it is *not* a mechanism for the evolution of female power or sexual control over males. As long as males continue to evolve mechanisms to advance their capacity for sexual coercion and violence, females may remain at some disadvantage. As I explained in the context of duck sex, the "war of the sexes" is highly asymmetrical—not really a war at all. Males evolve weapons and tools of control, while females are merely coevolving defenses of their freedom of choice. It's not a fair fight.

Although aesthetic remodeling in humans has provided great advances in female sexual autonomy, I think the subsequent evolution of human culture has resulted in the emergence of new *cultural mechanisms* of sexual conflict. In other words, I propose that cultural ideologies of male power, sexual domination, and social hierarchy—that is, patriarchy—developed to *reassert* male control over fertilization, reproduction, and parental investment as countermeasures to the evolutionary expansion of female sexual autonomy. The result is a new, human *sexual conflict arms race* being waged through the mechanisms of culture.

More specifically, I think that the advances in female sexual autonomy that occurred over millions of years since our common ancestry with the chimpanzees (evolutionary context 2) have been challenged by two relatively recent *cultural* innovations—agriculture and the market economy that developed along with agriculture (evolutionary context 4). These twin inventions came into being a scant six hundred human generations ago and created the first opportunity for wealth and the differential distribution of wealth. When males gained cultural control over these material resources, new opportunities were created for the cultural consolidation of male social power. The independent and parallel invention of patriarchy in many of the world's cultures has functioned to impose male control over nearly all aspects of female life, indeed human life. Thus, the cultural evolution of patriarchy has prevented modern women from fully consolidating the previous evolutionary gains in sexual autonomy.

This cultural sexual conflict theory poses a productive and exciting new intellectual interface between aesthetic evolution, sexual conflict, cultural evolution, and contemporary sexual and gender politics. From this perspective, for example, it is not an accident that patriarchal ideologies are focused so intently on the control of female sexuality and reproduction and also on the condemnation and prohibition of same-sex behavior. Female sexual autonomy and same-sex behavior have both evolved to be disruptive to male hierarchical power and control. These disruptive effects were likely the driving force behind the cultural invention and maintenance of the patriarchy itself.

Despite the near ubiquity of male culture dominance, this view implies that patriarchy is not inevitable, and it does not constitute human biological "destiny" (whatever that is). Patriarchy is a product not of our evolutionary history nor of human biology per se but of human culture. There is a tendency to respond to the many ills of male dominance—aggression, crime, sexual violence, rape, warfare, and so on—with weary inevitability: "Boys will be boys." However, such "boys" are more likely products of patriarchal culture than of human evolutionary history. Analysis of the history of sexual conflict in humans indicates that males have been evolutionarily deweaponized and only culturally rearmed. Remember that men have *less* physical advantage in body size over women than do the famously peaceful male bonobos. The social and sexual advantages men currently enjoy over women cannot be explained as the inevitable result of our biological, evolutionary history alone.

If patriarchy is part of a cultural sexual conflict arms race, then we should predict the emergence of cultural countermeasures to reassert and preserve female sexual and social autonomy, and so they have. Beginning with nineteenth-century feminist movements for women's suffrage, access to education, and rights to property and inheritance, there has been a culturally coevolved effort to counteract the control of patriarchy and to reassert and advance female sexual autonomy and freedom of choice. Although it took thousands of years to happen, the results of these efforts— legal recognition of women's suffrage, universal human rights, and the abolition of legal slavery—are demonstrations that it *is* possible to dismantle deeply ingrained components of patriarchy that are often, still, erroneously considered as biologically "natural."

The concept of an ongoing, culturally waged sexual conflict arms race also allows us to understand what is at stake in the battle between contemporary feminists and advocates of conservative, patriarchal views of human sexuality. After all, control over reproduction—including birth control and abortion—is at the very core of sexual conflict.

Like the evolved sexual autonomy of ducks, feminism is *not* an ideology of power or control over others; rather, it is an ideology of freedom of choice. This asymmetry of goals—the patriarchal aim of advancing male dominance versus the feminist commitment to freedom of choice—is inherent in all sexual conflict, from ducks to humans. But it gives the contemporary cultural struggle over universal sexual equal rights an especially frustrating quality.

As if to justify its use of power and privilege, the defenders of patriarchy often mischaracterize feminism as an ideology of power. Feminists, they claim, are attempting to take control of men's lives, deny them their natural, biological prerogatives, and put men in a subservient position. For example, one antifeminist legal scholar has even erroneously criticized the legal doctrine of "sexual autonomy"—which has become the basis of most rape and sex crimes law—as allegedly including the right to impose one's own personal sexual desires upon others. However, we can see that such views fundamentally misconstrue what sexual autonomy is and how it arises either biologically or culturally.

Watching recent political battles over birth control and reproductive rights in the United States, many experienced observers have remarked, "But I thought all these issues were settled *decades* ago!" Unfortunately, if these events are part of a cultural sexual conflict arms race, we can expect that the struggle for female sexual autonomy will continue as each side innovates new countermeasures to neutralize the previous advances by the other.

On the other hand, feminists themselves have often expressed discomfort with standards of beauty, sexual aesthetics, and discussions of desire. Beauty has been viewed as a punishing male standard that treats women and girls as sexual objects and persuades women to adopt the same self-destructive standard to judge themselves. Desire has been viewed as another route to finding themselves under the power of men. Yet aesthetic evolutionary theory reminds us that women are not only sexual *objects* but also sexual *subjects* with their own desires and the evolved agency to pursue them. Sexual desire and attraction are not just tools of subjugation but individual and collective instruments of social empowerment that can contribute to the expansion of sexual autonomy itself. Normative aesthetic agreement about *what is desirable* in a mate can be a powerful force to effect cultural change. The ancient lessons of *Lysistrata* are clear. Individuals can transform human society through their affirmative sexual choices.

This book has taken the concept of beauty from the humanities and applied it to the sciences by defining beauty as the result of a coevolutionary dance between desire and display. Now I would like to explore the opposite—take the coevolutionary view of beauty and see how it might apply to the humanities, specifically to the arts.

Indeed, progress in understanding aesthetic evolution in nature creates a whole new opportunity for intellectual exchange between evolutionary biology and aesthetic philosophy—the philosophy of art, aesthetic properties, art history, and art criticism—which I have been pursuing in new research. For centuries, the "aesthetics of nature" has consisted entirely of investigating *human* aesthetic experiences of nature—whether that be looking at a landscape, listening to the song of a Rose-Breasted Grosbeak, or contemplating the shape, color, and odor of an orchid flower. However, aesthetic evolution informs us that grosbeak songs and orchids (but not the landscape) have coevolved their aesthetic forms with the evaluations of nonhuman agents—female grosbeaks and insect pollinators, respectively. We humans can appreciate their beauty, but we have played no role in shaping it. Traditionally, aesthetic philosophy has failed to appreciate the aesthetic richness of the natural world, much of which has come into being through the subjective evaluations of animals. By viewing the beauties of nature through an exclusively *human gaze*, we have failed to comprehend the powerful aesthetic agency of many nonhuman animals. To be a more rigorous discipline, aesthetic philosophy must grapple with the full complexity of the biological world.

Another exciting implication of this aesthetic view of life is the realization that coevolutionary change is *the* fundamental feature underlying all aesthetic phenomena, including the *human arts*. As explained throughout this book, the evolution of sexual ornaments like the peacock's tail involves the corresponding coevolution of the peahen's cognitive aesthetic preferences. Changes in mating preferences have transformed the tail, and changes in the tail have transformed mating preferences. We can see a similar coevolutionary process at work in the fine arts. Mozart, for example, composed symphonies and operas that transformed his audiences' capacity to imagine what music could be and do. These new musical preferences then fed back upon future composers and performers to advance the classical style in Western music. Likewise, Manet, van Gogh, and Cézanne created paintings that pushed the genre of European painting beyond its previous bounds. The newly transformed aesthetic preferences of their audiences fed back upon new generations of artists, collectors, and museums, ultimately leading to Cubism, Dada, and other modernist art movements of the early twentieth century. These cultural mechanisms of aesthetic change in the human arts are inherently coevolutionary as well.

Once we understand that all art is the result of a coevolutionary historical process between audience and artist—a coevolutionary dance between display and desire, expression and taste—we must expand our conception of what art is and can be. We cannot define art by the objective qualities of an artwork nor by any special qualities of observer experience (that is, art is not merely in the eye of the beholder). Being an artwork means being the product of a historical process of aesthetic coevolution. In other words, *art is a form of communication that coevolves with its own evaluation.*

This coevolutionary definition of art implies that art necessarily emerges within an aesthetic community, or population of aesthetic producers and evaluators. In a now classic paper of aesthetic philosophy from 1964, Arthur Danto called this taste-making, aesthetic community "the artworld." This new, coevolutionary definition of art opens up an entirely new connection between evolutionary biology and the arts.

Perhaps the most revolutionary consequence of this definition of art is that it means that bird songs, sexual displays, animal-pollinated flowers, fruits, and so on are *art,* too. They are *biotic arts* that have emerged within myriad *biotic artworlds,* each of them a community that fostered the coevolution of animal aesthetic traits and preferences over time.

Of course, it could be argued that any definition of art should rest on the kind of cultural transmission of ideas that we see in human artworlds. The human arts are cultural phenomena that are transformed by aesthetic *ideas* that pass from person to person within a social network—a cultural mechanism of aesthetic innovation and influence. If we accept a cultural definition of art, that might seem to suggest that aesthetically coevolving, *genetic* entities cannot be art. However, this definition will not eliminate the biotic arts. For example, nearly half of all species of birds on the planet *learn* their songs from other members of their own species. These bird species have *avian cultures* that have persisted, thrived, and diversified for over forty million years. Consequently, learned bird songs have regional variations (that is, dialects), and cultural transmission can give rise to rapid and sometimes radical changes in these songs, just the way change sometimes occurs in the human arts. Similar aesthetic cultural processes occur in whales and bats.

In short, when we get out of the art museum and the library, and look closely at the aesthetic complexity of nature, and think about how it all came into being, we find that it is difficult to define the arts in any way that will include everything we recognize as human art but exclude the aesthetic productions of all nonhuman animals.

Some aesthetic philosophers, art historians, and artists may find the recognition of myriad new biotic art forms to be more of an annoyance, or even an outrage, than a contribution to their fields. But I think there is reason to welcome this more inclusive, "post-human" view of art as a real opportunity for progress in aesthetics. Originally, we humans conceived of ourselves as being at the center of all creation, with the sun and the stars revolving around *us*. Over the last five hundred years, however, scientific discoveries have demanded that we reframe our view of the cosmos and our place in it. With each discovery, humans have moved further and further from the organizing center of the universe. The reality is that we live in an entirely normal solar system, in the boring backwaters of a thoroughly vanilla galaxy—literally, a cosmic Nowheresville. Although the size of earth and its distance from the sun are indeed special, in every other way our position within the cosmos is profoundly random, unpredictable, and unimpressive. While many have found this intellectual change disconcerting, I think such knowledge can only enhance our appreciation of the astounding, unexpected richness of the biological world, human existence, our conscious experience, and our technological and cultural accomplishments.

In a similar way, I think that reframing aesthetic philosophy to remove humans from the organizing center of the discipline—to fully encompass the aesthetic productions of both human and nonhuman animals—can only enhance our appreciation of the marvelous diversity, complexity, aesthetic richness, and variable social functions of the human arts. By adopting a post-human aesthetic philosophy that places us, and our artworlds, in context with other animals, we will have a much deeper understanding of how we came to be and what is truly special about being human.

On a foggy late June morning in 1974, I stood in a large lobster boat eagerly gripping my binoculars as we pulled out of the harbor of West Jonesport, Maine. We were on our way to Machias Seal Island, at that time the southernmost nesting colony of the Atlantic Puffin (*Fratercula arctica*). The fog began to clear as we entered the deep water of the Bay of Fundy, and Captain Barna Norton was soon pointing out Greater Shearwaters, Sooty Shearwaters, and Wilson's Storm Petrels—smaller relatives of the great oceangoing albatrosses—as they skimmed over the gray water.

The sun broke through as we approached the grassy fifteen-acre island with its rocky shore and white postcard-ready lighthouse. Nesting in the grass along the boardwalks that crossed the island were thousands of Common Terns. Mixed in among them were a couple hundred Arctic Terns, distinguished from the Common Terns by their entirely bloodred beaks, silvery wings, shorter red legs, grayer breasts, and longer white tail plumes. In just six weeks, these Arctic Terns would begin their epic migration—the longest of any organism—down through the southern Atlantic to spend their winter in the Antarctic Ocean, only to return to breed here next summer. As we walked along the boardwalks through the tern colony, we instigated a traveling wave of consternation. Pairs of terns took turns screaming and diving down to attack our heads with their needle-sharp beaks. Being only twelve at the time, I was one of the shortest people in the group. So, the terns conveniently swooped down on the taller members of our party, and I escaped the brunt of these attacks.

From inside several blinds looking outward to the rocky shore, I saw dozens of Atlantic Puffins with their black-and-white tuxedo plumage and their big, clownishly colorful red, orange, and black beaks (color plate 21). The puffins sat sunning and socializing with each other on the granite boulders before flying back out to sea to feed. Occasionally, a new puffin would return from the sea, beak stuffed with a dozen or more thin little fishes that hung down on either side of the beak like the silvery walrus mustache so popular with rock stars and young men at that time. After landing on the rocks, the foraging puffin would descend between the boulders to his or her burrow to feed the single chick waiting hungrily below. Among the boulders were a few pairs of Common Murres (*Uria aalge*) and Razorbills (*Alca torda*), the closest living relative of the extinct, flightless Great Auk (*Alca impennis*), which plied these same waters centuries before.

The day flew by, and some hours later I returned to the boat sunburned, covered in smelly tern shit, and ecstatically happy. I remained vigilant all the way back to West Jonesport in hopes of catching yet one more view of a shearwater or a foraging Arctic Tern in our wake. Many events of that day—like waking up in my tent at dawn and identifying the song of my lifer Swainson's Thrush (*Catharus ustulatus*)—remain sealed in my memory over forty years later.

After I had spent months dreaming, planning, reading, and studying birds in my landlocked little hometown in southern Vermont, the experience of finally seeing the puffins and other seabirds had exceeded my wildest imaginings. The convergence of book learning and life experience—of *savoir* and *connaissance*—created a profound joy. It was an early and formative avian epiphany. In the years that followed, I would dedicate much of my life to reliving, expanding, and deepening the revelatory experience of natural history observation, scientific research, and discovery.

I realized along the way that bird-watching and science are both ways of exploring yourself in the world—parallel paths to find self-expression and meaning through engagement with the diversity and complexity of the natural world around us. But I am still astonished by the surprising new ways this continues to be true, how knowledge circles back and creates opportunity for richer, ever-deeper experiences and ever more stirring discoveries, and how that process enriches our lives.

I am still as excited for the next opportunity, the next discovery, the next new, beautiful bird, as I was on that expectant foggy morning in Maine.

Acknowledgments

I am indebted to many people for their insights, advice, assistance, and support during the writing and production of this book. Personally, I am grateful to my wife, Ann Johnson Prum, for her enthusiastic encouragement, helpful insights, editing advice, patience, and understanding along the way. I also thank my children, Gus, Owen, and Liam, for their open-hearted curiosity and interest. I thank my twin sister, Katherine, for her inspiration and understanding. Living shared, parallel lives as children had an immeasurable impact on me, my interest in feminism, and in the deep mystery of the subjective experiences of others. I would like to thank my parents, Bruce Prum and Joan Gahan Prum, who encouraged my interest in birds, science, and travel from my earliest days.

The writing of this book was supported by several fellowships. The book was begun in 2011–12 during an Ikerbasque Science Fellowship from the Ikerbasque Science Foundation and the Donostia International Physics Center (DIPC) in Donostia–San Sebastian, Spain. I am grateful to Pedro Miguel Echenique and Javier Aizpurua at the DIPC for their interest and support. The book was (nearly) completed during a fellowship at the Wissenschaftskolleg zu Berlin in 2015. The "Wiko" provided a marvelously productive, scholarly, and collegial environment, and I thank the many new friends I met there. The project was also supported by funds from the William

Robertson Coe Fund at Yale University and by a fellowship from the MacArthur Foundation.

I am thankful to Michael DiGiorgio and Rebecca Gelernter for their beautiful drawings and illustrations and to Juan José Arango, Brett Benz, Rafael Bessa, Marc Chrétien, Michael Dolittle, Ronan Donovan, Rodrigo Gavaria Obregón, Tim Laman, Kevin McCracken, Bryan Pfeiffer, João Quental, Ed Scholes, and Jim Zipp for permission to reproduce their lovely photographs.

The content and direction of the book were shaped and improved by many conversations and exchanges with and insights and comments from numerous colleagues and friends, including: Suzanne Alonzo, Ian Ayres, Dorit Bar-On, David Booth, Gerry Borgia, Brian Borovsky, Patricia Brennan, James Bundy, Tim Caro, Barbara Caspers, Innes Cuthill, Anne Dailey, Jared Diamond, Elizabeth Dillon, Michael Donoghue, Justin Eichenlaub, Teresa Feo, Michael Frame, Rich and Barbara Franke, Jennifer Friedmann, Jonathan Gilmore, Michael Gordin, Phil Gorski, Patty Gowaty, David Halperin, Brian Hare, Karsten Harries, Verity Harte, Geoff Hill, Dror Hawlena, Rebecca Helm, Geoff Hill, Jack Hitt, Rebecca Irwin, Susan Johnson Currier, Mark Kirkpatrick, Jonathan Kramnick, Susan Lindee, Pauline LeVen, Daniel Lieberman, Kevin McCracken, David McDonald, Erika Milam, Andrew Miranker, Michael Nachman, Barry Nalebuff, Tom Near, Daniel Osorio, Gail Patricelli, Robert B. Payne, Bryan Pfeiffer, Steven Pincus, Steven Pinker, Jeff Podos, Trevor Price, David Prum, Joanna Radin, Bill Rankin, Mark Robbins, Gil Rosenthal, David Rothenberg, Joan Roughgarden, Alexandre Roulin, Jed Rubenfeld, Dustin Rubenstein, Fred Rush, Bret Ryder, Lisa Sanders, Haun Saussy, Francis Sawyer, Sam See, Maria Servedio, Russ Shafer-Landau, Robert Shiller, Bryan Simmons, David Shuker, Bob Shulman, Stephen Stearns, Cassie Stoddard, Cordelia Swann, Gary Tomlinson, Chris Udry, Al Uy, Ralph Vetters, Michael Wade, Günter Wagner, David Watts, Mary Jane West-Eberhard, Tom Will, Catherine Wilson, Richard Wrangham, Marlene Zuk, and Kristof Zyskowski. I am sure there are others that I have forgotten!

Much of the research presented in the book was done in collaboration with my students and postdocs. I am very thankful for the creative input, discussions, and hard work of Marina Anciães, Jacob Berv, Kimberly Bostwick, Patricia Brennan, Chris Clark, Teresa

Feo, Todd Harvey, Jacob Musser, Vinod Saranathan, Ed Scholes, Sam Snow, Cassie Stoddard, and Kalliope Stournaras.

I am thankful to my editor at Doubleday, Kristine Puopolo, and her assistant, Daniel Meyer, who gave me encouragement, thoughtful insights, and excellent observations all along the way. Beth Rashbaum worked tirelessly on editing several drafts of the entire book, and she helped make the book clearer, more accessible, and easier to read. I am deeply thankful to Beth for her patience, persistence, and insights. Of course, I alone remain responsible for all errors, oversights, and omissions in the work.

I am very grateful to my agents, John Brockman and Katinka Matson, for their experience, advice, and guidance throughout the entire process.

Writing can be a lonely and uncertain process. Early on in the project, I had an e-mail correspondence with the poet Carter Revard about aesthetic evolution in birds, nature, and the arts. In closing, Carter shared with me Robert Frost's "The Tuft of Flowers," which concludes with the lines:

"Men work together," I told him from the heart,

"Whether they work together or apart."

The image from Frost's poem—of our many separate lives working in parallel in different ways, in isolation and perhaps even ignorance of each other, toward a shared goal of discovery, beauty, and justice—became an inspiration and encouragement throughout the project. Thus, I am grateful to all those working in parallel for scientific change and a new, more productive relationship between science and culture.

Notes

Introduction

4   Birding is about recognizing: Because the names of bird species are proper nouns, ornithologists always capitalize the common names of bird species. This is also the only way to distinguish between a Common Loon (*Gavia immer*) and a common loon, and a Ferruginous Hawk (*Buteo regalis*) and any hawk that is merely ferruginous.

4   functional magnetic resonance imaging studies: Gauthier et al. (2000), but for further debate on the neuroscience of visual expertise, see Harel et al. (2013) and other references therein.

4   when a birder identifies: Although bird-watching might be a neurological reutilization of this social part of the brain, it is also possible this part of the brain first evolved to recognize bird species, other wildlife, and plants that are potential food sources or predatory threats and that it was only later co-opted evolutionarily for its function in social recognition. Bird-watching might be among the very first functions of mind.

7   As Thomas Nagel has written: In the classic paper "What Is It Like to Be a Bat?," Nagel (1974) makes the claim that an organism is conscious if its sensory experience has specific qualities, that is, if "there is something it is like to be" that organism. While I have no stake in whether this is a productive definition of consciousness, I do think there is ample evidence that many organisms—including birds—have a flow of sensory and cognitive experience that varies in its qualities. These sensory and cognitive qualities ultimately give rise to ecological, social, and sexual decisions that are fundamental to aesthetic evolution.

9   the beaks of the Galápagos Finches: Research on the evolution of beaks of Galápagos Finches by Peter and Rosemary Grant has been summarized in Grant (1999) and in the classic book *The Beak of the Finch* by J. Weiner (1994).

9   the evolution of an avian ornament: Of course, beak shape can also be influenced by aesthetic sexual and social selection. The enormous and brilliant beaks of *Ramphastos* toucans and many hornbills are examples of complex social signals that have not evolved merely through natural selection on their ecological functions.

12   it has been nearly forgotten: I am indebted to Mary Jane West-Eberhard for both her classic work on sexual and social selection (1979, 1983) and her recent critiques of adaptive mate choice and advocacy for "Darwin's forgotten theory" (2014).

Chapter 1: Darwin's *Really* Dangerous Idea

18   I propose that Darwin's *really* dangerous idea: Darwin (1871).

18   "The sight of a feather in a peacock's tail": Darwin to Asa Gray, April 3 [1860], Darwin Correspondence Project, Letter 2743.

19   Charles Darwin, a member: For an excellent biography of Darwin, see Janet Browne's two-volume *Charles Darwin: Voyaging,* vol. 1 (2010), and *Charles Darwin: The Power of Place,* vol. 2 (2002).

20   "Much rubbish was talked there": Darwin (1887, 15).

20   "light will be thrown": Darwin (1859, 488).

22   "We thus learn that man is descended": Darwin (1871, 784).

22   "Courage, pugnacity, perseverance": Darwin (1871, 794–95).

23   *Descent* has never had: In Browne's biography of Darwin (2002), *Descent* is discussed in just a few pages, whereas over one hundred pages are devoted to the impact of *Origin.*

23   "With the great majority of animals": Darwin (1871, 61); sentence marked with * was added in the second edition.

24   "On the whole, birds appear": Darwin (1871, 466).

25   numerous experiments across the animal kingdom: A good, if somewhat out-of-date, summary of sexual selection theory and data is provided by Andersson (1994). A more recent review of the perceptual and cognitive nature of mate choice is Ryan and Cummings (2013).

25   "Amongst many animals": Darwin (1859, 127).

25   This view still prevails today: Contemporary evolutionary biologists often attempt to cover up their intellectual differences with Darwin by citing these early passages on sexual selection from the few paragraphs in *Origin* and entirely ignoring Darwin's explicitly aesthetic view of mate choice in the two volumes of *Descent.*

25   "The case of the male Argus Pheasant": Darwin (1871, 516).

26   "The male Argus Pheasant acquired his beauty": Darwin (1871, 793).

29   "Under the head of sexual selection": Mivart (1871, 53).

29   "The second process consists": Mivart (1871, 53).

29   "Even in Mr. Darwin's": Mivart (1871, 75–76).

30   "such is the instability": Mivart (1871, 59).

30   the word "vicious": "vicious, adj.," *OED* online, March 2016, Oxford University Press.

30   today "caprice" refers: "caprice, n.," *OED* online, March 2016, Oxford University Press.

30   "The display of the male": Mivart (1871, 62).

31   "The assignment of the law": Mivart (1871, 48).

32   But Darwin and Wallace never agreed: *The Ant and the Peacock* by Helena Cronin (1991) provides an excellent historical account of the Darwin-Wallace debate.

32   "I may perhaps be here permitted": Darwin (1882, 25). Darwin did make a single concession to the critics of sexual selection: "It is, however, probable that I may have extended it too far, as, for instance, in the case of the strangely formed horns and mandibles of male Lamellicorn beetles." In other words, before his death, Darwin barely gave an inch to the critics of mate choice—or about the length of the horns of a Lamellicorn beetle.

33   "The only way in which": Wallace (1895, 378–79).

33   "remember that physical beauty": Ben S. Bernanke, "The Ten Suggestions," June 2, 2013, Princeton University's 2013 Baccalaureate remarks.

33   "If there is (as I maintain)": Wallace (1895, 378–79).

34   "In rejecting that phase of sexual selection": Wallace (1889, xii).

34   "Natural selection acts": Wallace (1895, 379).

35   Wallace's hatchet job: For an interesting account of the various studies of mate choice during the early twentieth century, see Milam (2010).

35   Ronald A. Fisher proposed: Fisher (1915, 1930).

36   Fisher actually proposed: The two stages of Fisher's model have contributed to confusion about what "Fisherian" sexual selection is (1915, 1930). Does "Fisherian" refer to the first, adaptive stage or to the second, arbitrary stage? Or to a combination of the two? Throughout this book, "Fisherian" refers to Fisher's innovative description of the second stage of the sexual selection process.

37   "a runaway process": Fisher (1930, 137).

41   Around the centennial: This new awareness was marked by the publication of a volume of contributions edited by Campbell (1972) that included a highly influential paper on differential reproductive investment by Robert Trivers.

41   Russell Lande and Mark Kirkpatrick: Lande (1981); Kirkpatrick (1982).

44   Zahavi published his "handicap principle": Zahavi (1975).

45   "I suggest that sexual selection": Zahavi (1975, 207).

45   "Sexual selection is effective": Zahavi (1975, 207).

46   good genes are *different:* Some researchers have proposed that good genes and the Lande-Kirkpatrick mechanisms are merely points on a

continuum of indirect genetic benefits (Kokko et al. 2002). However, these mechanisms make diametrically opposite predictions about the evolved "meaning" of sexual ornaments and are still best understood as distinct evolutionary mechanisms (Prum 2010, 2012).

48   Alan Grafen at Oxford: Grafen (1990).

48   If a handicap is like a test: Another way to imagine the nonlinear costs of display traits is to think of these costs like money. The idea is that some individuals are quality impoverished and don't have enough, while others are quality rich and have plenty extra to spare. Just as a dollar is worth more to a poor man than to a rich man, the quality-impoverished individual will have to pay a larger relative cost for his ornaments than will a quality-rich individual. However, is variation in quality in natural populations distributed unequally like wealth? We don't know, because this vital assumption of the handicap principle has, as far as I know, never been explicitly tested in any animal species. After Grafen's (1990) proposal saved the handicap principle from its imminent intellectual demise, no one has apparently looked back to examine whether it is actually true.

48   "According to the handicap principle": Grafen (1990, 487).

49   "To believe in the Fisher-Lande process": Grafen (1990, 487).

49   "Fisher's idea is too clever": Grafen (1990, 487).

50   "The split between Fisher and Good-genes": Ridley (1993, 143).

52   the effect of redefining fitness: Ernst Mayr (1972) raised this exact issue in his chapter for the volume celebrating the centennial of *Descent*.

53   an authentically Darwinian view: Evolutionary biologists generally recognize four mechanisms of biological evolution: mutation, recombination, drift, and natural selection. This neo-Wallacean classification defines sexual selection as a form of adaptive natural selection. To restore a legitimately Darwinian framework to evolutionary biology, sexual selection should be added to this list as an independent, fifth evolutionary mechanism.

Chapter 2: Beauty Happens

54   "good evidence": Darwin (1871, 516).

58   the ornithologist G. W. H. Davison spent: Davison (1982).

58   most observations of Argus behavior: You can observe the display of the Great Argus by watching amateur videos on YouTube of captive individuals.

59   the shape of an inverted umbrella: Bierens de Haan (1926) cited in Davison (1982).

63   "There is no question": Beebe (1926, 2:185).

64   "ball and socket" designs: Campbell (1867, 202–3).

64   "it is undoubtedly": Darwin (1871, 516).

64   "Darwin's ideas": Beebe (1926, 2:185–86).

64   "It seems impossible to conceive": Beebe (1926, 2:187).

65 The paper discussed: Prum (1997).

68 "We may speak of this hypothesis": "null, adj.," *OED* online, March 2016, Oxford University Press.

70 "to guess better than the crowd": Keynes (1936, chap. 12).

71 in the 1950s, Ronald A. Fisher: Fisher (1957). For a detailed discussion of Fisher's advocacy of the safety of smoking, see Stolley (1991).

71 the Lande-Kirkpatrick sexual selection mechanism: For further details, see Prum (2010, 2012).

71 That is why it is the null model: Another famous null model in evolutionary biology is the Hardy-Weinberg law, which gives the frequency of genotypes in a population given the frequency of alleles, or gene variations. The Hardy-Weinberg law tells us what genotype frequencies we should expect in a population if nothing else is going on—including nonrandom mating, immigration, emigration, or selection. Biologists use observed deviations from Hardy-Weinberg to demonstrate that there is *something* special happening within a population. Interestingly, Fisher first proposed his mate choice theory in 1915, only seven years after the publication of Hardy-Weinberg. Like Hardy-Weinberg, Fisher's theory can best be understood as an attempt to describe the evolutionary consequences of the existence of genetic variation alone. In the case of mate choice, however, this variation is genetic variation in preference that selects on other genetic variation in the display trait. The Lande-Kirkpatrick models are mathematical realizations of that process.

71 Grafen's demand for "abundant proof": Grafen (1990, 487).

76 a recent "meta-analysis": Prokop et al. (2012).

78 Given free rein, mate choice: Pomiankowski and Iwasa (1993); Iwasa and Pomiankowski (1994).

79 the American Academy of Family Physicians: Mehrotra and Prochazka (2015).

79 it is very difficult to accurately assess: One could argue that annual physical exams are not cost-effective because the American population is so healthy, or that the human phenotype has specifically evolved to *conceal* genetic quality, health, and condition information from others, rather than to reveal it. But I doubt that either of these explanations is true.

80 the Food and Drug Administration: Alberto Gutierrez (director of FDA Office of In vitro Diagnostics and Radiological Health) to Anne Wojcicki (23andMe CEO), Nov. 22, 2013, FDA doc. GEN1300666. The FDA has subsequently given approval to 23andMe to market tests for specific genetic disorders.

81 "Unfortunately, I couldn't find the effect": Lehrer (2010).

81 meta-analyses of multiple data sets: Palmer (1999); Jennions and Møller (2002).

82 the "honesty of symmetry": One reason why the "honest symmetry" idea lives on in evolutionary psychology and neuroscience is that the

evolutionary biologists are so embarrassed by it that they no longer discuss it. This intellectual vacuum allows other disciplines to continue to cite this failed idea as if it were firmly established.

82   elaborate courtship displays: For example, Byers et al. (2010); Barske et al. (2011).

82   When Beauty Happens, costs will happen too: The vast majority of the papers on costly honest signaling assume that the existence of costly traits provides evidence for Zahavi's handicap principle. However, the Lande-Kirkpatrick null model also predicts the evolution of costly traits; the offset between the Lande-Kirkpatrick equilibrium and the natural selection optimum is an exact measure of the costs of being sexually attractive (see fig., page 42). To reject the Beauty Happens null, researchers need to demonstrate that the costly traits are specifically correlated with variation in direct benefits or good genes. This is much more rarely accomplished.

83   atonal twentieth-century concert music: This analogy to ballet and music may seem overwrought, but this same adaptive logic has been applied to explain the aesthetics of human art and performances. Denis Dutton (2009), for example, has proposed that human capacity for artistic creation and performance has evolved by mate choice for honest indicators of good genes and mental and physical capacity.

83   The value of a dollar was *extrinsic:* The further irony of the gold standard is that gold is itself assumed to have some intrinsic value. Although gold is a relatively inert metal and has plenty of useful physical properties, the establishment of gold as a "universal" standard of value is an arbitrary cultural phenomenon. This observation demonstrates how hard it is to establish any system of value that is not subject to arbitrary, aesthetic influences.

83   "social contrivance": This very apt phrase comes from Samuelson (1958).

84   Imagine that the next time you see a beautiful rainbow: I have momentarily violated my commitment to use beauty to mean *coevolved attraction.* Although we are obviously attracted to the rainbow, it has not, and cannot, coevolve with our evaluation of it (Prum 2013).

84   The burden of proof lies: The analogy between the value of beauty and money also provides an insight into the emotional energy used in defense of adaptive mate choice. Just as modern economics put goldbugs out of business, the Beauty Happens hypothesis poses an existential threat to the adaptationist worldview. Why? Because, to use St. George Mivart's phrase, adaptationism is based on its commitment to "the all-sufficiency of 'natural selection'" as an explanation of functional design in nature (1871, 48). Acknowledging any *intrinsic* evolutionary value to beauty would permit mate choice and aesthetic evolution to become unhinged from adaptation. The all sufficiency of adaptation would come tumbling down.

A further parallel between theories of the value of money and beauty comes from the observation that most currencies historically started with backing by an extrinsic commodity like gold. The social contrivance of value arises later once this currency creates a medium of economic exchange. This historical transformation from extrinsic to intrinsic value precisely parallels Fisher's two-phase model of the evolution of traits and preferences. The first phase begins with an adaptive indicator of some correlated extrinsic, adaptive benefit, but the origin of mating preference genes create a new opportunity for value—the indirect genetic benefit of having attractive offspring.

85   "The belief in the efficient market": Krugman (2009).

85   Shiller presented the case: Shiller (2015).

86   "To many economists, the mere existence": Conversation with Shiller, Sept. 16, 2013.

86   For the title of their 2009 book: Akerlof and Shiller (2009).

87   a team of economists published: Muchnik et al. (2013).

88   "Emperor wears no clothes": Prum (2010).

Chapter 3: Manakin Dances

90   research interest in phylogeny: The abandonment of the investigation of organismal phylogeny occurred during the first two-thirds of the twentieth century and was fostered by the notion that genetics and population genetics were the most appropriate and productive ways to investigate evolutionary questions. The result was that the mid-twentieth-century "New Synthesis" in evolutionary biology was a largely ahistorical science, based on population genetic machinery that aspired to emulate the ideal gas law—that is, $PV = nRT$; the pressure times the volume equals the temperature times the number of moles and the ideal gas constant. During the last decades of the twentieth century, it required a major intellectual battle to restore phylogeny and phylogenetics to their appropriate place in evolutionary biology, which provides a good ground plan for the future restoration of Darwinian aesthetic evolution. For a history of the early intellectual battle to restore phylogenetics to evolutionary biology, see Hull (1988).

90   Aesthetic radiation is the process: Aesthetic evolution can also proceed by various mechanisms of social selection. For example, when birds make choices about which baby bird mouth to feed, the plumages and mouth patterns may evolve to attract the attentions of parents. This process may result in the evolution of "cuteness"—attractive baby offspring.

91   Biogeography and Systematics Discussion Group: The group was run by the faculty adviser Bill Fink, an ichthyologist. The graduate students at the time included Michael Donoghue, a plant systematist, now a member of the U.S. National Academy of Sciences and one of my Yale colleagues; Wayne Maddison, a spider systematist and co-author with his identical twin brother, David Maddison, of MacClade, Mes-

quite, and other computer programs that made phylogenetic analysis of character evolution possible; Brent Mishler, a botanist and now curator at the herbarium at the University of California, Berkeley; and Jonathan Coddington, spider systematist, now at the Smithsonian.

91   observations of toucan plumage and skeletal characters: I published this research in Prum (1988) and Cracraft and Prum (1988).

91   Only, I don't smell like mothballs: Like all modern workplaces, natural history museums have had to respond to occupational health and safety regulations that limit workplace exposure to hazardous chemicals. In recent decades, museums stopped using paradichlorobenzene (mothballs) to control insect pests.

92   Jonathan Coddington's research: Coddington (1986).

92   a male Golden-headed Manakin: The basics of the behavior and reproduction of the Golden-headed Manakin have been described by Snow (1962b) and Lill (1976).

93   the tiny rapid steps: Kimberly Bostwick, my former doctoral student, was the first to describe the backward slide of *Ceratopipra* manakins as a moonwalk in an interview for the PBS *Nature* documentary *Deep Jungle* in 2005.

93   Lek breeding is a form of polygyny: For a review of the biology of leks, see Höglund and Alatalo (1995). Lek evolution is discussed in detail in chapter 7.

95   I also found the White-bearded Manakin: The display behavior and reproduction of the White-bearded Manakin have been described by Snow (1962a) and Lill (1974). The mechanism for the production of their mechanical wing *snap* sounds were established by Bostwick and Prum (2003).

96   all manakins evolved: The early, evolutionary origin of lekking in the common ancestor of the manakins was established in Prum (1994). The only non-lekking species in the family, the Helmeted Manakin (*Antilophia galeata*), is the sister group to the cooperatively lekking genus *Chiroxiphia* and is embedded deep in the phylogeny of the family. Thus, we can infer that the absence of lekking in *Antilophia* is an evolutionary loss, or reversal, in that species.

    The ages of the origins of living groups of birds are somewhat contested, but the most recent and well-supported estimates an age about fifteen million years for the manakins comes from Prum et al. (2015).

98   a land of milk and honey: Interestingly, like fruit, these icons of an easy life of pleasure—milk and honey—are both natural products that have specifically coevolved to be desirable and eaten.

98   Females used their capacity for mate choice: Snow's fruit-eating hypothesis for the evolution of polygyny is supported by the observation that many lekking birds are found tropical frugivores, including the manakins, the cotingas, the birds of paradise, and the bowerbirds. A similar ecological situation occurs in some obligate nectar feeders like hummingbirds. Like fruit, nectar *wants* to be eaten; it is a bribe

created by the plant to attract animal pollinators. Likewise, hummingbirds have entirely female parental care. Female-only parental care also occurs in birds with precocial young that can feed themselves immediately after hatching, including the pheasants, chickens, grouse, and their relatives. Because precocial young merely need to be watched and protected from predators, one parent can do the job as well as two. In the extreme case of complete brood parasitism, females lay their eggs in the nests of other species, and neither biological parent provides any parental care. In all these cases, uniparental care has resulted in the evolution of intense sexual selection through female mate choice and the evolution of male territorial display in arenas or leks.

98   Lekking birds feature so prominently: As the pioneering Yale ecologist George Evelyn Hutchinson (1965) wrote in the book *The Ecological Theater and the Evolutionary Play,* environmental conditions and ecological interactions create the setting in which evolutionary change takes place. Thus, a fruit-eating diet creates conditions that foster the evolution of polygynous breeding systems and extreme mate preferences. Other ecological conditions can give rise to other, *very* different breeding systems that have a big impact on patterns of aesthetic evolution. The vast majority of bird species have a pair bond in which males and females raise the young together. In many such species, like puffins and penguins, males and females have evolved identical sexual ornaments. Such ornaments evolve by *mutual mate choice* in which both sexes have the same traits and preferences, and both sexes are choosing. Some shorebirds exhibit polyandrous breeding systems with multiple male mates per female. For example, in the Plains Wanderer (*Pedionomus torquata*), the Painted Snipe (*Rostratula benghalensis*), and the long-toed, lily-trotting jacanas (*Jacana* species), females are larger, more brightly colored, sing the songs, and defend the territories against other females. If a female has a territory of high enough quality, she will be able to attract *multiple* males to nest with her. These smaller males each build a nest, incubate a clutch of eggs that she lays, and raise their young in her high-quality habitat. In these *polyandrous* species, mate choice is by *males.* However, the variation in reproductive success between the most successful and the least successful females is not nearly as great as the variation in sexual success among lekking male birds, so polyandrous birds do not evolve such aesthetic extremity as polygynous lekking birds.

99   David Snow and Alan Lill had already published: Snow (1962a, b); Lill (1974, 1976).

99   The species was so poorly known: Haverschmidt (1968); Mees (1974).

101   Marc Théry made later observations: Théry (1990).

101   had already been described: Snow (1961).

101   The courtship of the White-throated Manakin: Davis (1949).

105   Marc Théry in French Guiana: Théry (1990).

105   a spectacular *above-the-canopy* flight display: Davis (1982).

106 I came away with unique scientific observations: Prum (1985, 1986).

112 the other *Corapipo* manakins: There are three other *species* of *Cora-pipo* manakins from the Andes of Colombia and Venezuela (*C. leucor-rhoa*) and from the highlands of southern Central America (*C. altera* and *C. heteroleuca*).

115 Barbara and David Snow published: Snow and Snow (1985).

117 I developed a comprehensive hypothesis: Prum and Johnson (1987).

117 phylogeny of the entire manakin family: Prum (1990, 1992).

Chapter 4: Aesthetic Innovation and Decadence

122 the mechanical sounds of manakins: By examining manakin mechanical sounds across their phylogeny, we know that there have been multiple origins in manakins (Prum 1998).

123 adaptation provides at best an incomplete account: For an analysis of the limits of adaptation to explain morphological innovation, see Wagner (2015).

123 manakin display movements produced: Prum (1998); Clark and Prum (2015).

123 I first heard the wing songs: The wing songs of the Club-winged Manakin were briefly described by Edwin Willis (1966) from western Colombia. Willis hypothesized that the sound was produced by the "clapping of the thickened secondaries" but concluded that he could not eliminate the possibility that it was a vocalization. This was the only description of the song when my observations were made in 1985, and no recordings were available of the sounds.

124 The wing songs are a major component: Kimberly Bostwick (2000) followed up on the anecdotal observations of Willis (1966) with a full behavioral study of the display repertoire of the Club-winged Manakin in Ecuador.

124 Sclater's illustrations were reproduced: Sclater (1862); Darwin (1871, 491; fig. 35).

125 the first generation of field-worthy, high-speed video cameras: See Dalton (2002).

126 their Bronx cheer "roll snaps": Bostwick and Prum (2003). Interestingly, after the origin of mechanical sound in the genus, female *Manacus* were not satisfied merely with the firecracker-like *snap*. They continued to innovate through the addition of the mechanical sound repertoire with the roll-*snap* and the flight riffle (Bostwick and Prum 2003).

126 The tiny pumping movements: The fastest-contracting vertebrate muscles are all associated with sound production. For example, fast muscles produce rattlesnake rattles (around 90 Hz) and toadfish swim-bladder whistles (around 200 Hz) (Rome et al. 1996). However, each of these organisms makes a sound at the frequency of the muscle contractions. Club-winged Manakins link fast-cycling muscles to a

frequency-multiplying stridulatory organ to produce a much higher frequency communication sound.

128 Bostwick and other collaborators: Bostwick et al. (2009).

128 If vocal songs are already robust indicators: The quality of adaptive information of any sexual signal needs to evolve. This information is refined by the action of mate choice to become more and more closely correlated with quality. The problem is, if one mating display is already a robust indicator of quality, why should one ever change and abandon the adaptive advantages for new, untested ornament with initially worse-quality information? Honest signaling will constrain the evolution of display repertoires and aesthetic innovations.

129 Kim Bostwick has provided a definitive scientific answer: Bostwick et al. (2012).

129 Birds have only tinkered: See Chiappe (2007), Field et al. (2013), and Feo et al. (2015).

131 how the bizarre ulna morphology: Club-winged Manakins are uncommon birds that have rarely, if ever, been kept in captivity. It would be both logistically and legally challenging to bring them into a laboratory to make the necessary observations to measure their flight capabilities and physiology.

133 the morphological consistency in wing bone design: The argument that morphological stasis among species is evidence of maintenance by natural selection is an adaptationist view. Thus, in this example, adaptationists are trapped into either questioning this basic tenet or rejecting the hypothesis of adaptive mate choice in Club-winged Manakins.

133 the observation that *female* Club-wings: This "chooser decadence" evolves through the same genetic correlations that drive arbitrary aesthetic coevolution. Mate choice on display traits will result in genetic covariation between traits and preferences. This is why mate choice itself can drive the evolution of mating preferences. Similarly, as females select upon the male bodies through mate choices, they can also alter their *own* bodies in genetically correlated ways.

133 females will not be harming: Although it has been very difficult to generate direct evidence of preference coevolution between females and the display traits they prefer, the evolution of correlated female expression of male ornamental traits provides prima facie evidence that females do indeed coevolve through their mate choice on males.

134 the wing bones begin to develop: The cartilaginous precursors of the radius and ulna begin to develop on day 6 in chickens, and ossification begins on day 7 of incubation in ducks (Romanoff 1960, 1002). Sexual differentiation of the gonads begins on day 7 of incubation in chickens (Romanoff 1960, 822). However, sex hormones that are involved in sexual differentiation of non-gonadal body tissues begin to circulate around the body by day 10 of incubation when other sexually dimorphic organs begin to differentiate, like the syrinx (Romanoff 1960, 541, 842).

134 the costs of decadent display traits: Lande (1980).

135 female mate choice has resulted: This phenomenon is distinct from the evolution of genuine, female ornaments, such as in species with mutual mate choice, or in polyandrous species with male mate choice only. Rather, like Club-winged Manakins, females in these species exhibit features with purely ornamental functions that they will never use and, therefore, cannot benefit from having.

136 the distribution of feather follicles: Romanoff (1960, 1019); Lucas and Stettenheim (1972).

137 the crown feathers grow inward: The role of feather follicle orientation in the development of unusual bird crests has been demonstrated in domestic crested pigeons. Shapiro et al. (2013).

137 the critical *orientation* of the feather follicles: Shapiro et al. (2013).

139 the plumage colors of a black American Crow: For a review of avian feather melanins, see McGraw (2006).

140 we confirmed that the black stripes: Vinther et al. (2008).

141 Feathers first evolved: Prum (1999); Prum and Brush (2002, 2003).

141 the raptor-like dinosaur *Anchiornis huxleyi:* Li et al. (2010).

144 I proposed a model: Prum (1999).

144 the evo-devo theory: Prum and Brush (2002, 2003); Harris et al. (2002).

146 The advantage of this evo-devo approach: For further discussion, see Prum (2005).

147 *three* dinosaur lineages survived: Prum et al. (2015).

Chapter 5: Make Way for Duck Sex

149 "*Make Way for Ducklings*": McCloskey (1941).

153 Many ducks perform *sham preening* displays: Konrad Lorenz (1941, 1971) presented a detailed comparative analysis of the evolution of the courtship displays of ducks. The research was highly innovative and anticipated the future development of phylogenetic ethology. In this and other works, Lorenz proposed that one of the sources of novel communication signals is "displacement" behavior—random motor patterns that were originally performed at times of social or motivational tension. Accordingly, he proposed that sham preening displays evolved from such movements, just as some people might nervously play with their hair when they are on a first date. Over time, such behaviors could evolve to become communication displays through the process of ritualization, which involves exaggeration and reduced variation so that the display stands out from the rest of the bird's behavior.

156 Susan Brownmiller built a powerful: Brownmiller (1975).

157 "Because of the important differences": Gowaty (2010, 760). As Brownmiller (1975) has proposed in humans, Patty Gowaty has noted that sexual coercion in birds can foster the evolution of "convenience polyandry," in which a female accepts a male mate, or multiple male

mates, in order to protect herself from sexual violence by other males (Gowaty and Buschhaus 1998).

157 a desensitization to the social and evolutionary impact: For example, based on the idea that all female behaviors that bias male fertilization success are identical forms of adaptive sexual selection, Eberhard (1996, 2002), Eberhard and Cordero (2003), and others (for example, Adler 2009) have proposed that resistance to sexual attack is merely a form of mate choice. This "resistance as choice" mechanism leads to the proposal that rape is essentially adaptive for *females*. Accordingly, if a female resists all sexual attacks, then the male that is ultimately successful at fertilizing her offspring will inevitably be the best at sexual attack. Her male offspring will then inherit the capacity to excel at sexual attack, which produces an indirect, genetic benefit to her. The problem with this idea is that it ignores the direct costs to the female and the indirect costs to the female's *female* offspring. In other words, the benefits of having sons who are better rapists will have to balance against the disadvantages in the form of lower survival and fecundity of daughters who experienced sexual violence. I think the full, onerous intellectual implications of the "resistance as choice" hypothesis have been greatly obscured by the fact it has not been referred to, just as accurately, as the "rape as choice" hypothesis.

158 Females are often injured: For a recent review, see Brennan and Prum (2012).

159 Geoffrey Parker defined sexual conflict: Parker (1979).

159 Duck sex provides a premier example: For the evolution of sexual autonomy by this mechanism, it doesn't matter whether the indirect genetic benefits of mate choice are due to good genes or the arbitrary Beauty Happens mechanism; either will work.

160 the penis of the diminutive Argentine Lake Duck: McCracken et al. (2001).

163 Modern agriculture's answer: Artificial insemination is ubiquitous in modern agriculture. On the farm, mammals are never left to do it for themselves. Few people realize that nearly every bite of mammal flesh they eat—whether beef, pork, or lamb—begins with a prizewinning male animal, a farm employee, an artificial vagina, a liquid nitrogen tank, and a big syringe. However, for the most part, poultry are left to themselves, so this duck farm was a rare opportunity.

165 "Eversion of the 20 cm": Brennan et al. (2010).

167 Brennan showed that the longer and twistier the penis: Brennan et al. (2007).

167 A comparative analysis of penis and vaginal morphology: Brennan et al. (2007).

168 We hypothesized: The mechanism of sexually antagonistic coevolution in waterfowl is presented in full in Brennan and Prum (2012).

169 There have been both escalations *and* reductions: Brennan et al. (2007).

171 Conversely, we hypothesized: In her first attempt at these mechanical challenge experiments, Brennan made "artificial vaginas" out of silicone. As we predicted, when the straight or counterclockwise tubes were held up to the male cloaca at the moment of erection, the penis shot out completely unimpeded. However, when the hairpin turn or clockwise spiral was used, the penis became temporarily bottled up within the tube, after which it *burst* out the side of the tube by blowing a hole through the soft silicone. These tests provided a successful but anecdotal proof of concept. We were sure the penis had busted through the wall of the female-mimicking silicone tubes because its forward progress had been impeded, but we couldn't prove that it had been trapped. To get the data we needed, design improvements were necessary. That's when we moved to glass.

172 These observations confirmed: Brennan et al. (2010). High-speed video images of these experiments are available at the *Proceedings of the Royal Society B* and on YouTube.

172 expelling the unwanted sperm: Domestic chickens can eject sperm of males from unsolicited copulations (Pizzari and Birkhead 2000).

172 forced copulations are a stunning 40 percent: Evarts (1990). Discussed in Brennan and Prum (2012).

173 female ducks have indeed succeeded: Brennan and Prum (2012).

173 when female Muscovys were actively: Brennan et al. (2010).

174 Females do not, indeed cannot, evolve to assert power: The mechanism for the evolution of sexual autonomy functions through the shared, coevolved, normative agreement on what male traits are attractive and the cooperative advantages to *all* females of freedom of choice. Thus, unlike males, there is no available selection for females to take advantage of one another and assert their own individual desires over others. Thus, there is no selection for female power to directly confront male sexual aggression with a countervailing force for female sexual control.

174 Duckpenisgate: Asawin Suebsaeng, "The Latest Conservative Outrage Is About Duck Penis," *Mother Jones,* March 26, 2013. Suebsaeng reported, "The $16 muffin ain't got nothing on duck penis."

176 After Patricia Brennan wrote an awesome defense: Patricia Brennan, "Why I Study Duck Genitalia," Slate.com, April 3, 2013.

176 The *New York Post* headline read: S. A. Miller, "Government's Wasteful Spending Includes $385G Duck Penis Study," *New York Post,* Dec. 17, 2013. When I first read the headline, I asked myself, "What is G?" Someone had to point out to me that this stood for "grand." I assume that the (more logical to me) alternative $385K was considered an endorsement of the metric system, and perhaps of a one-world government under the United Nations.

176 If the pharmaceutical industry: The NSF-funded research program that was attacked in these news reports was specifically on the season-

ality of duck penis development and the effects of social environment and competition on duck penis size.

177   sexual violence is against the will: These insights into the evolution of sexual conflict contradict another, major, reductionist trend in contemporary evolutionary biology—the concept of the selfish gene. In *The Selfish Gene*, Richard Dawkins (2006) proposed that the gene is the essential level of selection and that individual organisms are merely "bags" that propagate their selfish genes. While gene selection can occur, duck sex teaches us that sexual conflict over fertilization *cannot* be reduced entirely to gene-level selection. Except for the tiny fraction of their genomes that controls sexual differentiation, male and female ducks have all the *same genes*. Female ducks have genes for long, spiky, harmful penises, and male ducks have genes for convoluted, clockwise vaginal morphologies. Genes for vaginal and penis morphologies are *not* competing with each other to propagate copies of themselves into future generations. These genes don't have a sex. Rather, it is only the *individuals* that have a sex, and it is only at the level of the individual organisms that the sexual conflict over fertilization can occur.

This observation is easily proven by looking at the evolution of sexual conflict in turtles, in which sexual determination is temperature dependent; warmer eggs become female, and colder eggs become male. There are *no genetic differences* between male and female turtles. Yet sexual conflict in turtles is rampant. Male tortoises sexually harass female tortoises by aggressively attempting to mount and copulate with them, and the costs of this harassment to females is significant. Selfish genes simply cannot explain the evolution of sexual conflict in a species that lacks genetic differences between the sexes. A similar analysis could be applied to hermaphroditic animals that simultaneously produce ova and sperm. In this case, selection is taking place at the emergent level of the organ, or gonad, and not at the level of the gene.

178   Birds originally inherited the penis: For a review, see Brennan et al. (2008). The penis first evolved in the exclusive common ancestor of mammals and reptiles. In living birds, the penis is present in all ratites and tinamous (that is, the ostrich and its kin) and in all waterfowl. The penis is also present in a few groups of game birds (Galliformes) that are most closely related to the waterfowl. These groups are descendants of the most ancient, independent lineages of the living birds, and they have inherited the reptilian penis from their dinosaur ancestors. The penis has been lost several times independently in tinamous, in various groups of game birds, and in the ancestor of all Neoaves—the group that includes 95 percent of species of birds of the world.

179   barnyard hens can eject sperm: Pizzari and Birkhead (2000).

179   female neoavian birds have: Interestingly, many Neoavian birds have evolved a cloacal protuberance: a short, button-shaped bump around

the cloaca that develops during the breeding season. This structure may have evolved as a male counter measure to the loss of the penis to allow a male to force open the female cloaca during forced copulations.

179 "On the whole, birds appear": Darwin (1871, 466).

Chapter 6: Beauty *from* the Beast

182 the aesthetic structures created by male bowerbirds: The biology and natural history of bowerbirds is beautifully surveyed by Frith and Frith (2004).

186 the word "bower" referred: "bower, n.1," *OED* online, March 2016, Oxford University Press.

187 much more elaborate avenue "bower-plans": No clarification is less called for than an explanation of a bad pun. But this one actually raises an interesting issue. In the field of developmental evolutionary biology, the concept of the body plan refers to the fundamental anatomical layout shared by members of the same higher groups, or phyla. Dating back to the Romantic poet, writer, and natural historian Johann von Goethe, the body plan concept was originally coined in German as *Bauplan*. Here, the question now becomes what term do we use to refer to the "body plan" of the extended phenotype? The *ExBauplan*? For the singular, aesthetic, extended phenotype of the male bowerbirds, the avenue and maypole architectures, and their variations, are perfect examples of the concept of the "bower-plan."

190 brilliant, iridescent fragments: We had the pleasure of describing the photonic crystal nanostructures in these extraordinary blue scales of *Entimus* weevils (Saranathan et al. 2015).

191 the male and the female stand on opposite sides: Frith and Frith (2004).

191 The bowerbird family (Ptilonorhynchidae): Frith and Frith (2004).

191 Coined by Richard Dawkins: Dawkins (1982).

192 As a confirmed neo-Wallacean: The only voice I know that has enthusiastically embraced the "neo-Wallacean" label is Richard Dawkins. In his book *The Ancestor's Tale,* Dawkins (2004) eagerly described the discoveries of Zahavi, Hamilton, and Grafen as "sophisticated neo-Wallacean" triumphs over Darwinian vagueness. Dawkins paints the following portrait of the Darwin-Wallace debate (2004, 265–66):

> For Darwin, the preferences that drove sexual selection were taken for granted—given. Men just prefer smooth women, and that's that. Alfred Russel Wallace, the co-discoverer of natural selection, hated the arbitrariness of Darwinian sexual selection. He wanted females to choose males not by whim but on merit . . . For Darwin, peahens choose peacocks simply because, in their eyes, they are pretty. Fisher's later mathematics put that Darwinian theory on a sounder mathematical footing. For Wallaceans, peahens choose

peacocks not because they are pretty but because their bright feathers are a token of their underlying health and fitness . . . Darwin did not try to explain female preference, but was content to postulate it to explain male appearance. Wallaceans seek evolutionary explanations of the preferences themselves.

Instead of taking Darwin's aesthetic language as a hypothesis about the evolutionary elaboration of traits and preferences, Dawkins confounds the arbitrariness of Darwinian sexual traits with a perceived ambiguity about his evolutionary mechanism for the origin of preferences. The anti-aesthetic Wallaceans are portrayed as scientifically progressive, while aesthetic Darwinism is portrayed as fuzzy, lazy, and incomplete. Although Dawkins admits Fisher's more solid theoretical grounding for the arbitrary, he does not entertain any modern Darwinian alternative to the Wallacean solution. Because the Fisherian answers don't provide the comforting "rhyme and reason" of the neo-Wallacean solutions, they aren't even entertained as scientific answers.

192 The earliest branch in the phylogeny: Though somewhat out of date in terms of data quality and quantity, the most current phylogeny of the bowerbirds is Kusmierski et al. (1997). The Australopapuan catbirds (*Ailuroedus*) are not related to the common North American Gray Catbird (*Dumetella carolinensis*), which is a member of the mockingbird family (Mimidae).

193 nest construction in catbirds: Frith and Frith (2001).

193 It's a stage set with props: Prior to the late twentieth-century revival of sexual selection theory, bowers were explained with an updated version of Mivart's idea of sensory stimulus as a form of adaptive physiological coordination between the sexes (see chapter 1). Jock Marshall (1954) proposed that in the absence of a pair bond, the ancestral female bowerbird needed extra sexual stimulus to be induced to copulate and reproduce. Marshall hypothesized that the bower evolved as a male method of reminding females of the sexually stimulating shared, ancestral nest of their evolutionary past. This would induce them to copulate and to build their own nest and continue with reproduction. This idea fails on so many levels that it is probably best left to the historical past, but it does document the intellectual contortions that evolutionary explanation achieved during the twentieth century in the absence of a theory of evolution by mate choice.

193 the females visited between 1 and 8 males: Uy et al. (2001).

194 the Tooth-Billed Bowerbird is a polygynous species: For the natural history of the Tooth-Billed Bowerbird, see Frith and Frith (2004).

195 pioneering work done by Jared Diamond: Diamond (1986).

195 Diamond did experiments: Diamond (1986).

196 Albert Uy repeated these ornament color choice experiments: Uy and Borgia (2000).

197 Jared Diamond established: Diamond (1986).

197 Joah Madden and Andrew Balmford conducted: Madden and Balmford (2004).

197 John Endler and colleagues: Endler et al. (2010); Kelly and Endler (2012).

198 an optical illusion known as forced perspective: This forced perspective illusion is the same phenomenon I propose to be at work in the array of three hundred golden spheres in the secondary feathers of the male Great Argus (see chapter 2).

198 Laura Kelley and John Endler: Kelley and Endler (2012).

198 this illusion could provide honest information: Unless the proposed "good brain genes" that make it possible for male bowerbirds to create these optical illusions are heritable by females, and useful to female survival or fecundity, then the optical illusions are *not* evolving as indicators of "good genes" per se. It is possible that female aesthetic selection on male display behavior could result in neural evolution and innovation, but if these neural advances are only used in aesthetic display and evaluation, then they are merely aesthetic innovations. Thus it is possible that when Beauty Happens, the artistic mind coevolves.

199 In a *New York Times* interview: Bhanoo (2012). I will return to the subject of animal art in chapter 12.

199 Gerry Borgia and Stephen and Melinda Pruett-Jones: Borgia et al. (1985).

199 Borgia proposed a compelling: Borgia (1995).

200 Borgia described the extremely abrupt courtship: Borgia (1995).

200 Borgia and Daven Presgraves investigated: Borgia and Presgraves (1998).

201 Borgia's threat reduction hypothesis: For a more detailed discussion, see Prum (2015).

201 females are frightened by the aggressive male displays: Although Gerry Borgia agrees that the source of the selection on female preference is an indirect, genetic benefit of control over paternity, he sees the outcome as an evolutionary negotiation between male and female interests. However, I think there is ample evidence that females have evolved to gain complete freedom of choice over fertilization and paternity identity. Males build bowers, gather and arrange decorations, sing, and display to visiting females because females have selected on them to do so. There is no other game in town. By controlling the standards of beauty and by evolving standards of beauty that empower female autonomy, females have gained near-complete control over the outcome of sexual selection.

201 the threat reduction response evolves: Prum (2015).

203 aesthetic remodeling proceeds through a correlation: To be easier on readers, I have described the association between the display trait and the phenotypic feature that contributes to female sexual autonomy as a correlation, but this is imprecise. The association is really a cova-

riance, in which specific genetic variations for the trait co-occur in the same individuals as specific genetic variations for the autonomy-enhancing phenotypic feature.

204 Gail Patricelli developed: Patricelli et al. (2002, 2003, 2004).

205 Patricelli was able to confirm: Patricelli et al. (2004) also found that the females' tolerance of intense display behavior was not related to the order of visitation or their familiarity with particular males from previous breeding seasons. Rather, the biggest predictor of female tolerance of intense display was the actual attractiveness of the male, the quality of his decorations, and the quality of his bower.

Chapter 7: Bromance Before Romance

207 leks have evolved in a wide variety: For a review of lek diversity and evolution, see Höglund and Alatalo (1995).

207 In "The Law of Battle" section: Darwin (1871, 468–77).

207 in the "Vocal Music" and "Love-Antics and Dances" sections: Darwin (1871, 477–95).

208 *The Life of Birds:* Welty (1982, 304).

208 "a forum for male-male competition": Emlen and Oring (1977).

208 males cannot actually gain: Bradbury (1981). Bradbury showed that the increases in the number of males would provide a linear increase in the volume of a group advertisement. But the power of the signal is inversely proportional to the square of the distance away from the source. So these linear increases in volume are not enough to increase the active area of the lek per male and provide each male with a proportional increase in female visits.

209 the "hotspot" model: Bradbury et al. (1986).

209 the "hotshot" model: Beehler and Foster (1988).

209 some Blue-crowned Manakin: Durães et al. (2007).

209 Durães captured and analyzed: Durães et al. (2009).

209 Bradbury proposed the revolutionary hypothesis: Bradbury (1981). Bradbury's female choice model is an adaptive model involving natural selection on female preferences to minimize the costs of searching for a mate.

210 David Queller went even further: Queller's model (1987) was a simple adaptation of Kirkpatrick's haploid model (1982) of the Fisher process in which lek size was treated as a male display trait. Of course, an individual male cannot display his genes for lek size all by himself, but in a large enough population of mobile individuals groups of males with genes for greater sociality might establish themselves in clusters and receive reproductive benefits if females preferred to mate in aggregations.

214 coordinated and cooperative behavior: Coordinated and cooperative male display has evolved multiple times independently within the family (Prum 1994).

215   males perform a coordinated version: Prum and Johnson (1987).

215   descriptions of coordinated displays: These behaviors have been described by Snow (1963b), Schwartz and Snow (1978), Robbins (1983), and Ryder et al. (2008, 2009).

217   *Chiroxiphia* males engage: General descriptions of *Chiroxiphia* manakin display behavior and breeding system can be found in Snow (1963a, 1976), Foster (1977, 1981, 1987), McDonald (1989), and DuVal (2007a).

217   perform coordinated displays: In some *Chiroxiphia* species, some copulations take place without the group displays. For example, Emily DuVal (2007b) has documented that nearly 50 percent of copulations in Lance-tailed Manakins (*Chiroxiphia lanceolata*) occur after a bout of solitary display by a single male without an immediate coordinated display preceding. However, it is unknown whether these females had previously observed a bout of group display by the same males. Furthermore, this copulating male was always an alpha male territory owner with a beta partner; his beta partner was merely missing when this female visited. Apparently, there is no route to sexual success by males that never display in partnerships or groups. So, coordinated display is obligate, at least at the level of the breeding system if not individual female visits.

219   the timing of the vocal coordination: Trainer and McDonald (1993).

219   each year they molt: For descriptions of molt in *Chiroxiphia,* see Foster (1987) and DuVal (2005).

220   males typically spend several more years: As DuVal has documented, some male Lance-tailed Manakins become alphas as soon as they acquire mature plumage, without serving as a beta to any other male. Apparently, these particular guys really have whatever it takes to succeed in this social competition.

220   a very few males may obtain: McDonald (1989).

220   By using the DNA fingerprints of chicks: DuVal and Kempenaers (2008).

221   even by me in the 1980s: Prum (1985, 1986).

221   males within displaying partnerships: McDonald and Potts (1994).

222   another example of the evolutionary cascade: Recall from chapter 3 that an ancestor of the Pin-tailed Manakin first evolved the tail-pointing posture, which created the opportunity for the evolution of elongate tail feathers featured in this display in the Pin-tailed Manakin. However, we can see that such patterns are not deterministic, because while the Golden-winged Manakin shares the same tail-pointing display, it never evolved a pointy tail. Similarly, although the origin of coordinated display ultimately transformed in the obligate coordinated male courtship display in the *Chiroxiphia* manakin, no such evolutionary change has occurred in other manakin species.

223   McDonald has pioneered: McDonald (2007).

223 the best predictor of a young male's future sexual success: McDonald (2007).

223 social connectedness among young: Ryder et al. (2008, 2009).

Chapter 8: Human Beauty Happens Too

227 a field called evolutionary psychology: Contemporary evolutionary psychology is an intellectual descendant of the science of sociobiology, which was championed by E. O. Wilson and others in the 1970s and 1980s. Sociobiology was based on the hypothesis that the social and sexual behaviors we see in both humans and other animals could be explained through adaptive evolution by natural selection. In recent decades, human sociobiology has been succeeded by evolutionary psychology, which shares the same adaptationist goals. But it has gone much further, incorporating neo-Wallacean ideas, which preclude the possibility of an authentically Darwinian mechanism of aesthetic mate choice.

Of course, the study of the evolutionary history of the human species and how it has shaped our sexuality, psychology, cognition, linguistics, personality, and so forth is a profoundly fascinating and fertile discipline. In fact, all of the work in these chapters about human evolution could be considered speculative new theory in the field of evolutionary psychology in this broad sense. The problem is that the field of evolutionary psychology as it is currently construed is *not* that discipline.

227 Evolutionary psychologists view: Two brief examples provide a flavor of evolutionary psychology research on mate choice. Aki Sinkkonen (2009) proposed that the "umbilicus" (yes, that's the belly button) evolved as an honest signal of mate quality in bipedal, furless humans, even though the belly button has existed for 200 million years in all placental mammals, long before bipedality (or furlessness) appeared. Hobbs and Gallup (2011) also "discovered" that 92 percent of the lyrics from popular hit songs from the *Billboard* charts include "embedded reproductive messages." Who knew? Of course, the existence of these messages about fidelity, commitment, rejection, arousal, and body parts supports their hypothesis that popular music has "adaptive value."

227 There is never any doubt: For broader critiques of the intellectual and empirical problems of evolutionary psychology, see Bolhuis et al. (2011), Buller (2005), Richardson (2010), and Zuk (2013).

227 evolutionary psychology is bad science: A prominent example of a failing zombie idea in evolutionary psychology is the continued interest in the hypothesis that deviations from body symmetry communicate individual genetic or developmental flaws and that humans have evolved adaptive mating preferences for symmetrical faces and bodies as a result. This "fluctuating asymmetry" hypothesis originated in the

study of birds in the early 1990s, but was soon soundly rejected and became a famous example of a failed theory (see chapter 2, "Beauty Happens"). Yet, twenty years later, the idea still thrives in evolutionary psychology.

Even evolutionary psychologists admit that there is no evidence that human facial symmetry is associated with superior genes or development (Gangestad and Scheyd 2005). There is also no consistent evidence that people actually prefer symmetrical faces. In fact, human facial diversity (including asymmetry) is not an accident. Rather, diversity in human facial appearance has likely evolved under strong social selection for indicators of individuality (Sheehan and Nachman 2014). Complex social interactions are based on recognizing others as individuals, and then treating them accordingly. Faces are diverse because there are evolutionary advantages to being recognizable as *you*. One of the primary features that makes faces recognizable is facial *a*symmetry. Given our neural mechanisms of facial cognition, symmetrical faces are simply harder to register, to recognize, and to remember. Humans are highly evolved to recognize and remember the features of individual faces, and therefore to find some asymmetry more appealing than symmetry. This phenomenon is not limited to people. For example, some highly social paper wasps have evolved distinctive, asymmetrical face patterns and the ability to learn and recognize them (Sheehan and Tibbets 2011).

Symmetrical faces are not especially beautiful, because symmetry is bland. Bland is not beautiful, and facial symmetry can be the ultimate in bland. By contrast, asymmetry is actually *attractive,* in part because it is recognizable. This is why three of the twentieth century's most glamorous and sexually idolized American women—Marilyn Monroe, Madonna, and Cindy Crawford—came to fame with prominent, symmetry-defying facial moles. It's also why the majority of hairstyles—like side parts—create and enhance facial asymmetries. Of course, monstrous asymmetries are unattractive, but so are monstrous symmetries. Think Cyrano de Bergerac.

The adaptationist hypothesis that we have evolved a preference for symmetry because it is an indicator of genetic quality is a zombie idea that refuses to die despite all the evidence to the contrary, because people are ideologically committed to believing it. Researchers will go to practically any length to keep the zombie alive, no matter how dubious the kinds of evidence they have to turn to for support. For example, a team of evolutionary psychologists from Rutgers University, including the famous sociobiologist Robert Trivers, published a study of symmetry in 185 Jamaican men and women in *Nature* (Brown et al. 2005). Their paper claimed that human dancing ability is an indicator of underlying body symmetry, and therefore an honest signal of genetic quality, which is why we have evolved to admire good dancing and to consider it sexy. The paper was featured on the cover of *Nature* and was covered in newspaper stories and media reports around the world. Unfortunately, the data were too good to be true. Several years after publication, Trivers himself uncovered irregularities in the data set and began to discredit the paper as a fraud perpetrated by one of his co-authors. Ultimately, a full investigation by Rutgers University concluded that there was "clear and convincing evidence" of data fabrication by the postdoc and lead author on the study. The paper was finally retracted by *Nature* in December 2013. See Reich (2013).

228 humans have evolved bones: An excellent discussion of evolutionary context 1 is presented by Neil Shubin (2008) *Your Inner Fish*.

233 better body-cooling efficiency: See Bramble and Lieberman (2004) and Lieberman (2013).

234 Elizabeth Grice and colleagues have written: Grice et al. (2009, 1190).

234 average lifetime numbers of sex partners: Accurate data on lifetime numbers of human sexual partners are difficult to obtain. There is an entire literature studying how men and women distort their self-reported numbers of sexual partners—men up and women down—to meet cultural expectations. Terri Fisher (2013) has shown that young American women reported higher numbers of lifetime sexual partners when they were attached to a bogus lie detector than not, but young men gave *lower* numbers of sexual partners with the bogus lie detector. No such pattern was found for reporting nonsexual behavior. Interestingly, among the nonrepresentative sample of men and women enrolled in a psychology course at a major American university, women reported more lifetime sexual partners. Unsurprisingly, distortions in reporting number of sexual partners are biased in the direction that conforms with culturally accepted norms for men's and women's sexual behavior and for the predictions of evolutionary psychology.

    Some comprehensive data on lifetime sexual partners come from a study of sexual behavior in Sweden. Lewin et al. (2000) reported that the differences between lifetime numbers of sexual partners increased considerably between 1967 and 1996 for both sexes, but the median number of partners were not that different (1967: 1.4 for women, 4.7 for men; 1996: 4.6 for women; 7.1 for men). The differences between men and women were mostly the result of the activities of a small subset of the most sexually active males. The differences between men and women are smaller than the differences between 1967 and 1996.

235 males should be expected to evolve: Male sexual choosiness is not unique to humans and is quite common among insects that have no male intromittent organ. Like Neoavian birds, in those insects that have evolutionarily lost any male "penis" or intromittent organ, there is an associated advance by females in sexual conflict over fertilization and reproductive investment. To induce a female to accept his sperm, these male insects present a "nuptial gift" to the female before mating that consists of either a nutritious bug or an especially calorie-rich,

edible spermatophore. Nuptial gifts greatly increase a female's fecundity because she can turn those nutrients directly into more eggs. Consequently, females evolve to demand greater male investment (direct benefits) as part of reproduction. Predictably, many female insects have evolved to mate multiply in order to acquire multiple gifts. But these nuptial gifts are expensive for males to produce, and males of many insects have evolved to be quite choosy about mating. For example, in some species of dance flies, females have entirely lost their feeding mouth parts and must rely entirely on male nuptial gifts. As a result of male mating preferences, females have coevolved inflatable abdominal display sacs. Funk and Tallamy (2000) interpreted the exaggerated, swollen female body ornaments of long-tailed dance flies as "deceptive" manipulations of male preferences for indicators of high female quality—large bodies swollen with eggs. But their data are entirely consistent with Fisher's original model of an initially informative display trait leading to the coevolution of preferences for an entirely arbitrary and meaningless trait—a big beautiful abdomen.

237 there *must be* something of greater value: Mating value is a great example of how a cultural imperative—the need to see the sexually and socially successful as objectively better—has become reified into a scientific concept that excludes any other possibility. Once the concept of mating value exists, all questions of mate choice and sexual success are then framed to produce only adaptive answers.

237 one well-known study looked at a sample: Jasienska et al. (2004).

238 attractive, arbitrary traits: Once males evolve preferences for younger, more fertile females with more "feminine" facial features, it is possible that any subsequent exaggerations in facial femininity that arise in a population would become especially attractive. But these variations would only add noise to the original correlation between the "feminine" features and age, or "reproductive value." The result would be the arbitrary elaboration of these new distracting and dishonest femininity traits away from the adaptive origins of the preference for honest information about actual age. This is exactly the scenario that Ronald A. Fisher proposed in his "runaway" model of popular, arbitrary traits evolving from initially honest, adaptive traits (see chapter 1). This is also a great example of why it is so hard to maintain an honest sexual signal.

239 "Scant research has addressed": Gangestad and Scheyd (2005, 537).

240 Numerous studies have shown: Gangestad and Scheyd (2005) cite two studies that support female preferences for somewhat masculine features, two supporting preferences for somewhat feminine features, and three that find no particular pattern.

240 females prefer a light stubble: Neave and Shields (2008).

240 the "male gaze": The phrase "the male gaze" was coined by Laura Mulvey (1975) in her essay "Visual Pleasure and Narrative Cinema." Since that time, the term has expanded to refer not just to cinematic

or artistic depictions of women but to the power of chauvinistic and patriarchal attitudes to women and women's bodies, and to the self-concepts of female attractiveness internalized by women themselves to accommodate these expectations as a consequence.

241 how social interactions alter perceptions: Eastwick and Finkel (2008); Eastwick and Hunt (2014).

241 "This idiosyncrasy will prove fortuitous": Eastwick and Hunt (2014, 745).

243 Darwin himself struggled: Darwin (1871, 248–49).

244 *The Third Chimpanzee:* Diamond (1992).

245 Gallup and colleagues tested this hypothesis: Gallup et al. (2003).

247 "the most variable of all bones": Romer (1955, 192).

247 PRICC: Two mammalogists that I queried about *baculum* evolution told me about the PRICC mnemonic, but they both cautioned me to remember that insectivores are no longer considered monophyletic. But PRICC lives on in mammalogy class notes because some intellectual conveniences are worth tolerating a little polyphyly.

249 "Genesis 2:21 contains": Gilbert and Zevit (2001, 284). The mammalian and reptilian penises are homologous. The original vertebrate penis had an external groove, or sulcus, for the transport of semen, which is still retained in the penises of birds, crocodiles, lizards, and snakes. The mammalian penis evolved an enclosed urethra by fusing the two edges of the sulcus to create a new tube from this groove.

249 "A female who behaves": Dawkins (2006, 305–8).

251 his penile handicap hypothesis: Cellerino and Janini (2005).

251 The convergence of these various features: Is it a coincidence that the only other primates that have lost the *baculum*—the spider (*Ateles*) and woolly (*Lagothrix*) monkeys—also have prominently dangling genitalia? Interestingly, however, the genital dangle is displayed by the pendulous clitoris of *female* spider and woolly monkeys. The function and evolution of this female genital display is not well understood. However, some mammals have a homologous clitoris bone called the *os clitoridis*. Perhaps social selection for the loss of the *os clitoridis* and clitoral dangle led to the loss of the homologous bone in the penises of spider and woolly monkeys.

252 the evolution of bipedality: Maxine Sheets-Johnstone (1989) has also proposed that human penis size, shape, and display have evolved for an aesthetic display that includes visual and tactile components. She goes further to suggest that bipedality itself might have evolved partly through sexual selection to enhance the genital display.

Jared Diamond (1997) rejected the aesthetic dangle hypothesis because of anecdotal evidence that many women do not find men's penises particularly attractive. However, I think these responses from contemporary women are likely to be highly influenced by the fact that penises are mostly covered up in the modern world by clothing. Because they are rarely seen, women have little opportunity to evaluate them comparatively. By comparison, I think that noses would look pretty weird and unattractive as well if they were rarely visible and only revealed immediately before kissing began.

252 An aesthetic function: Smith (1984).

254 mate choice does not have to end: William Eberhard (1985, 1996) established that mate choice can act on features evaluated during copulation in those species that mate multiply.

257 girls of normal body weight: See Haworth (2011).

257 an anonymous man wrote a piece: The way in which this anonymous sexual liaison story was used during the campaign against the political candidate was so unfair and irresponsible that I hesitated to mention it. But the man's story provides an extraordinarily vivid example of the power of cultural fashion to shape human sexual behavior, so I have left out the details of the politician's name and so on.

257 "the waxing trend": The increasing prevalence and recent rapid increase in extreme forms of pubic hair grooming by American women have recently been documented by Rowen et al. (2016).

258 men are quite sexually picky: As this anonymous report documents, many cultures police sexual practices by deploying the powerful emotion of disgust. Although disgust is a deeply biologically structured emotion, the specific things that *elicit* disgust—foods, odors, or sexual practices—can be extremely variable and highly culturally determined. Sexual practices can be particularly effectively regulated through cultural stories that recruit the emotion of disgust. The disgust with pubic hair reported by this anonymous blogger is an example of how fast these cultural mechanisms can change.

258 the evolution of lactose tolerance in adults: Studies of the coevolution of genes, culture, and human diversity were pioneered by William Durham (1991), who introduced lactase expression evolution as an example of a cultural top-down effect. Genetic and genomic research on the evolution of lactase expression in adult humans has been reviewed by Curry (2013).

In the absence of lactase, lactose ingested into the digestive system is broken down by bacteria in the large intestine, causing bloating, pain, and gas.

Because this evolutionary process has been far too recent to have resulted in the fixation of genes for adult lactase production, many people on the planet are lactose intolerant. Recent genomic studies of this phenomenon have discovered strong evidence of natural selection for mutations at several sites upstream of the lactase gene in a region that is known to be involved in the regulation of lactase enzyme expression. This source of natural selection has not been strong enough, or universal enough, to result in the complete fixation of this genetic novelty in all human populations. There are still many populations on the planet—especially east Asian and many African populations that have not had a history of dairy culture—who have not evolved to produce lactase as adults.

258 cultural ideas about beauty: Similar ideas have previously been discussed by Charles Darwin (1872), Jared Diamond (1992), and Jerry Coyne (2009, 235).

260 strong natural selection for darker skin: Jablonski (2006); Jablonski and Chapin (2010). Diamond (1992) questions whether skin color has any adaptive basis and hypothesizes that all variations in human skin color are the result of arbitrary social and sexual selection.

261 cultural preference for this kind of female body shape: Cultural top-down effects may also influence the evolutionary future of human beings. The distribution of underarm and pubic hair strongly indicates that body odors produced by an interaction of secreted pheromones, sweat, and the microbiota of the skin have coevolved as sexual communications. Many of us can identify the body odors of specific individuals and have experienced the particular attraction to the body odors of our partners. Yet the culture of hygiene—that is, frequent washing of the body with soap and application of deodorants to eliminate body odors and the removal of body hair—likely influences what body odors people think are culturally acceptable and sexually attractive. Furthermore, hygienic cultural concern about the risk posed by the bacteria lurking in human bodies, body parts, body cavities, and bodily fluids can also influence people's sexual behavior. Ultimately, the culture of hygiene could disrupt millions of years of human intersexual chemical communication and aesthetic coevolution. Mate choices by generations of people practicing modern hygiene could contribute to the loss of human pheromone specificity and sensitivity. The culture of hygiene could eliminate an entire sensory dimension of human sexual beauty. Of course, people would still smell; body odors would just cease to be beautiful.

261 cultural mating preferences: Bailey and Moore (2012).

Chapter 9: Pleasure Happens

264 woman's sexual pleasure is a nonlinear, exponential increase: A mathematician colleague, Michael Frame, expressed mystification with my logic. It is true that two numbers—1 and 9—cannot by themselves imply any correlation other than a straight line, a linear relationship. But I am asking us to think poetically about numbers in a way that I imagine was instinctive for the Greeks. The strongest association for the number 9 is, I think, as 3^2, which implies a pleasure difference that is squared, more expansive rather than merely larger.

265 Perhaps no topic in human sexual evolution: Elisabeth Lloyd's *Case of the Female Orgasm* (2005) provides an excellent review of the literature on this fascinating question. Pavličev and Wagner (2016) pro-

vide a new hypothesis for the original, ancient evolutionary origin of orgasm in placental mammals. They propose that female orgasm originally evolved as the sensory signal for ovulation when ovulation itself was induced by copulation.

265 sexual pleasure as merely: Freud's theory of orgasm in women was also an "adaptive" theory of sexual function, but from a psychological rather than an evolutionary perspective. For Freud, the move from clitoral to vaginal orgasm was necessary for the development of a woman's full sexual and emotional maturity. The "right" kind of orgasm provided the *direct benefit* of helping women overcome the psychological challenges of moving from their infantile mother attachment to mature fitness-enhancing, heterosexual relations. In this sense, both evolutionary and psychological "adaptation" involve an appropriate, functional fit between the phenotype and the environment.

266 "vicious feminine caprice": Mivart (1871, 59).

266 Freud's failed theory: Freud's theory had a devastating toll on educated and privileged women throughout Europe and the United States. As Alfred Kinsey (1953) wrote in *Sexual Behavior in the Human Female,* "This question is one of considerable importance because much of the literature and many of the clinicians, including psychoanalysts and some of the clinical psychologists and marriage counselors, have expended considerable effort trying to teach their patients to transfer 'clitoral responses' into 'vaginal responses.' Some hundreds of women in our own study and many thousands of the patients of certain clinicians have consequently been much disturbed by their failure to accomplish this biological impossibility."

266 Symons's by-product hypothesis: Gould (1987); Lloyd (2005).

266 "Male and female both have the same": Sutherland (2005).

267 female orgasm is broadly distributed: The criteria for female orgasm defined by Masters and Johnson (1966) include increased heart rate and rapid vaginal and uterine contractions. These variables have been measured in captive female stump-tailed macaques (*Macaca arctoides*) (Goldfoot et al. 1980). Although captive female stump-tailed macaques can apparently experience orgasm during male-female copulation, it is much more frequent during female-female mountings (Chevalier-Sklonikoff 1974).

268 Lloyd goes on to document: Lloyd (2005) pays particular attention to a monumentally flawed but highly cited paper by Baker and Bellis (1998). Furthermore, she points out that several influential studies correlating female orgasm during intercourse to the attractiveness or symmetry of the women's male partners are flawed because they fail to test the sperm competition hypothesis. No published studies have actually tested the upsuck hypothesis that women orgasm more frequently during intercourse with genetically superior men when they have multiple sexual partners during the same estrous cycle.

269 "propitiousness of [their] mating circumstances": Puts (2007, 338).

269 A fundamental problem with the upsuck hypothesis: For example, Baker and Bellis (1998) propose that variation in orgasmability indicates the strategic variation in orgasm among women and their mating circumstances. However, this gambit is a surefire method for preventing falsification because every variation in the data can be reinterpreted ad hoc as yet another example of a specific variation in adaptive strategy.

269 frequency of orgasm during copulation: Wallen and Lloyd (2011).

270 female orgasm apparently rarely occurs: Allen and Lemon (1981).

271 Anthropological data from a range of cultures: Davenport (1977).

271 A 2000 survey found: Qidwai (2000).

272 the by-product hypothesis marginalizes: Lloyd (2005, 139–43) summarizes feminists' objections to Symons's original proposal of the by-product account. She correctly points out that the cultural status of women's sexual pleasure is not determined by whether or not orgasm is an adaptation; that is, "adaptive value" does not determine cultural or personal value. But she fails to counter the critique by Fausto-Sterling et al. (1997) that in the upsuck theory "women have much more agency than they do" in the by-product account.

275 extending copulation duration: The lack of correlation of copulation duration and sperm competition in primates is documented by the very short copulation duration in chimpanzees despite strong sperm competition. Dogs and some other mammals have evolved to extend copulation duration—the well-known copulatory lock—which may enhance male success in sperm competition by preventing the female from mating with another individual for an extended period of time. However, this mechanism extends post-ejaculatory copulation duration and is quite unlike human sexual behavior, which extends pre-ejaculatory copulation duration.

277 male orgasm is more pleasurable: This conclusion is further supported by the fact that male fishes and birds pursue copulations fervently, even though there is no intromission in most species and therefore no opportunity for tactile genital sensory experience or pleasure during mating.

279 a role for a Fisherian "runaway process": Miller (2004, 240).

Chapter 10: The *Lysistrata* Effect

282 In colonial White-fronted Bee-Eaters: Emlen and Wrege (1992).

284 there is no evolutionary advantage: The strong limit on the number of eggs and offspring per female means that there is less variance between the most reproductively successful female and the average, whereas variance in male reproductive success can be very high. As a result, males may be able to gain substantially from trying to monopolize many reproductive opportunities, but females have little or nothing to gain by exercising social control in this way.

284 The "average" female old-world monkey: Given the many variations in social structure and reproductive biology within and among primate species. there really is no "average" old-world monkey. Thus, my shorthand description of these breeding systems falls short. However, I think this summary remains an essentially accurate summary of the ancestral condition of sexual conflict in the old-world primate clade.

284 females make all the reproductive investment: Reproductive investment encompasses the total energy, time, and resources that an individual commits to the production, health, and survival of its offspring. The combination of exclusively female parental care and complete lack of sexual autonomy found in many old-world primates is completely unknown in birds. By contrast, manakin and bowerbird females do all the parental care, but they have evolved complete sexual autonomy as a result.

285 infanticide by males accounts: Palombit (2009, 380). Sometimes, infanticidal attacks are part of a broader disruptive male strategy to obtain social dominance in the first place.

287 gorilla females who are newcomers: Robbins (2009).

287 male infanticide frequently occurs: I use the term "male infanticide" to mean "infanticide by males," and not infanticide of males. Group fission provides female gorillas with a rare opportunity for mate choice, because the female may be able to decide which group she joins. Of course, she is also choosing to join the other females that go with that group, so it may not be purely dictated by mate choice.

287 one-third of all infant mortality: David Watts, personal communication.

288 alpha males achieve about 50 percent: A general summary of chimpanzee breeding behavior comes from Muller and Mitani (2005). Paternity estimate is from Boesch et al. (2006).

288 During the consortship: Muller et al. (2009).

288 males that have been most aggressive: Muller et al. (2009).

288 sexual coercion over fertilization: Poali (2009). Paternity in bonobos is nonrandom and biased toward males of high social rank, which is determined in part by the rank of a male's mother within the group (Gerloff et al. 1999).

289 Although sexual conflict and coercion: The nature of human sexual violence has also been qualitatively transformed during our evolution from our ape ancestors. Shannon Novak and Mallorie Hatch (2009) conducted a fascinating forensic study of cranial injuries inflicted by violent encounters between individual chimpanzees and humans. They discovered that female chimpanzees exhibit significantly more injuries to the top and back of the skull, whereas male chimpanzees have more injuries directly to the face. This is because male chimpanzees face their attackers and females are more likely to flee or huddle down during an attack. In contrast, women sustain more facial than cranial

injuries in male partner violence, matching the pattern of male chimpanzees. Despite the devastating impact of sexual violence on women, these data indicate that women have evolved a new, confrontational, frontal orientation toward male violence since common ancestry with chimpanzees.

289 human males simply *do not* murder young children: In the United States, the three most frequent causes of infant mortality—congenital malformations, prematurity and low birth weight, and sudden infant death syndrome—account together for a total of 44 percent of all infant deaths (CDC 2007, 1115). Although children are one hundred times more likely to be murdered or fatally abused by an unrelated stepparent than they are by their genetic parents (Daly and Wilson 1988), infanticide accounts for less than 1 in 100,000 infant deaths.

289 infanticide by mothers: For example, see Scrimshaw (2008).

290 "self-domestication": Hare et al. (2012).

291 the social temperament of humans: Hare and Tomasello (2005). By "historically independent," I mean that the evolution of a more tolerant social temperament occurred separately, at different places and times, in each of the ancestors of modern humans and bonobos.

294 our evolutionary history: Gordon (2006).

295 sexual differences in body size: Rensch's rule (Rensch 1950) is basically a null model of sexual size dimorphism evolution with body size evolution based on many independent observations from nature. If bodies evolve to be larger, and nothing else special occurs to influence that process, then the size differences between males and females will become proportionally even *greater*. The fact that the opposite has occurred in humans—that as we have gotten larger, the sexual size difference has decreased instead—indicates that we can reject the null model and that something special *has* happened during human evolution (evolutionary context 2). That special something is likely selection for reduced sexual size dimorphism. Now, what kind of selection—natural or sexual—is the question. I propose that it's a sexual selection in the form of female mate choice for reduced sexual size dimorphism—that is, females' preference for males who are more similar in body size to themselves.

295 a decades-long, Soviet-era experiment: Trut (2001); the implications of this study are discussed extensively by Hare and Tomasello (2005).

295 reduction in aggression in bonobos: Hare et al. (2012).

296 elongate canines are kept razor sharp: Walker (1984). The enamel on the inner surface of the upper canines is thinner than the enamel on the outer surface of the third premolars so that the canine teeth are constantly sharpened by chewing motions.

296 The canines of *Sahelanthropus tchadensis:* Lieberman (2011).

297 the reduction of male canines: See Jolly (1970), Hylander (2013), and Lieberman (2011).

297 A hamadryas baboon uses his extremely large canines: Swedell and Schreier (2009).

297 Male mountain gorillas use canine teeth: Robbins (2009).

297 In chimps, the repertoire: Muller et al. (2009).

298 *aesthetic* expansion of female social and sexual autonomy: The aesthetic deweaponization hypothesis implies that smiling might have evolved through female mating preferences for a positive, nonaggressive social signal that directly facilitates the aesthetic evaluation of canine size. Previous theories of the origins of smiling, dating back to Darwin, have proposed that the human smile evolved from any of various teeth-baring displays of our primate ancestors, which can signal either dominance/aggression or fear/submissiveness. However, none of these narratives actually address the "content" of smiling explicitly, nor why these other tooth-baring signals would evolve new meanings. In fact, a smile is not merely baring one's teeth (like a grimace). A smile is an efficient and explicit display of one's canines and positive, nonviolent intent. The novel evolutionary association of canine display with nonaggressive, positive social and seductive messages seems more likely to have evolved from selection for aesthetic display of canine size.

298 a mathematical, genetic model: Snow et al. (forthcoming).

298 those traits that are associated: Gangestad and Scheyd (2005); Neave and Shields (2008).

301 Male investment in parenting: Among other primates, paternal care is found in some gibbons, tamarins, and owl monkeys (Fernandez-Duque et al. 2009).

302 this second major evolutionary transformation: In an analysis of the biology of *Ardipithecus ramidus,* the human paleontologist C. Owen Lovejoy (2009) proposed that female choice for less aggressive males, reduced male-to-male violence, canine reduction, and loss of the canine-premolar honing complex all occurred in hominin evolution by the late Pliocene. Lovejoy envisioned an evolutionary process that was driven by natural selection for a new "adaptive suite" of morphological, behavioral, and life history characteristics related to cooperative behavior, male reproductive investment, and mate choice. For example, Lovejoy proposed that the evolution of bipedalism was facilitated by the food-carrying behavior of males providing food in exchange for sex. However, Lovejoy did not outline any specific ecological, life history, or selective explanation for the concurrent reduction of male social dominance, origin of male investment, and male and female mate choice. Lovejoy's evolutionary scenarios show that the challenges raised in this chapter to the evolution of human reproduction are broadly recognized in evolutionary anthropology as critical to the explanation of human origins but that the field has yet to establish a clear evolutionary mechanism to achieve these changes in the absence

of theories of sexual conflict, aesthetic mate choice, and sexual autonomy.

Chapter 11: The Queering of *Homo sapiens*

304   we tend to think that sexual identity categories: Like racial identities,
the cultural categories of sexual identity have been imposed upon a
biological phenomenon that is much richer, more variable, continuous, and complex than the cultural categories that we use to tidy up
this reality. Categories of sexual identity have been vital, progressive
political tools in the struggle for political and social recognition of
the rights of lesbian, gay, bisexual, and transgender individuals. But
these categories can also become a burden, because they obscure the
fact that the variation and diversity in human sexual preferences and
behavior exist on a continuum.

304   ample scientific literature: An excellent exception to this trend is Bailey and Zuk (2009).

305   same-sex behavior is still *sex:* Same-sex behavior is well-known
in a wide variety of animals (Bagemihl 1999; Roughgarden 2009).
Throughout most of the twentieth century, biologists largely ignored
same-sex behavior as an aberration or struggled to reinterpret it as
a form of nonsexual, social behavior. For example, George Murray
Levick was a Victorian explorer and natural historian who published a
book on the natural history and behavior of the Adelie Penguin (*Pygos-
celis adeliae*) and other Antarctic penguins (Levick 1914). He made
numerous observations of same-sex behavior, but he did not publish
them. They remained unpublished in his original notebooks, where he
recorded them in ancient Greek to keep these salacious details secret
from any but the most educated readers. The notes were recently redis-
covered, translated, and published (Russell et al. 2012). It is important
to emphasize, however, that same-sex behavior—whether in nonhuman animals or in humans—is an extremely diverse class of phenomena, which does not have a single, unified causal explanation. I do not
think that it will be possible to make *any* broad scientific generalizations about this diverse phenomenon beyond its definition.

305   the combined effects of these many small genetic influences: Ironically,
substantial continuous variation in sexual preference implies that some
cultural opinions and judgments about whether same-sex behavior is a
personal "choice" are correct for many individuals. Same-sex behavior
is not a "choice" for those minority of individuals that are toward the
ends of the continuous variation in sexual preference. However, for the
majority of people within the distribution, same-sex behavior *may be*
one possible choice among a variety of sexual attractions.

306   The problem with the "Helpful Uncle" hypothesis: The notion that
individuals with exclusively same-sex preferences will have the spare

time and energy to raise the younger generations of their family (or the interest in doing so) because they have no offspring of their own is just another cultural construction. Actually, this idea seems more like a homophobic cultural solution about how to make practical use of such people, who have been prevented from pursuing their own sexual autonomy, than an evolutionary mechanism to explain their existence.

306 there is no evidence that same-sex behavior: In some cultures, men with culturally variant gender presentations may identify with and adopt female gender roles, often including child care. But it is not clear that this is a biological phenomenon or a top-down cultural effect in which individuals conform to the limited available cultural roles for gender presentation variance.

306 Kin selection arguments fail: Another recent hypothesis proposes that specific genes that advance the reproductive success of one sex may result in maladaptive behavior in the other sex (Camperio Ciani et al. 2008). If natural selection for some reproductive trait in one sex, such as mothers, is strong enough, then the evolutionary advantages of that trait may outweigh the losses of reproductive success in some offspring that inherit these same genetic variations—that is, sons with preferences. This mechanism could work because, on average, gene copies spend half of the time in females and half of the time in males. A big enough advantage in one context could overcome a smaller disadvantage in the other, leading such genes to evolve.

Although evolutionarily plausible, this mechanism remains entirely speculative in that there is no specific hypothesis about what kinds of genes and traits could contribute to advancement of reproductive success in mothers but to altered sexual preference in sons. This mechanism treats variation in sexual preference as an accidental and unintended by-product of adaptation in the opposite sex. Same-sex behavior is hypothesized to result merely from a breakdown in the efficiency of natural selection to produce adapted individuals of both sexes from the same gene pool. Like the kin selection hypotheses, this idea fails to specifically address the evolution of subjective experience of sexual desire itself that is the core of the issue.

More recently, Rice et al. (2012) proposed that homosexuality is a consequence of the accidental intergenerational inheritance of epigenetic modifications to the genome that occur during individual sexual development. These modifications are hypothesized to regulate the sensitivity of developing embryos to maternal androgens in utero, and they are proposed to be "turned off" or reset during later development. When this reset does not occur, these epigenetic modifications could be passed on to the next generation and could cause androgen hypersensitivity or desensitivity in offspring of the opposite sex.

Although this evolutionary mechanism is also theoretically plausible, it erroneously equates same-sex preference and behavior with developmental "feminization" or "masculinization" of men and women, respectively. The authors define "homosexuality" as any non-opposite-sex sexual attraction or experience—that is, any Kinsey score greater than 0. I think these authors seek to find a solution to theoretical fitness costs that have never been demonstrated. Did individuals with a Kinsey score greater than 0 have lower fitness before the invention of cultural sexual identity categories that the authors embrace as biologically real? This is unknown. Further, the idea that same-sex attraction involves sexual "inversion" explicitly pathologizes it, and has long been rejected as a relevant explanation of the variety of same-sex preferences.

307 human same-sex behavior: Qazi Rahman and Glen Wilson (2003; Wilson and Rahman 2008) have articulated a similar proposal but without the explicit recognition of the role of aesthetic mate choice and sexual conflict. Without these elements, they cannot elaborate the numerous testable predictions that establish that mechanism as being a more consistent explanation.

307 have evolved the opposite pattern: Greenwood (1980); Sterk et al. (1997); Kappeler and van Schaik (2002).

307 male friends help protect the females' offspring: Smuts (1985).

307 female-female friendships contribute to protection: Silk et al. (2009).

310 In baboon society, male-female friendships function: Palombit (2009).

310 social alliances between males: There are several reasons why I think that this proposed evolved social function for variation in sexual preference is more plausible than the kin selection, or "Helpful Uncle," hypothesis. First, this selective advantage is just one advantage of the proposed mechanism of selection, not the only one. Second, there is good evidence that very similar nonsexual friendships between males and females advance female fitness in nonhuman primates, which is entirely outside the human cultural context. Third, I think that there is more contemporary evidence of gay male–straight female relationships in human societies than evidence that variation in sexual desire contributes to raising nieces and nephews.

311 there is good evidence: From various identical and fraternal twin comparisons, Pillard and Bailey (1998) have reported heritability estimates of self-identified homosexuality as high as 0.74.

312 Bonobos are notable for the nearly complete absence: Paoli (2009).

314 exclusive homosexual identity: Pillard and Bailey (1998) review this literature.

314 women generally prefer men: Reviewed in Gangestead and Scheyd (2005).

314 Alfred Kinsey found: Kinsey et al. (1948, 650); Kinsey et al. (1953, 475).

315 The biological capacity: The near ubiquity of a capacity for same-sex attraction likely fuels the anxiety over same-sex desire in societies where it is condemned, thus exacerbating homophobia and violence against sexual minorities.

315 fascinating work: Wekker (1999).

316 all categories of sexual partner violence: The reported lifetime incidence of each class of same-sex partner violence for heterosexual
women and gay men, respectively, were as follows: rape, 9.1 percent
versus about 0 percent; physical violence, 33.2 percent versus 28.7 percent; stalking, 10.2 percent versus about 0 percent; overall, 35 percent
versus 29 percent (Walters et al. 2010). Unfortunately, the CDC data
were reported only in terms of the sexual orientation of the victims,
not of their intimate partners. So we do not yet know whether bisexual
men are less likely to engage in sexual coercion, partner violence, or
rape of their female sexual partners than are exclusively heterosexual
men.

317 82 percent of the simplest DNA sequence variations: Keinan and Clark
(2012). The primary reason why humans have so many rare genomic
variations is the explosively rapid expansion of human population
sizes over the last fifteen thousand years. Keinan and Clark describe
this condition as an "excess" of rare genetic variants, but they are only
in excess in relation to the assumption of stable or equilibrium evolutionary conditions that are irrelevant to the history of contemporary
humans.

317 a cultural mechanism to co-opt same-sex behavior: Patriarchal co-
option of same-sex desire may be one of the reasons that very hierarchical, traditionally male-dominated institutions—like the military,
some traditional religious institutions, or boarding schools—have
a particularly hard time controlling or eliminating sexual coercion,
sexual violence, and abuse both within and between sexes. The inherently hierarchical structure of these organizations facilitates and insti-
tutionalizes the sexual misuse of hierarchical power.

318 This opinion, well represented: Warner (1999); Halperin (2012).

Chapter 12: This *Aesthetic* View of Life

320 the most succinct and memorable articulation: The closing lines of
Keats's ode are remarkable both for the stringent synonymy of beauty
and truth and for their vigorous insistence that this view is an *all-
sufficient* explanation of the world. In both ways, Keats anticipates
the Wallacean worldview on sexual ornament.

320 *Hamlet, Prince of Denmark:* In the spring of 2013, the Yale Repertory
Theatre put on a production of Shakespeare's *Hamlet* starring Paul
Giamatti as the troubled Danish prince. The show was a blockbuster
hit, and tickets were sold out completely. For a month, the entire city
of New Haven was abuzz with *Hamlet.* Even our weekly lab meetings, which involve presentations of current research by students and
postdocs, or discussions of recent scientific papers on evolution and
ornithology, became conversations about *Hamlet.* During this time,
Jennifer Friedmann, a Yale class of '13 cognitive science major who

was doing a senior research project in my lab on avian aesthetic evolution, brought to my attention this astounding passage from *Hamlet*'s act 3. She was struck by the similarities to our discussions of Fisherian and Wallacean views of sexual selection, and I am grateful for her insightful suggestion to analyze this passage.

321 "Ha, ha! Are you honest?": When I first read this passage (for the first time since high school), my head reeled! Here, Shakespeare was obviously grappling with beauty and honesty in a surprisingly resonant way, but he packs so much into these dense lines that I needed help to figure out how to unravel it.

I sought expert assistance from my friend and Yale colleague James Bundy, dean of the Yale School of Drama and the director of the 2013 Yale Rep production of *Hamlet*. Over lunch, James gave me a quick course in dramatic analysis for ornithologists. With James's encouragement, I have embarked on my own evo-ornithological analysis of this passage from *Hamlet*. Of course, I remain *solely* responsible for any errors, omissions, overextensions, or oversights.

321 Beauty, he says, can transform truth: Following Hamlet's suggestion of "discourse," Ophelia characterizes the relationship between beauty and honesty as "commerce." But then Hamlet subverts Ophelia's usage by implying a more degraded transaction—the purely carnal business of a brothel.

321 *power of beauty* that actually subverts honesty: Like Fisher, Hamlet understands that the combination of beauty and truth lies unstably on a knife's edge because the very existence of beauty creates a seductive power that can degrade its own honesty.

Hamlet's personal realization that Ophelia's beauty is not an indicator of her honesty follows the same course as Fisher's two-stage model of evolution by mate choice. Hamlet begins his relationship with Ophelia in a rosy state of Wallacean contentment, in which her beauty *is* an honest indicator of the inner quality of her soul and her commitment to him. Yet this inherently unstable relationship cannot endure, just as Fisher proposed that correlation between display traits and quality would be eroded by the emerging advantages of attraction—the power of beauty.

In defense of Ophelia, however, she is not acting with sexual autonomy. She has shunned and lied to Hamlet under the coercive instructions of her father. (I haven't focused a lot on the sexual coercion of offspring by parents, but this is a great example from literature.) In the final act when Ophelia goes mad, she finally expresses some of her true, autonomous sexual desires. She sings a *bawdy*(!) tale of her own Valentine's Day deflowering by a deceptive rogue (perhaps Hamlet?). She then imagines herself as Hamlet's queen, addresses her wise counselors and fine courtiers, and orders the servants to bring around her carriage. In her madness, Ophelia can finally reveal her real desires and fantasy. Constrained in life from realizing her sexual self because of her father's coercion, Ophelia is only liberated and self-realized through madness and death. This is, perhaps, Shakespeare's cautionary tale about the social risks of the pursuit of female sexual autonomy in Elizabethan society. Indeed, Ophelia's demise is the second tragedy of *Hamlet*.

322 "The fox knows many things": Berlin (1953).

323 dominated, indeed hijacked, by adaptationist Hedgehogs: See David Hull's *Science as a Process* (1988) and Ron Amundson's *The Changing Role of the Embryo in Evolutionary Thought* (2005).

325 the painful history of political and ethical abuse: For an authoritative social history of eugenics, see Kevles (1985).

330 Sexual autonomy is not a mythical: In "The Riddle of Rape-by-Deception and the Myth of Sexual Autonomy," Yale law professor Jed Rubenfeld (2013) argues that the concept of sexual autonomy underlying U.S. rape laws is an unsupportable myth. Rubenfeld conceives of sexual autonomy broadly as including the right to *assert* one's individual desires over the desires of others. Obviously, this concept of sexual autonomy is designed to fail, because the desires of different individuals will inevitably diverge and conflict. In his view, sexual autonomy is unrealizable and therefore mythical. Rubenfeld briefly entertains a "thinner" concept of a sexual autonomy that is basically congruent with my definition—freedom to pursue one's sexual desires without coercion. But he dismisses this idea as conceptually muddled with a single odd example. He asks how we could describe a lonely, disabled, homeless beggar as sexually autonomous? The answer, of course, is that this unfortunate person's multiple miseries have nothing to do with violations of his sexual autonomy. So, yes, this person is sexually autonomous; the fact that his autonomy gains him no pleasures is entirely irrelevant to the issue. Autonomy is freedom from coercion, not power to assert your desires. This conclusion is illustrated precisely in the observation that sexual autonomy in animals does not involve the imposition of sexual desire on others. Female ducks can still be turned down by prospective mates, even though they have evolved anatomical structures to protect their sexual autonomy in the face of forced copulation.

Evolutionary biology demonstrates that sexual autonomy is not a myth. Although the evolution of sexual autonomy in animals is not a justification for a legal theory of rape based on this definition, it is proof that the concept is not specious but a natural consequence of individuality, preference, choice, and complex social interaction. I leave it to legal scholars to pursue whether this scientific result is an appropriate basis for the establishment of law, but it is clear that these biological phenomena involve exactly the kind of complex social conflicts that law was invented to resolve.

332 the cultural evolution of patriarchy: The near ubiquity of patriarchy in contemporary human cultures also has obscured the role of female

mate choice in the evolution of humans. By adopting an aesthetic view, we are able to see that human evolution required a transformation of male physical and social phenotype and that female sexual autonomy provides a mechanism to achieve that change.

333 control over reproduction: The traditional patriarchal insistence that women be stay-at-home mothers is yet another manifestation of sexual conflict over parental investment. These cultural ideas function to prevent women from gaining sexual, economic, and social independence through the pursuit of their own independent, nonreproductive social and economic activities.

333 criticized the legal doctrine of "sexual autonomy": Rubenfeld (2013).

335 opportunity for intellectual exchange: I have published the basic framework for a coevolutionary aesthetic philosophy in the journal *Biology and Philosophy* (Prum 2013).

335 an exclusively *human gaze:* The "human gaze" refers to a power relation between the human and the natural that places human sensory and material gratification as the objective purpose of nature. Analogous to the "male gaze," this anthropocentric perspective prevents the recognition of organismal agency and the autonomous aesthetic ends of other species.

336 *art is a form of communication:* Prum (2013).

336 In a now classic paper: Danto (1964).

337 nearly half of all species: Song-learning birds include oscine passerines, parrots, hummingbirds, and *Procnias* bellbirds (Cotingidae). For an introduction to bird song learning and its cultural consequences, see Kroodsma (2005).

337 Similar aesthetic cultural processes: A dramatic case of aesthetic cultural revolution in Australian populations of humpback whales has been documented by Noad et al. (2000).

337 it is difficult to define the arts: In Prum (2013), I provide a detailed analysis of the impact of various definitions of art on whether there are nonhuman arts.

338 the harbor of West Jonesport, Maine: For this lovely trip to the Bay of Fundy all those years ago, I am deeply indebted to Mary and Richard Burton-Beinecke, with whom I have sadly lost contact. Mary was a Unitarian minister in nearby Arlington, Vermont, and we met the previous spring in a bird-watching course organized by the Vermont Institute of Natural Science and taught by my (now lifelong) friend Tom Will. Mary and Richard were kind enough to take me along on their trip to Machais Seal Island and thereby contributed substantially to my growing obsession with birds.

Bibliography

Adler, M. 2009. "Sexual Conflict in Waterfowl: Why Do Females Resist Extra-pair Copulations?" *Behavioral Ecology* 21:182–92.

Akerlof, G. A., and R. J. Shiller. 2009. *Animal Spirits: How Human Psychology Drives the Economy, and Why It Matters for Global Capitalism.* Princeton, N.J.: Princeton University Press.

Allen, M. L., and W. B. Lemmon. 1981. "Orgasm in Female Primates." *American Journal of Primatology* 1:15–34.

Amundson, R. 2005. *The Changing Role of the Embryo in Evolutionary Thought: Roots of Evo-Devo.* Cambridge, U.K.: Cambridge University Press.

Andersson, M. 1994. *Sexual Selection.* Princeton, N.J.: Princeton University Press.

Bagemihl, B. 1999. *Biological Exuberance: Animal Homosexuality and Natural Diversity.* New York: St. Martin's Press.

Bailey, N. W., and A. J. Moore. 2012. "Runaway Sexual Selection Without Genetic Correlations: Social Environments and Flexible Mate Choice Initiate and Enhance the Fisher Process." *Evolution* 66:2674–84.

Bailey, N. W., and M. Zuk. 2009. "Same-Sex Sexual Behavior and Evolution." *Trends in Ecology & Evolution* 24:439–46.

Baker, R. R., and M. A. Bellis. 1993. "Human Sperm Competition: Ejaculate Manipulation by Females and a Function for the Female Orgasm." *Animal Behaviour* 46:887–909.

Barkse, J., B. A. Schlinger, M. Wikelski, and L. Fusani. 2011. "Female

Choice for Male Motor Skills." *Proceedings of the Royal Society of London B* 278:3523–28.

Beebe, W. 1926. *Pheasants: Their Lives and Homes.* 2 vols. New York: New York Zoological Garden and Doubleday.

Beehler, B. M., and M. S. Foster. 1988. "Hotshots, Hotspots, and Female Preference in the Organization of Lek Mating Systems." *American Naturalist* 131:203–19.

Berlin, I. 1953. *The Hedgehog and the Fox: An Essay on Tolstoy's View of History.* London: Weidenfeld & Nicolson.

Bhanoo, S. N. 2012. "Observatory: Design and Illusion, to Impress the Ladies." *New York Times,* Jan. 24, D3.

Bierens de Haan, J. A. 1926. "Die Balz des Argusfasans." *Biologische Zentralblatt* 46:428–35.

Boesch, C., G. Kohou, H. Néné, and L. Vigilant. 2006. "Male Competition and Paternity in Wild Chimpanzees of the Taï Forest." *American Journal of Physical Anthropology* 130:103–15.

Bolhuis, J. J., G. R. Brown, R. C. Richardson, and K. N. Laland. 2011. "Darwin in Mind: New Opportunities for Evolutionary Psychology." *PLoS Biology* 9:e1001109.

Borgia, G. 1995. "Why Do Bowerbirds Build Bowers?" *American Scientist* 83:542–47.

Borgia, G., and D. C. Presgraves. 1998. "Coevolution of Elaborated Male Display Traits in the Spotted Bowerbird: An Experimental Test of the Threat Reduction Hypothesis." *Animal Behaviour* 56:1121–28.

Borgia, G., S. G. Pruett-Jones, and M. A. Pruett-Jones. 1985. "The Evolution of Bower-Building and the Assessment of Male Quality." *Zeitschrift für Tierpsychology* 67:225–36.

Bostwick, Kimberly S. 2000. "Display behaviors, mechanical sounds, and their implications for evolutionary relationships of the Club-winged Manakin (*Machaeropterus deliciosus*)." *Auk* 117 (2):465–78.

Bostwick, K. S., D. O. Elias, A. Mason, and F. Montealegre-Z. 2009. "Resonating Feathers Produce Courtship Song." *Proceedings of the Royal Society of London B.*

Bostwick, K. S., and R. O. Prum. 2003. "High-Speed Video Analysis of Wing-Snapping in Two Manakin Clades (Pipridae: Aves)." *Journal of Experimental Biology* 206 (20): 3693–706.

Bostwick, K. S., M. L. Riccio, and J. M. Humphries. 2012. "Massive, Solidified Bone in the Wing of a Volant Courting Bird." *Biology Letters* 8:760–63.

Bradbury, J. W. 1981. "The Evolution of Leks." In *Natural Selection and Social Behavior: Recent Research and Theory,* edited by R. D. Alexander and D. W. Tinkle, 138–69. New York: Chiron Press.

Bradbury, J. W., R. M. Gibson, and I. M. Tsai. 1986. "Hotspots and the Dispersion of Leks." *Animal Behaviour* 34:1694–709.

Bramble, D. M., and D. E. Lieberman. 2004. "Endurance Running and the Evolution of *Homo.*" *Nature* 432:345–52.

Brennan, P. L. R., T. R. Birkhead, K. Zyskowski, J. Van Der Waag, and R. O. Prum. 2008. "Independent Evolutionary Reductions of the Phallus in Basal Birds." *Journal of Avian Biology* 39:487–92.

Brennan, P. L. R., C. J. Clark, and R. O. Prum. 2010. "Explosive Eversion and Functional Morphology of the Duck Penis Supports Sexual Conflict in Waterfowl Genitalia." *Proceedings of the Royal Society of London B* 277:1309–14.

Brennan, P. L. R., and R. O. Prum. 2012. "The Limits of Sexual Conflict in the Narrow Sense: New Insights from Waterfowl Biology." *Philosophical Transactions of the Royal Society of London B* 367:2324–38.

Brennan, P. L. R., R. O. Prum, K. G. McCracken, M. D. Sorenson, R. E. Wilson, and T. R Birkhead. 2007. "Coevolution of Male and Female Genital Morphology in Waterfowl." *PLoS One* 2:e418.

Brown, W. M., L. Cronk, K. Grochow, A. Jacobson, C. K. Liu, Z. Popovic, and R. Trivers. 2005. "Dance Reveals Symmetry Especially in Young Men." *Nature* 438:1148–50.

Browne, J. 2002. *Charles Darwin: The Power of Place.* Vol. 2. Princeton, N.J.: Princeton University Press.

———. 2010. *Charles Darwin: Voyaging.* Vol. 1. New York: Random House.

Brownmiller, S. 1975. *Against Our Will: Men, Women, and Rape.* New York: Simon & Schuster.

Buller, D. J. 2005. *Adapting Minds: Evolutionary Psychology and the Persistent Quest for Human Nature.* Cambridge, Mass.: MIT Press.

Byers, J., E. Hebets, and J. Podos. 2010. "Female Mate Choice Based upon Male Motor Performance." *Animal Behaviour* 79:771–78.

Campbell, B. 1972. *Sexual Selection and the Descent of Man, 1871–1971.* Chicago: Aldine.

Campbell, G. D., Duke of Argyll. 1867. *The Reign of Law.* London: Strahan.

Camperio Ciani, A., P. Cermelli, and G. Zanzotto. 2008. "Sexually Antagonistic Selection in Human Male Homosexuality." *PLoS One* 3:e2282.

CDC, Morbidity and Mortality Weekly Report. 2007. "QuickStats: Infant Mortality Rates for 10 Leading Causes of Infant Death—United States, 2005," edited by Centers for Disease Control and Prevention. Atlanta.

Cellerino, A., and E. A. Jannini. 2005. "Male Reproductive Physiology as a Sexually Selected Handicap? Erectile Dysfunction Is Correlated with General Health and Health Prognosis and May Have Evolved as a Marker of Poor Phenotypic Quality." *Medical Hypotheses* 65:179–84.

Chevalier-Skolnikoff, S. 1974. "Male-Female, Female-Female, and Male-Male Sexual Behavior in the Stump-tailed Monkey, with Special Attention to Female Orgasm." *Archives of Sexual Behavior* 3:95–116.

Chiappe, L. M. 2007. *Glorified Dinosaurs: The Origin and Early Evolution of Birds.* Hoboken, N.J.: Wiley & Sons.

Clark, C. J., and R. O. Prum. 2015. "Aeroelastic Flutter of Feathers, Flight, and the Evolution of Non-vocal Communication in Birds." *Journal of Experimental Biology* 218:3520–27.

Coddington, J. A. 1986. "The Monophyletic Origin of the Orb Web." In *Spiders: Webs, Behavior, and Evolution,* edited by W. A. Shear, 319–63. Palo Alto, Calif.: Stanford University Press.

Coyne, J. A. 2009. *Why Evolution Is True.* Oxford: Oxford University Press.

Cracraft, J., and R. O. Prum. 1988. "Patterns and Processes of Diversification: Speciation and Historical Congruence in Some Neotropical Birds." *Evolution* 42:603–20.

Cronin, H. 1991. *The Ant and the Peacock.* Cambridge, U.K.: Cambridge University Press.

Curry, A. 2013. "The Milk Revolution." *Nature* 500:20–22.

Dalton, R. "High Speed Biomechanics Caught on Camera." *Nature* 418: 721–22.

Daly, M., and M. Wilson. 1988. "Evolutionary Social Psychology and Family Homicide." *Science* 242:519–24.

Danto, A. 1964. "The Artworld." *Journal of Philosophy* 61:571–84.

Darwin, C. 1859. *On the Origin of Species.* London: John Murray.

———. 1871. *The Descent of Man, and Selection in Relation to Sex.* London: John Murray.

———. 1882. "A Preliminary Notice to 'On the Modification of the Race of Syrian Street Dog by Means of Sexual Selection' by Dr. Van Dyck." *Proceedings of the Zoological Society of London* 25:367–69.

———. 1887. *The Autobiography of Charles Darwin.* New York: Barnes & Noble Reprint.

Davenport, W. H. 1977. "Sex in Cross-cultural Perspective." In *Human Sexuality in Four Perspectives,* edited by F. Beach, 115–63. Baltimore: Johns Hopkins University Press.

Davis, T. A. W. 1949. "Display of White-throated Manakins *Corapipo gutturalis.*" *Ibis* 91:146–47.

Davis, T. H. 1982. "A Flight-Song Display of the White-throated Manakin." *Wilson Bulletin* 94:594–95.

Davison, G. W. H. 1982. "Sexual Displays of the Great Argus Pheasant *Argusianus argus.*" *Zeitschrift für Tierpsychologie* 58:185–202.

Dawkins, R. 1982. *The Extended Phenotype: The Long Reach of the Gene.* Oxford: Oxford University Press.

——. 2004. *The Ancestor's Tale.* New York: Houghton Mifflin.

——. 2006. *The Selfish Gene.* 30th anniversary ed. New York: Oxford University Press.

Diamond, J. M. 1986. "Animal Art: Variation in Bower Decorating Style among Male Bowerbirds *Amblyornis inornatus.*" *Proceedings of the National Academy of Sciences* 83:3402–06.

——. 1992. *The Third Chimpanzee.* New York: HarperCollins.

——. 1997. *Why Is Sex Fun?* New York: Basic Books.

Durães, R., B. A. Loiselle, and J. G. Blake. 2007. "Intersexual Spatial Relationships in a Lekking Species: Blue-crowned Manakins and Female Hotspots." *Behavioral Ecology* 18:1029–39.

Durães, R., B. A. Loiselle, P. G. Parker, and J. G. Blake. 2009. "Female Mate Choice Across Spatial Scales: Influence of Lek and Male Attributes on Mating Success of Blue-crowned Manakins." *Proceedings of the Royal Society of London B* 276:1875–81.

Durham, W. H. 1991. *Coevolution: Genes, Culture, and Human Diversity.* Palo Alto, Calif.: Stanford University Press.

Dutton, D. 2009. *The Art Instinct.* New York: Bloomsbury Press.

DuVal, E. H. 2005. "Age-Based Plumage Changes in the Lance-tailed Manakin: A Two-Year Delay in Plumage Maturation." *Condor* 107:915–20.

——. 2007a. "Cooperative Display and Lekking Behavior of the Lance-tailed Manakin (*Chiroxiphia lanceolata*)." *Auk* 124:1168–85.

——. 2007b. "Social Organization and Variation in Cooperative Alliances Among Male Lance-tailed Manakins." *Animal Behaviour* 73:391–401.

DuVal, E. H., and B. Kempenaers. 2008. "Sexual Selection in a Lekking Bird: The Relative Opportunity for Selection by Female Choice and Male Competition." *Proceedings of the Royal Society of London B* 275:1995–2003.

Eastwick, P. W., and E. J. Finkel. 2008. "Sex Differences in Mate Preferences Revisited: Do People Know What They Initially Desire in a Romantic Partner?" *Journal of Personality and Social Psychology* 94:245–64.

Eastwick, P. W., and L. L. Hunt. 2014. "Relational Mate Value: Consensus and Uniqueness in Romantic Evaluations." *Journal of Personality and Social Psychology* 106:728–51.

Eberhard, W. G. 1985. *Sexual Selection and Animal Genitalia.* Cambridge, Mass.: Harvard University Press.

——. 1996. *Female Control: Sexual Selection by Cryptic Female Choice.* Princeton, N.J.: Princeton University Press.

——. 2002. "The Function of Female Resistance Behavior: Intromission by Coercion vs. Female Cooperation in Sepsid Flies (Diptera)." *Revista de Biologia Tropical* 50:485–505.

Eberhard, W. G., and C. Cordero. 2003. "Sexual Conflict and Female Choice." *Trends in Ecology and Evolution* 18:439–40.

Emlen, S. T., and L. W. Oring. 1977. "Ecology, Sexual Selection, and the Evolution of Mating Systems." *Science* 197:215–23.

Emlen, S. T., and P. H. Wrege. 1992. "Parent-Offspring Conflict and the Recruitment of Helpers Among Bee-Eaters." *Nature* 356:331–33.

Endler, J. A., L. C. Endler, and N. R. Doerr. 2010. "Great Bowerbirds Create Theaters with Forced Perspective When Seen by the Audience." *Current Biology* 20:1679–84.

Evarts, S. 1990. "Male Reproductive Strategies in a Wild Population of Mallards (*Anas plathyrhynchos*)." Ph.D. diss., University of Minnesota.

Fausto-Sterling, A., P. A. Gowaty, and M. Zuk. 1997. "Evolutionary Psychology and Darwinian Feminism." *Feminist Studies* 23:402–17.

Feo, Teresa J., D. J. Field, and R. O. Prum. 2015. "Barb Geometry of Asymmetrical Feathers Reveals a Transitional Morphology in the Evolution of Avian Flight." *Proceedings of the Royal Society of London B: Biological Sciences* 282 (1803: 20142864).

Fernandez-Duque, E., C. R. Valeggia, and S. P. Mendoza. 2009. "The Biology of Paternal Care in Human and Nonhuman Primates." *Annual Review of Anthropology* 38:115–30.

Field, D. J., C. Lynner, C. Brown, and S. A. F. Darroch. 2013. "Skeletal Correlates for Body Mass Estimation in Modern and Fossil Flying Birds." *PLoS One* 8:e82000.

Fisher, R. A. 1915. "The Evolution of Sexual Preference." *Eugenics Review* 7:184–91.

———. 1930. "The Genetical Theory of Natural Selection." Oxford: Clarendon Press.

Fisher, R. A. 1957. "The Alleged Dangers of Cigarette Smoking." *British Medical Journal* 2:1518.

Fisher, T. D. 2013. "Gender Roles and Pressure to Be Truthful: The Bogus Pipeline Modifies Gender Differences in Sexual but Not Non-sexual Behavior." *Sex Roles* 68:401–14.

Foster, M. S. 1977. "Odd Couples in Manakins: A Study of Social Organization and Cooperative Breeding in *Chiroxiphia linearis*." *American Naturalist* 11:845–53.

———. 1981. "Cooperative Behavior and Social Organization of the Swallow-tailed Manakin (*Chiroxiphia caudata*)." *Behavioral Ecology and Sociobiology* 9:167–77.

———. 1987. "Delayed Plumage Maturation, Neoteny, and Social System Differences in Two Manakins of the Genus *Chiroxiphia*." *Evolution* 41:547–58.

Frith, C. B., and D. W. Frith. 2001. "Nesting Biology of the Spotted Catbird, *Ailuruedus melanotis,* a Monogamous Bowerbird (Ptilonorhynchidae), in Australian Wet Tropical Upland Rainforests." *Australian Journal of Zoology* 49:279–310.

———. 2004. *The Bowerbirds.* Oxford: Oxford University Press.

Funk, D. H., and D. W. Tallamy. 2000. "Courtship Role Reversal and Deceptive Signals in the Long-tailed Dance Fly, *Rhamphomyia longicauda.*" *Animal Behaviour* 59:411–21.

Gallup, G. G., R. L. Burch, M. L. Zappieri, R. A. Parvez, M. L. Stockwell, and J. A. Davis. 2003. "The Human Penis as a Semen Displacement Device." *Evolution and Human Behavior* 24:277–89.

Gangestad, S. W., and G. J. Scheyd. 2005. "The Evolution of Human Physical Attractiveness." *Annual Review of Anthropology* 34:523–48.

Gauthier, I., P. Skudlarski, J. C. Gore, and A. W. Anderson. 2000. "Expertise for Cars and Birds Recruits Brain Areas Involved in Face Recognition." *Nature Neuroscience* 3:191–97.

Gerloff, U., B. Hartung, B. Fruth, G. Hohmann, and D. Tautz. 1999. "Intracommunity Relationships, Dispersal Pattern, and Paternity Success in a Wild Living Community of Bonobos (*Pan paniscus*) Determined from DNA Analysis of Faecal Samples." *Proceedings of the Royal Society of London B* 266:1189–95.

Gilbert, S. F., and Z. Zevit. 2001. "Congenital Baculum Deficiency in the Human Male." *American Journal of Medical Genetics* 101:284–85.

Goldfoot, D. A., H. Westerborg-van Loon, W. Groeneveld, and A. Koos Slob. 1980. "Behavioral and Physiological Evidence of Sexual Climax in the Female Stump-tailed Macaque (*Macaca arctoides*)." *Science* 208:1477–79.

Gordon, A. D. 2006. "Scaling of Size and Dimorphism in Primates II: Macroevolution." *International Journal of Primatology* 27:63–105.

Gould, S. J. 1987. "Freudian Slip." *Natural History* 87 (2): 14–21.

Gowaty, P. A. 2010. "Forced or Aggressively Coerced Copulation." In *Encyclopedia of Animal Behavior,* edited by M. D. Breed and J. Moore, 759–63. Burlington, Mass.: Elsevier.

Gowaty, P. A., and N. Buschhaus. 1998. "Ultimate Causation of Aggressive and Forced Copulation in Birds: Female Resistance, the CODE Hypothesis, and Social Monogamy." *American Zoologist* 38:207–25.

Grafen, A. 1990. "Sexual Selection Unhandicapped by the Fisher Process." *Journal of Theoretical Biology* 144:473–516.

Grant, P. R. 1999. *Ecology and Evolution of Darwin's Finches.* Princeton, N.J.: Princeton University Press.

Greenwood, P. J. 1980. "Mating Systems, Philopatry, and Dispersal in Birds and Mammals." *Animal Behaviour* 28:1140–62.

Grice, E. A., H. H. Kong, S. Conlan, C. B. Deming, J. Davis, A. C. Young,

NISC Comparative Sequencing Program, G. G. Bouffard, R. W. Blakesley, P. R. Murray, E. D. Green, M. L. Turner, and J. A. Segre. 2009. "Topographical and Temporal Diversity of the Human Skin Microbiome." *Science* 324:1190–92.

Halperin, D. M. 2012. *How to Be Gay.* Cambridge, Mass.: Belknap Press.

Hare, B., and M. Tomasello. 2005. "Human-Like Social Skills in Dogs?" *Trends in Cognitive Sciences* 9:439–44.

Hare, B., V. Wobber, and R. W. Wrangham. 2012. "The Self-Domestication Hypothesis: Evolution of Bonobo Psychology Is Due to Selection Against Aggression." *Animal Behaviour* 83:573–85.

Harel, A., D. Kravitz, and C. I. Baker. 2013. "Beyond Perceptual Expertise: Revisiting the Neural Substrates of Expert Object Recognition." *Frontiers in Human Neuroscience* 7 (885): 1–11.

Harris, M. K., J. F. Fallon, and R. O. Prum. 2002. "Shh-Bmp2 Signaling Module and the Evolutionary Origin and Diversification of Feathers." *Journal of Experimental Zoology (Molecular and Developmental Evolution)* 294:160–76.

Haverschmidt, F. 1968. *Birds of Surinam.* Edinburgh: Oliver & Boyd.

Haworth, A. 2011. "Forced to Be Fat." *Marie Claire,* July 20.

Hobbs, D. R., and G. G. Gallup. 2011. "Songs as a Medium for Embedded Reproductive Messages." *Evolutionary Psychology* 9:390–416.

Höglund, J., and R. V. Alatalo. 1995. *Leks.* Princeton, N.J.: Princeton University Press.

Hrdy, S. B. 1981. *The Woman That Never Evolved.* Cambridge, Mass.: Harvard University Press.

Hull, D. L. 1988. *Science as a Process.* Chicago: University of Chicago Press.

Hutchinson, G. E. 1965. *The Ecological Theater and the Evolutionary Play.* New Haven, Conn.: Yale University Press.

Hylander, W. L. 2013. "Functional Links Between Canine Height and Jaw Gape in Catarrhines with Special Reference to Early Hominins." *American Journal of Physical Anthropology* 150:247–59.

Iwasa, Y., and A. Pomiankowski. 1994. "The Evolution of Mate Preferences for Multiple Sexual Ornaments." *Evolution* 48:853–67.

Jablonski, N. G. 2006. *Skin: A Natural History.* Berkeley: University of California Press.

Jablonski, N. G., and G. Chaplin. 2010. "Human Skin Pigmentation as an Adaptation to UV Radiation." *Proceedings of the National Academy of Science* 107:8962–68.

Jasienska, G., A. Ziomkiewicz, P. T. Ellison, S. F. Lipson, and I. Thune. 2004. "Large Breasts and Narrow Waists Indicate High Reproductive Potential in Women." *Proceedings of the Royal Society of London B* 271:1213–17.

Jennions, M. D., and A. P. Møller. 2002. "Relationships Fade with Time: A Meta-analysis of Temporal Trends in Publication in Ecology and Evolution." *Proceedings of the Royal Society of London B* 269:43–48.

Jolly, C. T. 1970. "The Seed-Eaters: A New Model of Hominid Differentiation Based on a Baboon Analogy." *Man* 5:5–26.

Kappeler, P. M., and C. P. van Schaik. 2002. "Evolution of Primate Social Systems." *International Journal of Primatology* 23:707–40.

Keinen, A., and A. G. Clark. 2012. "Recent Explosive Human Population Growth Has Resulted in an Excess of Rare Genetic Variants." *Science* 336:740–43.

Kelley, L. A., and J. A. Endler. 2012. "Illusions Promote Mating Success in Great Bowerbirds." *Science* 335:335–38.

Kevles, D. J. 1985. *In the Name of Eugenics*. New York: Alfred A. Knopf.

Keynes, J. M. 1936. *The General Theory of Employment, Interest, and Money*. New York: Harcourt Brace.

Kinsey, A. C. 1953. *Sexual Behavior in the Human Female*. Bloomington: Indiana University Press.

Kinsey, A. C., W. B. Pomeroy, and C. E. Martin. 1948. *Sexual Behavior in the Human Male*. Bloomington: Indiana University Press.

Kirkpatrick, M. 1982. "Sexual Selection and the Evolution of Female Choice." *Evolution* 82:1–12.

———. 1986. "The Handicap Mechanism of Sexual Selection Does Not Work." *American Naturalist* 127:222–40.

Kokko, H., R. Brooks, J. M. McNamara, and A. I. Houston. 2002. "The Sexual Selection Continuum." *Proceedings of the Royal Society of London B* 269:1331–40.

Kroodsma, D. 2005. *The Singing Life of Birds: The Art and Science of Listening to Birdsong*. New York: Houghton Mifflin.

Krugman, P. 2009. "How Did Economists Get It So Wrong?" *New York Times Sunday Magazine,* Sept. 6, 36–43.

Kusmierski, R., G. Borgia, J. A. Uy, and R. H. Corzier. 1997. "Labile Evolution of Display Traits in Bowerbirds Inidicate Reduced Effects of Phylogenetic Constraint." *Proceedings of the Royal Society of London B* 264:307–13.

Lande, R. 1980. "Sexual Dimorphism, Sexual Selection, and Adaptation in Polygenic Characters." *Evolution* 34 (2): 292–305.

———. 1981. "Models of Speciation by Sexual Selection on Polygenic Traits." *Proceedings of the National Academy of Sciences of the United States of America* 78 (6): 3721–25.

Lehrer, J. 2010. "The Truth Wears Off." *The New Yorker,* Dec. 6.

Levick, G. M. 1914. "Antarctic Penguins–a Study of Their Social Habits." London: William Heinemann.

Lewin, B. F., K. Helmius, G. Lalos, and S. A. Månsson. 2000. *Sex in Sweden: On the Swedish Sexual Life*. Stockholm: Swedish National Institute of Public Health.

Li, Q., K.-Q. Gao, J. Vinther, M. D. Shawkey, J. Clarke, L. D'Alba, Q. Meng, D. E. G. Briggs, and R. O. Prum. 2010. "Plumage Color Patterns of an Extinct Dinosaur." *Science* 327:1369–72.

Lieberman, D. E. 2011. *The Evolution of the Human Head*. Cambridge, Mass.: Belknap Press.

———. 2013. *The Story of the Human Body*. New York: Vintage.

Lill, A. 1974. "Social Organization and Space Utilization in the Lek-Forming White-bearded Manakin, *M. manacus trinitatus* Hartert." *Zeitschrift für Tierpsychologie* 36:513–30.

———. 1976. "Lek Behavior in the Golden-headed Manakin, *Pipra erythrocephala,* in Trinidad (West Indies)." *Advances in Ethology* 18:1–83.

Lloyd, E. A. 2005. *The Case of the Female Orgasm: Bias in the Science of Evolution*. Cambridge, Mass.: Harvard University Press.

Lorenz, K. 1941. "Vergleichende Bewegungsstudien an Anatiden." *Journal für Ornithologie* 89:194–293.

———. 1971. "Comparative Studies of the Motor Patterns of Anatinae (1941)." In *Studies in Animal and Human Behaviour,* edited by K. Lorenz, 14–114. Cambridge, Mass.: Harvard University Press.

Lovejoy, C. O. 2009. "Reexamining Human Origins in Light of *Ardipithecus ramidus*." *Science* 326:74e1–8.

Lucas, A. M., and P. R. Stettenheim. 1972. *Avian Anatomy: Integument*. Washington, D.C.: U.S. Department of Agriculture.

Madden, J. R., and A. Balmford. 2004. "Spotted Bowerbirds *Chlamydera maculata* Do Not Prefer Rare or Costly Bower Decorations." *Behavioral Ecology and Sociobiology* 55:589–95.

Marshall, A. J. 1954. *Bower-Birds: Their Displays and Breeding Cycles*. Oxford: Clarendon Press.

Mayr, E. 1972. "Sexual Selection and Natural Selection." In *Sexual Selection and the Descent of Man, 1871–1971,* edited by B. Campbell, 87–104. Chicago: Aldine.

Masters, W. H., and V. E. Johnson. 1966. *Human Sexual Response*. New York: Little, Brown.

McCloskey, R. 1941. *Make Way for Ducklings*. New York: Viking.

McCracken, K. G., R. E. Wilson, P. J. McCracken, and K. P. Johnson. 2001. "Are Ducks Impressed by Drakes' Display?" *Nature* 413:128.

McDonald, D. B. 1989. "Cooperation Under Sexual Selection: Age Graded Changes in a Lekking Bird." *American Naturalist* 134:709–30.

———. 2007. "Predicting Fate from Early Connectivity in a Social Network." *Proceedings of the National Academy of Sciences of the United States of America* 104:10910–14.

McDonald, D. B., and W. K. Potts. 1994. "Cooperative Display and Relatedness Among Males in a Lek-Breeding Bird." *Science* 266:1030–32.

McGraw, K. J. 2006. "Mechanics of Melanin-Based Coloration in Birds." In *Bird Coloration,* vol. 1, *Mechanisms and Measurements,* edited by G. E. Hill and K. J. McGraw, 243–94. Cambridge, Mass.: Harvard University Press.

Mees, G. F. 1974. "Additions to the Avifauna of Suriname." *Zoologische Mededelingen* 38:55–68.

Mehrotra, A., and A. Prochazka. 2015. "Improving Value in Health Care—Against the Annual Physical." *New England Journal of Medicine* 373: 1485–87.

Milam, E. K. 2010. *Looking for a Few Good Males: Female Choice in Evolutionary Biology.* Baltimore: Johns Hopkins University Press.

Miller, G. 2000. *The Mating Mind: How Sexual Choice Shaped the Evolution of Human Nature.* New York: Doubleday.

Mivart, St. G. 1871. Review of *The Descent of Man,* by Charles Darwin. *Quarterly Review* 131:47–90.

Møller, A. P. 1990. "Fluctuating Asymmetry in Male Sexual Ornaments May Reliably Reveal Male Quality." *Animal Behaviour* 40:1185–87.

———. 1992. "Female Swallow Preference for Symmetrical Male Sexual Ornaments." *Nature* 357:238–40.

Muchnik, L., S. Aral, and S. J. Taylor. 2013. "Social Influence Bias: A Randomized Experiment." *Science* 341:647–51.

Muller, M. N., S. M. Kahlenberg, and R. W. Wrangham. 2009. "Male Aggression Against Females and Sexual Coercion in Chimpanzees." In *Sexual Coercion in Primates and Humans: An Evolutionary Perspective on Male Aggression Against Females,* edited by M. N. Muller and R. W. Wrangham, 184–217. Cambridge, Mass.: Harvard University Press.

Muller, M. N., and J. C. Mitani. 2005. "Conflict and Cooperation in Wild Chimpanzees." *Advances in the Study of Behavior* 35:275–331.

Mulvey, L. 1975. "Visual Pleasure and Narrative Cinema." *Screen* 16:6–18.

Nagel, T. 1974. "What Is It Like to Be a Bat?" *Philosophical Review* 83: 435–50.

Neave, N., and K. Shields. 2008. "The Effects of Facial Hair Manipulation on Female Perceptions of Attractiveness, Masculinity, and Dominance in Male Faces." *Personality and Individual Differences* 45:373–77.

Noad, M. J., D. H. Cato, M. M. Bryden, M.-N. Jenner, and K. C. S. Jenner. 2000. "Cultural Revolution in Whale Songs." *Nature* 408:537.

Novak, S. A., and M. A. Hatch. 2009. "Intimate Wounds: Craniofacial Trauma in Women and Female Chimpanzees." In *Sexual Coercion in Pri-*

mates and Humans: An Evolutionary Perspective on Male Aggression Against Females, edited by M. N. Muller and R. W. Wrangham, 322–45. Cambridge, Mass.: Harvard University Press.

Palmer, A. R. 1999. "Detecting Publication Bias in Meta-analyses: A Case Study of Fluctuating Asymmetry and Sexual Selection." *American Naturalist* 154:220–33.

Palombit, R. 2009. "'Friendship' with Males: A Female Counterstrategy to Infanticide in Chacma Baboons of the Okavango Delta." In *Sexual Coercion in Primates and Humans: An Evolutionary Perspective on Male Aggression Against Females,* edited by M. N. Muller and R. W. Wrangham, 377–409. Cambridge, Mass.: Harvard University Press.

Paoli, T. 2009. "The Absence of Sexual Conflict in Bonobos." In *Sexual Coercion in Primates and Humans: An Evolutionary Perspective on Male Aggression Against Females,* edited by M. N. Muller and R. W. Wrangham, 410–23. Cambridge, Mass.: Harvard University Press.

Parker, G. A. 1979. "Sexual Selection and Sexual Conflict." In *Sexual Selection and Reproductive Competition in Insects,* edited by M. S. Blum and N. B. Blum, 123–66. New York: Academic Press.

Patricelli, G. L., J. A. Uy, and G. Borgia. 2003. "Multiple Male Traits Interact: Attractive Bower Decorations Facilitate Attractive Behavioural Displays in Satin Bowerbirds." *Proceedings of the Royal Society of London B* 270:2389–95.

———. 2004. "Female Signals Enhance the Efficiency of Mate Assessment in Satin Bowerbirds (*Ptilonorhynchus violaceus*)." *Behavioral Ecology* 15:297–304.

Patricelli, G. L., J. A. Uy, G. Walsh, and G. Borgia. 2002. "Male Displays Adjusted to Female's Response." *Nature* 415:279–80.

Pavlicev, M., and G. Wagner. 2016. "The Evolutionary Origin of Female Orgasm." *Journal of Experimental Zoology B: Molecular and Developmental Evolution* 326:326–37.

Pillard, R. C., and J. M. Bailey. 1998. "Human Sexual Orientation Has a Heritable Component." *Human Biology* 70:347–65.

Pizzari, T., and T. R. Birkhead. 2000. "Female Feral Fowl Eject Sperm of Subdominant Males." *Nature* 405:787–89.

Pomiankowski, A., and Y. Iwasa. 1993. "The Evolution of Multiple Sexual Ornaments by Fisher's Process of Sexual Selection." *Proceedings of the Royal Society of London B* 253:173–81.

Prokop, Z. M., L. Michalczyk, S. Drobniak, and M. Herdegen. 2012. "Meta-analysis Suggests Choosy Females Get Sexy Sons More Than 'Good Genes.'" *Evolution* 66:2665–73.

Prum, R. O. 1985. "Observations of the White-fronted Manakin (*Pipra serena*) in Suriname." *Auk* 102:384–87.

————. 1986. "The Displays of the White-throated Manakin *Corapipo gutturalis* in Suriname." *Ibis* 128:91–102.

————. 1988. "Phylogenetic Interrelationships of the Barbets (Capitonidae) and Toucans (Ramphastidae) Based on Morphology with Comparisons to DNA-DNA Hybridization." *Zoological Journal of the Linnean Society* 92:313–43.

————. 1990. "Phylogenetic Analysis of the Evolution of Display Behavior in the Neotropical Manakins (Aves: Pipridae)." *Ethology* 84:202–31.

————. 1992. "Syringeal Morphology, Phylogeny, and Evolution of the Neotropical Manakins (Aves: Pipridae)." *American Museum Novitates* 3043: 1–65.

————. 1994. "Phylogenetic Analysis of the Evolution of Alternative Social Behavior in the Manakins (Aves: Pipridae)." *Evolution* (48): 1657–75.

————. 1997. "Phylogenetic Tests of Alternative Intersexual Selection Mechanisms: Macroevolution of Male Traits in a Polygynous Clade (Aves: Pipridae)." *American Naturalist* 149:668–92.

————. 1998. "Sexual Selection and the Evolution of Mechanical Sound Production in Manakins (Aves: Pipridae)." *Animal Behaviour* 55:977–94.

————. 1999. "Development and Evolutionary Origin of Feathers." *Journal of Experimental Zoology (Molecular and Developmental Evolution)* 285:291–306.

————. 2005. "Evolution of the Morphological Innovations of Feathers." *Journal of Experimental Zoology: Part B, Molecular and Developmental Evolution* 304B (6): 570–79.

————. 2010. "The Lande-Kirkpatrick Mechanism Is the Null Model of Evolution by Intersexual Selection: Implications for Meaning, Honesty, and Design in Intersexual Signals." *Evolution* 64:3085–100.

————. 2012. "Aesthetic Evolution by Mate Choice: Darwin's *Really* Dangerous Idea." *Philosophical Transactions of the Royal Society of London B* 367:2253–65.

————. 2013. "Coevolutionary Aesthetics in Human and Biotic Artworlds." *Biology and Philosophy* 28:811–32.

————. 2015. "The Role of Sexual Autonomy in Evolution by Mate Choice." In *Current Perspectives in Sexual Selection,* edited by T. Hoquet, 237–62. New York: Springer.

Prum, R. O., J. S. Berv, A. Dornburg, D. J. Field, J. P. Townsend, E. M. Lemmon, and A. R. Lemmon. 2015. "A Comprehensive Phylogeny of Birds (Aves) Using Targeted Next-Generation DNA Sequencing." *Nature* 526:569–73.

Prum, R. O., and A. H. Brush. 2002. "The Evolutionary Origin and Diversification of Feathers." *Quarterly Review of Biology* 77:261–95.

————. 2003. "Which Came First, the Feather or the Bird?" *Scientific American,* March, 60–69.

Prum, R. O., and A. E. Johnson. 1987. "Display Behavior, Foraging Ecology, and Systematics of the Golden-winged Manakin (*Masius chrysopterus*)." *Wilson Bulletin* 87:521–39.

Puts, D. A. 2007. "Of Bugs and Boojums: Female Orgasm as a Facultative Adaptation." *Archives of Sexual Behavior* 36:337–39.

Qidwai, W. 2000. "Perceptions About Female Sexuality Among Young Pakistani Men Presenting to Family Physicians at a Teaching Hospital in Karachi." *Journal of the Pakistan Medical Association* 50 (2): 74–77.

Queller, D. C. 1987. "The Evolution of Leks Through Female Choice." *Animal Behaviour* 35:1424–32.

Rahman, Q., and G. D. Wilson. 2003. "Born Gay? The Psychobiology of Human Sexual Orientation." *Personality and Individual Differences* 34:1337–82.

Reich, E. S. 2013. "Symmetry Study Deemed a Fraud." *Nature* 497:170–71.

Rensch, B. 1950. "Die Abhängigkeit der relativen Sexualdifferenz von der Körpergrösse." *Bonner Zoologische Beitrage* 1:58–69.

Rice, W. R., U. Friberg, and S. Gavrilets. 2012. "Homosexuality as a Consequence of Epigenetically Canalized Sexual Development." *Quarterly Review of Biology* 87:343–68.

Richardson, R. C. 2010. *Evolutionary Psychology as Maladapted Psychology*. Cambridge, Mass.: MIT Press.

Ridley, M. 1993. *The Red Queen: Sex and the Evolution of Human Nature*. London: Viking.

Robbins, M. B. 1983. "The Display Repertoire of the Band-tailed Manakin (*Pipra fasciicauda*)." *Wilson Bulletin* 95:321–42.

Robbins, M. M. 2009. "Male Aggression Against Females in Mountain Gorillas: Courtship or Coercion?" In *Sexual Coercion in Primates and Humans: An Evolutionary Perspective on Male Aggression Against Females*, edited by M. N. Muller and R. W. Wrangham, 112–27. Cambridge, Mass.: Harvard University Press.

Romanoff, A. L. 1960. *The Avian Embryo: Structural and Functional Development*. New York: Macmillan.

Rome, L. C., D. A. Syme, S. Hollingworth, and S. L. Lindstedt. 1996. "The Whistle and the Rattle: The Design of Sound Producing Muscles." *Proceedings of the National Academy of Science* 93:8095–100.

Romer, A. S. 1955. *The Vertebrate Body*. New York: Saunders.

Roughgarden, J. 2009. *Evolution's Rainbow: Diversity, Gender, and Sexuality in Nature and People*. Berkeley: University of California Press.

Rowen, T. S., T. W. Gaither, M. A. Awad, E. C. Osterberg, A. W. Shindel, and B. N. Breyer. 2016. "Pubic Hair Grooming Prevalence and Motivation Among Women in the United States." *JAMA Dermatology* 2016:2154.

Rubenfeld, J. 2013. "The Riddle of Rape-by-Deception and the Myth of Sexual Autonomy." *Yale Law Journal* 122:1372–443.

Russell, D. G. D., W. J. L. Sladen, and D. G. Ainley. 2012. "Dr. George Murray Levick (1876–1956): Unpublished Notes on the Sexual Habits of the Adélie Penguin." *Polar Record* 48:387–93.

Ryan, M. J., and M. E. Cummings. 2013. "Perceptual Biases and Mate Choice." *Annual Review of Ecology, Evolution, and Systematics* 44:437–59.

Ryder, T. B., D. B. MacDonald, J. G. Blake, P. G. Parker, and B. A. Loiselle. 2008. "Social Networks in Lek-Mating Wire-tailed Manakin (*Pipra filicauda*)." *Proceedings of the Royal Society of London B* 275:1367–74.

Ryder, T. B., P. G. Parker, J. G. Blake, and B. A. Loiselle. 2009. "It Takes Two to Tango: Reproductive Skew and Social Correlates of Male Mating Success in a Lek-Breeding Bird." *Proceedings of the Royal Society of London B* 276:2377–84.

Samuelson, P. A. 1958. "An Exact Consumption-Loan Model of Interest With or Without the Social Contrivance of Money." *Journal of Political Economy* 66:467–82.

Saranathan, V., A. E. Seago, A. Sandy, S. Narayanan, S. G. J. Mochrie, E. R. Dufresne, H. Cao, C. Osuji, and R. O. Prum. 2015. "Structural Diversity of Arthropod Biophotonic Nanostructures Spans Amphiphilic Phase-Space." *Nanoletters* 15:3735–42.

Scheinfeld, A. 1939. *You and Heredity.* New York: Frederick A. Stokes.

Schwartz, P., and D. W. Snow. 1978. "Display and Related Behavior of the Wire-tailed Manakin." *Living Bird* 17:51–78.

Sclater, P. L. 1862. "Notes on *Pipra deliciosa*." *Ibis* 4:175–78.

Scrimshaw, S. C. M. 1984. "Infanticide in Human Populations: Societal and Individual Concerns." In *Infanticide: Comparative and Evolutionary Perspectives,* edited by G. Hausfater and S. B. Hrdy, 439–62. London: Aldine Transaction.

Shapiro, M. D., Z. Kronenberg, C. Li, E. T. Domyan, H. Pan, M. Campbell, H. Tan, C. D. Huff, H. Hu, A. I. Vickery, S. C. A. Nielsen, S. A. Stringham, H. Hu, E. Willerslev, M. Thomas, P. Gilbert, M. Yandell, G. Zhang, and J. Wang. 2013. "Genomic Diversity and Evolution of the Head Crest in the Rock Pigeon." *Science* 339:1063–67.

Sheehan, M. J., and M. W. Nachman. 2014. "Morphological and Population Genomic Evidence That Human Faces Have Evolved to Signal Individual Identity." *Nature Communications* 5:4800.

Sheehan, M. J., and E. A. Tibbets. 2011. "Specialized Face Learning Is Associated with Individual Recognition in Paper Wasps." *Science* 334:1271–75.

Sheets-Johnstone, M. 1989. "Hominid Bipedality and Sexual Selection Theory." *Evolutionary Theory* 9:57–70.

Shiller, R. J. 2015. *Irrational Exuberance*. 3rd ed. Princeton, N.J.: Princeton University Press.

Shubin, N. 2008. *Your Inner Fish: A Journey into the 3.5 Billion-Year History of the Human Body*. New York: Pantheon.

Silk, J. B., J. C. Beehner, T. J. Bergman, C. Crockford, A. L. Engh, L. R. Moscovice, R. M. Wittig, R. M. Seyfarth, and D. L. Cheney. 2009. "The Benefits of Social Capital: Close Social Bonds Among Female Baboons Enhance Offspring Survival." *Proceedings of the Royal Society of London B* 276:3099–104.

Sinkkonen, A. 2009. "Umbilicus as a Fitness Signal in Humans." *FASEB Journal* 23:10–12.

Smith, R. L. 1984. "Human Sperm Competition." In *Sperm Competition and the Evolution of Animal Mating Systems*, edited by R. L. Smith, 601–59. New York: Academic Press.

Smuts, B. 1985. *Sex and Friendship in Baboons*. Cambridge, Mass.: Harvard University Press.

Snow, B. K., and D. W. Snow. 1985. "Display and Related Behavior of Male Pin-tailed Manakins." *Wilson Bulletin* 97:273–82.

Snow, D. W. 1961. "The Displays of Manakins *Pipra pipra* and *Tyranneutes virescens*." *Ibis* 103:110–13.

———. 1962a. "A Field Study of the Black-and-white Manakin, *Manacus manacus*, in Trinidad, W.I." *Zoologica* 47:65–104.

———. 1962b. "A Field Study of the Golden-headed Manakin, *Pipra erythrocephala*, in Trinidad, W.I." *Zoologica* 47:183–98.

———. 1963a. "The Display of the Blue-backed Manakin, *Chiroxiphia pareola*, in Tobago, W.I." *Zoologica* 48:167–76.

———. 1963b. "The Display of the Orange-headed Manakin." *Condor* 65:44–48.

———. 1976. *The Web of Adaptation*. Ithaca, N.Y.: Cornell University Press.

Snow, S. S., S. H. Alonzo, M. R. Servedio, and R. O. Prum. Forthcoming. "Evolution of Resistance to Sexual Coercion Through the Indirect Benefits of Mate Choice."

Sterck, E. H. M., D. P. Watts, and C. O. van Schaik. 1997. "The Evolution of Female Social Relationships in Nonhuman Primates." *Behavioral Ecology and Sociobiology* 41:291–309.

Stolley, P. D. 1991. "When Genius Errs: R. A. Fisher and the Lung Cancer Controversy." *American Journal of Epidemiology* 133:416–25.

Sullivan, A. 1995. *Virtually Normal*. New York: Vintage Books.

Sutherland, J. 2005. "The Ideas Interview: Elisabeth Lloyd." *Guardian*, Sept. 26.

Swedell, L., and A. Schreier. 2009. "Male Aggression Toward Females in Hamadryas Baboons: Conditioning, Coercion, and Control." In *Sexual Coercion in Primates and Humans: An Evolutionary Perspective on Male Aggression Against Females,* edited by M. N. Muller and R. W. Wrangham, 244–68. Cambridge, Mass.: Harvard University Press.

Symons, D. 1979. *The Evolution of Human Sexuality.* Oxford: Oxford University Press.

Théry, M. 1990. "Display Repertoire and Social Organization of the White-fronted and White-throated Manakins." *Wilson Bulletin* 102:123–30.

Trainer, J. M., and D. B. McDonald. 1993. "Vocal Repertoire of the Long-tailed Manakin and Its Relation to Male-Male Cooperation." *Condor* 95:769–81.

Trut, L. N. 2001. "Experimental Studies of Early Canid Domestication." In *The Genetics of the Dog,* edited by A. Ruvinsky and J. Sampson, 15–41. Wallingford, U.K.: CABI.

Uy, J. A., and G. Borgia. 2000. "Sexual Selection Drives Rapid Divergence in Bowerbird Display Traits." *Evolution* 54:273–78.

Uy, J. A., G. L. Patricelli, and G. Borgia. 2001. "Complex Mate Searching in the Satin Bowerbird *Ptilonorhynchus violaceus." American Naturalist* 158:530–42.

Vinther, J., D. E. G. Briggs, R. O. Prum, and V. Saranathan. 2008. "The Colour of Fossil Feathers." *Biology Letters* 4:522–25.

Wagner, G. P. 2015. *Homology, Genes, and Evolutionary Innovation.* Princeton, N.J.: Princeton University Press.

Walker, A. 1984. "Mechanisms of Honing in the Male Baboon Canine." *American Journal of Physical Anthropology* 65:47–60.

Wallace, A. R. 1889. *Darwinism.* London: Macmillan.

———. 1895. *Natural Selection and Tropical Nature.* 2nd ed. London: Macmillan.

Wallen, K., and E. A. Lloyd. 2011. "Female Sexual Arousal: Genital Anatomy and Orgasm in Intercourse." *Hormones and Behavior* 59:780–92.

Warner, M. 1999. *The Trouble with Normal: Sex, Politics, and the Ethics of Queer Life.* Cambridge, Mass.: Harvard University Press.

Weiner, J. 1994. *The Beak of the Finch.* New York: Alfred A. Knopf.

Wekker, G. 1999. " 'What's Identity Got to Do with It?': Rethinking Identity in Light of the Mati Work in Suriname." In *Female Desires: Same-Sex and Transgender Practices Across Cultures,* edited by E. Blackwood and S. E. Wieringa, 119–38. New York: Columbia University Press.

Welty, J. C. 1982. *The Life of Birds.* 2nd ed. New York: Saunders.

West-Eberhard, M. J. 1979. "Sexual Selection, Social Competition, and Evolution." *Proceedings of the American Philosophical Society* 123:222–34.

———. 1983. "Sexual Selection, Social Competition, and Speciation." *Quarterly Review of Biology* 58:155–83.

———. 2014. "Darwin's Forgotten Idea: The Social Essence of Sexual Selection." *Neuroscience and Biobehavioral Reviews* 46:501–8.

Willis, E. O. 1966. "Notes on a Display and Nest of the Club-winged Manakin." *Auk* 83:475–76.

Wilson, G., and Q. Rahman. 2008. *Born Gay: The Psychobiology of Sex Orientation.* London: Peter Owen.

Zahavi, A. 1975. "Mate Selection—a Selection for a Handicap." *Journal of Theoretical Biology* 53:205–14.

Zuk, M. 2013. *Paleofantasy: What Evolution Really Tells Us About Sex, Diet, and How We Live.* New York: Norton.

Index

Page numbers in *italics* refer to illustrations.
Page numbers beginning with 347 refer to end notes.